D0999628

CRITICISM AND THE GROWTH OF KNOWLEDGE

CRITICISM AND THE GROWTH OF KNOWLEDGE

Proceedings of the International Colloquium in the Philosophy of Science, London, 1965, volume 4

Edited by

IMRE LAKATOS
Professor of Logic, University of London

ALAN MUSGRAVE
Professor of Philosophy, University of Otago

CAMBRIDGE
AT THE UNIVERSITY PRESS
1970

Published by the Syndics of the Cambridge University Press
Bentley House, 200 Euston Road, London N.W.1
American Branch: 32 East 57th Street, New York, N.Y. 10022

Library of Congress Catalogue Card Number 78–105496

Standard Book Number: 521 07826 1

Printed in Great Britain
at the University Press, Aberdeen

CONTENTS

PREFACE

This book constitutes the fourth volume of the Proceedings of the 1965 International Colloquium in the Philosophy of Science held at Bedford College, Regent's Park, London, from 11 to 17 July 1965. The Colloquium was organized jointly by the British Society for the Philosophy of Science and the London School of Economics and Political Science, under the auspices of the Division of Logic, Methodology and Philosophy of Science of the International Union of History and Philosophy of Science.

The Colloquium and the Proceedings were generously subsidized by the sponsoring institutions, and by the Leverhulme Foundation and the Alfred P. Sloan Foundation.

The members of the Organizing Committee were: W. C. Kneale (Chairman), I. Lakatos (Honorary Secretary), J. W. N. Watkins (Honorary Joint Secretary), S. Körner, Sir Karl Popper, H. R. Post and J. O. Wisdom.

The first three volumes of the *Proceedings* were published by the North-Holland Publishing Company, Amsterdam, under the following titles:

Lakatos (*ed.*): *Problems in the Philosophy of Mathematics*, 1967.

Lakatos (*ed.*): *The Problem of Inductive Logic*, 1968.

Lakatos and Musgrave (*eds.*): *Problems in the Philosophy of Science*, 1968.

The full programme of the Colloquium is printed in the first volume of the *Proceedings*.

This fourth volume follows the editorial policy pursued in the first three volumes: it is a rational reconstruction and expansion rather than a faithful report of the actual discussion. The whole volume arises from one symposium, the one held on 13 July on *Criticism and the Growth of Knowledge*. Originally, Professor Kuhn, Professor Feyerabend and Dr Lakatos were to be the main speakers, but for different reasons (see *below*, p. 25) Professor Feyerabend's and Dr Lakatos's contributions arrived only after the Colloquium. Professor Watkins kindly agreed to step in in their stead. Professor Sir Karl Popper took the chair of the lively discussion in which, among others, Professor Stephen Toulmin, Professor Pearce Williams, Miss Margaret Masterman and the Chairman participated.

The texts of the papers as here printed were finished at different times. Professor Kuhn's paper is printed essentially in the form in which it was first read. The papers by Professors John Watkins, Stephen Toulmin, Pearce Williams and Sir Karl Popper are slightly amended versions of their original contributions. On the other hand, Miss Masterman's paper was finished only in 1966; while Dr Lakatos's and Professor Feyerabend's papers, together with Professor Kuhn's final reply, were finished in 1969.

The Editors—greatly assisted by Peter Clark and John Worrall—wish to thank all the contributors for their kind cooperation. They are also grateful to Mrs Christine Jones and to Miss Mary McCormick for their conscientious and careful work in preparing the manuscripts for publication.

THE EDITORS

London, August 1969

Logic of Discovery or Psychology of Research?[1]

THOMAS S. KUHN
Princeton University

My object in these pages is to juxtapose the view of scientific development outlined in my book, *The Structure of Scientific Revolutions*, with the better known views of our chairman, Sir Karl Popper.[2] Ordinarily I should decline such an undertaking, for I am not so sanguine as Sir Karl about the utility of confrontations. Besides, I have admired his work for too long to turn critic easily at this date. Nevertheless, I am persuaded that for this occasion the attempt must be made. Even before my book was published two and a half years ago, I had begun to discover special and often puzzling characteristics of the relation between my views and his. That relation and the divergent reactions I have encountered to it suggest that a disciplined comparison of the two may produce peculiar enlightenment. Let me say why I think this could occur.

On almost all the occasions when we turn explicitly to the same problems, Sir Karl's view of science and my own are very nearly identical.[3] We are both concerned with the dynamic process by which scientific knowledge is acquired rather than with the logical structure of the products of scientific research. Given that concern, both of us emphasize, as legitimate data, the facts and also the spirit of actual scientific life, and both of us turn often to history to find them. From this pool of shared data, we draw many of the same conclusions. Both of us reject the view that science progresses

[1] This paper was initially prepared at the invitation of P. A. Schilpp for his forthcoming volume, *The Philosophy of Karl R. Popper*, to be published by The Open Court Publishing Company, La Salle, Ill., in The Library of Living Philosophers. I am most grateful to both Professor Schilpp and the publishers for permission to print it as part of the proceedings of this symposium before its appearance in the volume for which it was first solicited.

[2] For purposes of the following discussion I have reviewed Sir Karl Popper's [1959], his [1963], and his [1957]. I have also occasionally referred to his original [1935] and his [1945]. My own [1962] provides a more extended account of many of the issues discussed below.

[3] More than coincidence is presumably responsible for this extensive overlap. Though I had read none of Sir Karl's work before the appearance in 1959 of the English translation of his [1935] (by which time my book was in draft), I had repeatedly heard a number of his main ideas discussed. In particular, I had heard him discuss some of them as William James Lecturer at Harvard in the spring of 1950. These circumstances do not permit me to specify an intellectual debt to Sir Karl, but there must be one.

by accretion; both emphasize instead the revolutionary process by which an older theory is rejected and replaced by an incompatible new one[1]; and both deeply underscore the role played in this process by the older theory's occasional failure to meet challenges posed by logic, experiment, or oberva-tion. Finally, Sir Karl and I are united in opposition to a number of classical positivism's most characteristic theses. We both emphasize, for example, the intimate and inevitable entanglement of scientific observa-tion with scientific theory; we are correspondingly sceptical of efforts to produce any neutral observation language; and we both insist that scien-tists may properly aim to invent theories that *explain* observed phenomena and that do so in terms of *real* objects, whatever the latter phrase may mean.

That list, though it by no means exhausts the issues about which Sir Karl and I agree,[2] is already extensive enough to place us in the same minority among contemporary philosophers of science. Presumably that is why Sir Karl's followers have with some regularity provided my most sympathetic philosophical audience, one for which I continue to be grateful. But my gratitude is not unmixed. The same agreement that evokes the sympathy of this group too often misdirects its interest. Apparently Sir Karl's followers can often read much of my book as chapters from a late (and, for some, a drastic) revision of his classic, *The Logic of Scientific Discovery*. One of them asks whether the view of science outlined in my *Scientific Revolutions* has not long been common knowledge. A second, more charitably, isolates my originality as the demonstration that dis-coveries-of-fact have a life cycle very like that displayed by innovations-of theory. Still others express general pleasure in the book but will discuss only the two comparatively secondary issues about which my disagreement with Sir Karl is most nearly explicit: my emphasis on the importance of deep commitment to tradition and my discontent with the implications of the term 'falsification'. All these men, in short, read my book through a quite special pair of spectacles, and there is another way to read it. The view through those spectacles is not wrong—my agreement with Sir Karl is real and substantial. Yet readers outside of the Popperian circle almost

[1] Elsewhere I use the term 'paradigm' rather than 'theory' to denote what is rejected and replaced during scientific revolutions. Some reasons for the change of term will emerge below.

[2] Underlining one additional area of agreement about which there has been much misunderstanding may further highlight what I take to be the real differences between Sir Karl's views and mine. We both insist that adherence to a tradition has an essential role in scientific development. He has written, for example, 'Quantitatively and qualitatively by far the most important source of our knowledge—apart from inborn knowledge—is tradition' (Popper [1963], p. 27). Even more to the point, as early as 1948 Sir Karl wrote, 'I do not think that we could ever free ourselves entirely from the bonds of tradition. The so-called freeing is really only a change from one tradition to another' ([1963], p. 122).

invariably fail even to notice that the agreement exists, and it is these readers who most often recognize (not necessarily with sympathy) what seem to me the central issues. I conclude that a gestalt switch divides readers of my book into two or more groups. What one of these sees as striking parallelism is virtually invisible to the others. The desire to understand how this can be so motivates the present comparison of my view with Sir Karl's.

The comparison must not, however, be a mere point by point juxtaposition. What demands attention is not so much the peripheral area in which our occasional secondary disagreements are to be isolated but the central region in which we appear to agree. Sir Karl and I do appeal to the same data; to an uncommon extent we are seeing the same lines on the same paper; asked about those lines and those data, we often give virtually identical responses, or at least responses that inevitably seem identical in the isolation enforced by the question-and-answer mode. Nevertheless, experiences like those mentioned above convince me that our intentions are often quite different when we say the same things. Though the lines are the same, the figures which emerge from them are not. That is why I call what separates us a gestalt switch rather than a disagreement and also why I am at once perplexed and intrigued about how best to explore the separation. How am I to persuade Sir Karl, who knows everything I know about scientific development and who has somewhere or other said it, that what he calls a duck can be seen as a rabbit? How am I to show him what it would be like to wear my spectacles when he has already learned to look at everything I can point to through his own?

In this situation a change in strategy is called for, and the following suggests itself. Reading over once more a number of Sir Karl's principal books and essays, I encounter again a series of recurrent phrases which, though I understand them and do not quite disagree, are locutions that *I* could never have used in the same places. Undoubtedly they are most often intended as metaphors applied rhetorically to situations for which Sir Karl has elsewhere provided unexceptionable descriptions. Nevertheless, for present purposes these metaphors, which strike me as patently inappropriate, may prove more useful than straightforward descriptions. They may that is, be symptomatic of contextual differences that a careful literal expression hides. If that is so, then these locutions may function not as the lines-on-paper but as the rabbit-ear, the shawl, or the ribbon-at-the-throat which one isolates when teaching a friend to transform his way of seeing a gestalt diagram. That, at least, is my hope for them. I have four such differences of locutions in mind and shall treat them *seriatim*.

I

Among the most fundamental issues on which Sir Karl and I agree is our insistence that an analysis of the development of scientific knowledge must take account of the way science has actually been practiced. That being so, a few of his recurrent generalizations startle me. One of these provides the opening sentences of the first chapter of the *Logic of Scientific Discovery:* 'A scientist', writes Sir Karl, 'whether theorist or experimenter, puts forward statements, or systems of statements, and tests them step by step. In the field of the empirical sciences, more particularly, he constructs hypotheses, or systems of theories, and tests them against experience by observation and experiment.'[1] The statement is virtually a cliché, yet in application it presents three problems. It is ambiguous in its failure to specify which of two sorts of 'statements' or 'theories' are being tested. That ambiguity can, it is true, be eliminated by reference to other passages in Sir Karl's writings, but the generalization that results is historically mistaken. Furthermore, the mistake proves important, for the unambiguous form of the description misses just that characteristic of scientific practice which most nearly distinguishes the sciences from other creative pursuits.

There is one sort of 'statement' or 'hypothesis' that scientists do repeatedly subject to systematic test. I have in mind statements of an individual's best guesses about the proper way to connect his own research problem with the corpus of accepted scientific knowledge. He may, for example, conjecture that a given chemical unknown contains the salt of a rare earth, that the obesity of his experimental rats is due to a specified component in their diet, or that a newly discovered spectral pattern is to be understood as an effect of nuclear spin. In each case, the next steps in his research are intended to try out or test the conjecture or hypothesis. If it passes enough or stringent enough tests, the scientist has made a discovery or has at least resolved the puzzle he had been set. If not, he must either abandon the puzzle entirely or attempt to solve it with the aid of some other hypothesis. Many research problems, though by no means all, take this form. Tests of this sort are a standard component of what I have elsewhere labelled 'normal science' or 'normal research', an enterprise which accounts for the overwhelming majority of the work done in basic science. In no usual sense, however, are such tests directed to current theory. On the contrary, when engaged with a normal research problem, the scientist must *premise* current theory as the rules of his game. His object is to solve a puzzle, preferably one at which others have failed, and current theory is required to

[1] Popper [1959], p. 27.

define that puzzle and to guarantee that, given sufficient brilliance, it can be solved.[1] Of course the practitioner of such an enterprise must often test the conjectural puzzle solution that his ingenuity suggests. But only his personal conjecture is tested. If it fails the test, only his own ability not the corpus of current science is impugned. In short, though tests occur frequently in normal science, these tests are of a peculiar sort, for in the final analysis it is the individual scientist rather than current theory which is tested.

This is not, however, the sort of test Sir Karl has in mind. He is above all concerned with the procedures through which science grows, and he is convinced that 'growth' occurs not primarily by accretion but by the revolutionary overthrow of an accepted theory and its replacement by a better one.[2] (The subsumption under 'growth' of 'repeated overthrow' is itself a linguistic oddity whose *raison d'être* may become more visible as we proceed.) Taking this view, the tests which Sir Karl emphasizes are those which were performed to explore the limitations of accepted theory or to subject a current theory to maximum strain. Among his favourite examples, all of them startling and destructive in their outcome, are Lavoisier's experiments on calcination, the eclipse expedition of 1919, and the recent experiments on parity conservation.[3] All, of course, are classic tests, but in using them to characterize scientific activity Sir Karl misses something terribly important about them. Episodes like these are very rare in the development of science. When they occur, they are generally called forth either by a prior crisis in the relevant field (Lavoisier's experiments or Lee and Yang's[4]) or by the existence of a theory which competes with the existing canons of research (Einstein's general relativity). These are, however, aspects of or occasions for what I have elsewhere called 'extraordinary research', an enterprise in which scientists do display

[1] For an extended discussion of normal science, the activity which practitioners are trained to carry on, see my [1962], pp. 23–42, and 135–42. It is important to notice that when I describe the scientist as a puzzle solver and Sir Karl describes him as a problem solver (e.g. in his [1963], pp. 67, 222), the similarity of our terms disguises a fundamental divergence. Sir Karl writes (the italics are his), 'Admittedly, our expectations, and thus our theories, may precede, historically, even our problems. *Yet science starts only with problems.* Problems crop up especially when we are disappointed in our expectations, or when our theories involve us in difficulties, in contradictions'. I use the term 'puzzle' in order to emphasize that the difficulties which *ordinarily* confront even the very best scientists are, like crossword puzzles or chess puzzles, challenges only to his ingenuity. *He* is in difficulty, not current theory. My point is almost the converse of Sir Karl's.

[2] Cf. Popper [1963], pp. 129, 215 and 221, for particularly forceful statements of this position.

[3] For example, Popper [1963], p. 220.

[4] For the work on calcination see, Guerlac [1961]. For the background of the parity experiments see, Hafner and Presswood [1965].

very many of the characteristics Sir Karl emphasizes, but one which, at least in the past, has arisen only intermittently and under quite special circumstances in any scientific speciality.[1]

I suggest then that Sir Karl has characterized the entire scientific enterprise in terms that apply only to its occasional revolutionary parts. His emphasis is natural and common: the exploits of a Copernicus or Einstein make better reading than those of a Brahe or Lorentz; Sir Karl would not be the first if he mistook what I call normal science for an intrinsically uninteresting enterprise. Nevertheless, neither science nor the development of knowledge is likely to be understood if research is viewed exclusively through the revolutions it occasionally produces. For example, though testing of basic commitments occurs only in extraordinary science, it is normal science that discloses both the points to test and the manner of testing. Or again, it is for the normal, not the extraordinary practice of science that professionals are trained; if they are nevertheless eminently successful in displacing and replacing the theories on which normal practice depends, that is an oddity which must be explained. Finally, and this is for now my main point, a careful look at the scientific enterprise suggests that it is normal science, in which Sir Karl's sort of testing does not occur, rather than extraordinary science which most nearly distinguishes science from other enterprises. If a demarcation criterion exists (we must not, I think, seek a sharp or decisive one), it may lie just in that part of science which Sir Karl ignores.

In one of his most evocative essays, Sir Karl traces the origin of 'the tradition of critical discussion [which] represents the only practicable way of expanding our knowledge' to the Greek philosophers between Thales and Plato, the men who, as he sees it, encouraged critical discussion both between schools and within individual schools.[2] The accompanying description of Presocratic discourse is most apt, but what is described does not at all resemble science. Rather it is the tradition of claims, counterclaims, and debates over fundamentals which, except perhaps during the Middle Ages, have characterized philosophy and much of social science ever since. Already by the Hellenistic period mathematics, astronomy, statics and the geometric parts of optics had abandoned this mode of discourse in favour of puzzle solving. Other sciences, in increasing numbers, have undergone the same transition since. In a sense, to turn Sir Karl's view on its head, it is precisely the abandonment of critical discourse that marks the transition to a science. Once a field has made that transition, critical discourse recurs only at moments of crisis when the bases of the

[1] The point is argued at length in my [1962], pp. 52–97.

[2] Popper [1963], chapter 5, especially pp. 148–52.

field are again in jeopardy.[1] Only when they must choose between competing theories do scientists behave like philosophers. That, I think, is why Sir Karl's brilliant description of the reasons for the choice between metaphysical systems so closely resembles my description of the reasons for choosing between scientific theories.[2] In neither choice, as I shall shortly try to show, can testing play a quite decisive role.

There is, however, good reason why testing has seemed to do so, and in exploring it Sir Karl's duck may at last become my rabbit. No puzzle-solving enterprise can exist unless its practitioners share criteria which, for that group and for that time, determine when a particular puzzle has been solved. The same criteria necessarily determine failure to achieve a solution, and anyone who chooses may view that failure as the failure of a theory to pass a test. Normally, as I have already insisted, it is not viewed that way. Only the practitioner is blamed, not his tools. But under the special circumstances which induce a crisis in the profession (e.g. gross failure, or repeated failure by the most brilliant professionals) the group's opinion may change. A failure that had previously been personal may then come to seem the failure of a theory under test. Thereafter, because the test arose from a puzzle and thus carried settled criteria of solution, it proves both more severe and harder to evade than the tests available within a tradition whose normal mode is critical discourse rather than puzzle solving.

In a sense, therefore, severity of test-criteria is simply one side of the coin whose other face is a puzzle-solving tradition. That is why Sir Karl's line of demarcation and my own so frequently coincide. That coincidence is, however, only in their *outcome*; the *process* of applying them is very different, and it isolates distinct aspects of the activity about which the decision—science or non-science—is to be made. Examining the vexing cases, for example, psychoanalysis or Marxist historiography, for which Sir Karl tells us his criterion was initially designed,[3] I concur that they cannot now properly be labelled 'science'. But I reach that conclusion by a route far surer and more direct than his. One brief example may suggest that of the two criteria, testing and puzzle solving, the latter is at once the less equivocal and the more fundamental.

To avoid irrelevant contemporary controversies, I consider astrology rather than, say, psychoanalysis. Astrology is Sir Karl's most frequently cited example of a 'pseudo-science'.[4] He says: 'By making their interpretations and prophecies sufficiently vague they [astrologers] were able to

[1] Though I was not then seeking a demarcation criterion, just these points are argued at length in my [1962], pp. 10–22 and 87–90.

[2] Cf. Popper [1963], pp. 192–200, with my [1962], pp. 143–58. [3] Popper [1963], p. 34.

[4] The index to Popper [1963] has eight entries under the heading 'astrology as a typical pseudo science'.

explain away anything that might have been a refutation of the theory had the theory and the prophecies been more precise. In order to escape falsification they destroyed the testability of the theory.'[1] Those generalizations catch something of the spirit of the astrological enterprise. But taken at all literally, as they must be if they are to provide a demarcation criterion, they are impossible to support. The history of astrology during the centuries when it was intellectually reputable records many predictions that categorically failed.[2] Not even astrology's most convinced and vehement exponents doubted the recurrence of such failures. Astrology cannot be barred from the sciences because of the form in which its predictions were cast.

Nor can it be barred because of the way its practitioners explained failure. Astrologers pointed out, for example, that, unlike general predictions about, say, an individual's propensities or a natural calamity, the forecast of an individual's future was an immensely complex task, demanding the utmost skill, and extremely sensitive to minor errors in relevant data. The configuration of the stars and eight planets was constantly changing; the astronomical tables used to compute the configuration at an individual's birth were notoriously imperfect; few men knew the instant of their birth with the requisite precision.[3] No wonder, then, that forecasts often failed. Only after astrology itself became implausible did these arguments come to seem question-begging.[4] Similar arguments are regularly used today when explaining, for example, failures in medicine or meteorology. In times of trouble they are also deployed in the exact sciences, fields like physics, chemistry, and astronomy.[5] There was nothing unscientific about the astrologer's explanation of failure.

Nevertheless, astrology was not a science. Instead it was a craft, one of the practical arts, with close resemblances to engineering, meteorology, and medicine as these fields were practised until little more than a century ago. The parallels to an older medicine and to contemporary psychoanalysis are, I think, particularly close. In each of these fields shared theory was adequate only to establish the plausibility of the discipline and to provide a rationale for the various craft-rules which governed practice. These rules had proved their use in the past, but no practitioner supposed they were sufficient to prevent recurrent failure. A more articulated theory and more powerful rules were desired, but it would have been absurd to

[1] Popper [1963], p. 37.

[2] For examples see, Thorndike [1923–58], **5**, pp. 225 ff.; **6**, pp. 71, 101, 114.

[3] For reiterated explanations of failure see, *ibid.* **1**, pp. 11 and 514 f.; **4**, 368; **5**, 279.

[4] A perceptive account of some reasons for astrology's loss of plausibility is included in Stahlman [1956]. For an explanation of astrology's previous appeal see, Thorndike [1955].

[5] Cf. my [1962], pp. 66–76.

abandon a plausible and badly needed discipline with a tradition of limited success simply because these desiderata were not yet at hand. In their absence, however, neither the astrologer nor the doctor could do research. Though they had rules to apply, they had no puzzles to solve and therefore no science to practise.[1]

Compare the situations of the astronomer and the astrologer. If an astronomer's prediction failed and his calculations checked, he could hope to set the situation right. Perhaps the data were at fault: old observations could be re-examined and new measurements made, tasks which posed a host of calculational and instrumental puzzles. Or perhaps theory needed adjustment, either by the manipulation of epicycles, eccentrics, equants, etc., or by more fundamental reforms of astronomical technique. For more than a millennium these were the theoretical and mathematical puzzles around which, together with their instrumental counterparts, the astronomical research tradition was constituted. The astrologer, by contrast, had no such puzzles. The occurrence of failures could be explained, but particular failures did not give rise to research puzzles, for no man, however skilled, could make use of them in a constructive attempt to revise the astrological tradition. There were too many possible sources of difficulty, most of them beyond the astrologer's knowledge, control, or responsibility. Individual failures were correspondingly uninformative, and they did not reflect on the competence of the prognosticator in the eyes of his professional compeers.[2] Though astronomy and astrology were regularly practised by the same people, including Ptolemy, Kepler, and Tycho Brahe, there was never an astrological equivalent of the puzzle-solving astronomical tradition. And without puzzles, able first to challenge and then to attest the ingenuity of the individual practitioner, astrology could

[1] This formulation suggests that Sir Karl's criterion of demarcation might be saved by a minor restatement entirely in keeping with his apparent intent. For a field to be a science its conclusions must be *logically derivable* from *shared premises*. On this view astrology is to be barred not because its forecasts were not testable but because only the most general and least testable ones could be derived from accepted theory. Since any field that did satisfy this condition *might* support a puzzle solving tradition, the suggestion is clearly helpful. It comes close to supplying a sufficient condition for a field's being a science. But in this form, at least, it is not even quite a sufficient condition, and it is surely not a necessary one. It would, for example, admit surveying and navigation as sciences, and it would bar taxonomy, historical geology, and the theory of evolution. The conclusions of a science may be both precise and binding without being fully derivable by logic from accepted premises. Cf. my [1962], pp. 35–51, and also the discussion in Section III, *below*.

[2] This is not to suggest that astrologers did not criticize each other. On the contrary, like practitioners of philosophy and some social sciences, they belonged to a variety of different schools, and the inter-school strife was sometimes bitter. But these debates ordinarily revolved about the *implausibility* of the particular theory employed by one or another school. Failures of individual predictions played very little role. Compare Thorndike [1923–58], 5, p. 233.

not have become a science even if the stars had, in fact, controlled human destiny.

In short, though astrologers made testable predictions and recognized that these predictions sometimes failed, they did not and could not engage in the sorts of activities that normally characterize all recognized sciences. Sir Karl is right to exclude astrology from the sciences, but his over-concentration on science's occasional revolutions prevents his seeing the surest reason for doing so.

That fact, in turn, may explain another oddity of Sir Karl's historiography. Though he repeatedly underlines the role of tests in the replacement of scientific theories, he is also constrained to recognize that many theories, for example the Ptolemaic, were replaced before they had in fact been tested.[1] On some occasions, at least, tests are not requisite to the revolutions through which science advances. But that is not true of puzzles. Though the theories Sir Karl cites had not been put to the test before their displacement, none of these was replaced before it had ceased adequately to support a puzzle-solving tradition. The state of astronomy was a scandal in the early sixteenth century. Most astronomers nevertheless felt that normal adjustments of a basically Ptolemaic model would set the situation right. In this sense the theory had not failed a test. But a few astronomers, Copernicus among them, felt that the difficulties must lie in the Ptolemaic approach itself rather than in the particular versions of Ptolemaic theory so far developed, and the results of that conviction are already recorded. The situation is typical.[2] With or without tests, a puzzle-solving tradition can prepare the way for its own displacement. To rely on testing as the mark of a science is to miss what scientists mostly do and, with it, the most characteristic feature of their enterprise.

II

With the background supplied by the preceding remarks we can quickly discover the occasion and consequences of another of Sir Karl's favourite locutions. The preface to *Conjectures and Refutations* opens with the sentence: 'The essays and lectures of which this book is composed, are variations upon one very simple theme—the thesis that *we can learn from our mistakes*.' The emphasis is Sir Karl's; the thesis recurs in his writing from an early date[3]; taken in isolation, it inevitably commands assent. Everyone

[1] Cf. Popper [1963], p. 246.

[2] Cf. my [1962], pp. 77–87.

[3] The quotation is from Popper [1963], p. vii, in a preface dated 1962. Earlier Sir Karl had equated 'learning from our mistakes' with 'learning by trial and error' ([1963], p. 216), and the trial-and-error formulation dates from at least 1937 ([1963], p. 312) and is in spirit older than that. Much of what is said below about Sir Karl's notion of 'mistake' applies equally to his concept of 'error'.

can and does learn from his mistakes; isolating and correcting them is an essential technique in teaching children. Sir Karl's rhetoric has roots in everyday experience. Nevertheless, in the contexts for which he invokes this familiar imperative, its applications seems decisively askew. I am not sure a mistake has been made, at least not a mistake to learn from.

One need not confront the deeper philosophical problems presented by mistakes to see what is presently at issue. It is a mistake to add three plus three and get five, or to conclude from 'All men are mortal' to 'All mortals are men'. For different reasons, it is a mistake to say, 'He is my sister', or to report the presence of a strong electric field when test charges fail to indicate it. Presumably there are still other sorts of mistakes, but all the normal ones are likely to share the following characteristics. A mistake is made, or is committed, at a specifiable time and place by a particular individual. That individual has failed to obey some established rule of logic, or of language, or of the relations between one of these and experience. Or he may instead have failed to recognize the consequences of a particular choice among the alternatives which the rules allow him. The individual can learn from his mistake only because the group whose practice embodies these rules can isolate the individual's failure in applying them. In short, the sorts of mistakes to which Sir Karl's imperative most obviously applies are in individual's failure of understanding or of recognition within an activity governed by pre-established rules. In the sciences such mistakes occur most frequently and perhaps exclusively within the practice of normal puzzle-solving research.

That is not, however, where Sir Karl seeks them, for his concept of science obscures even the existence of normal research. Instead, he looks to the extraordinary or revolutionary episodes in scientific development. The mistakes to which he points are not usually acts at all but rather out-of-date scientific theories: Ptolemaic astronomy, the phlogiston theory, or Newtonian dynamics, and 'learning from our mistakes' is, correspondingly, what occurs when a scientific community rejects one of these theories and replaces it with another.[1] If this does not immediately seem an odd usage,

[1] Popper [1963], pp. 215 and 220. In these pages Sir Karl outlines and illustrates his thesis that science grows through revolutions. He does not, in the process, ever juxtapose the term 'mistake' with the name of an out-of-date scientific theory, presumably because his sound historic instinct inhibits so gross an anachronism. Yet the anachronism is fundamental to Sir Karl's rhetoric, which does repeatedly provide clues to more substantial differences between us. Unless out-of-date theories are mistakes, there is no way to reconcile, say, the opening paragraph of Sir Karl's preface ([1963], p. vii: 'learn from our mistakes'; 'our often mistaken attempts to solve our problems'; 'tests which may help us in the discovery of our mistakes') with the view ([1963], p. 215) that 'the growth of scientific knowledge . . . [consists in] the repeated overthrow of scientific theories and their replacement by better or more satisfactory ones'.

that is mainly because it appeals to the residual inductivist in us all. Believing that valid theories are the product of correct inductions from facts, the inductivist must also hold that a false theory is the result of a mistake in induction. In principle, at least, he is prepared to answer the questions: what mistake was made, what rule broken, when and by whom, in arriving at, say, the Ptolemaic system? To the man for whom those are sensible questions and to him alone, Sir Karl's locution presents no problems.

But neither Sir Karl nor I is an inductivist. We do not believe that there are rules for inducing correct theories from facts, or even that theories, correct or incorrect, are induced at all. Instead we view them as imaginative posits, invented in one piece for application to nature. And though we point out that such posits can and usually do at last encounter puzzles they cannot solve, we also recognize that those troublesome confrontations rarely occur for some time after a theory has been both invented and accepted. In our view, then, no mistake was made in arriving at the Ptolemaic system, and it is therefore difficult for me to understand what Sir Karl has in mind when he calls that system, or any other out-of-date theory, a mistake. At most one may wish to say that a theory which was not previously a mistake has become one or that a scientist has made the mistake of clinging to a theory for too long. And even these locutions, of which at least the first is extremely awkward, do not return us to the sense of mistake with which we are most familiar. Those mistakes are the normal ones which a Ptolemaic (or a Copernican) astronomer makes within his system, perhaps in observation, calculation, or the analysis of data. They are, that is, the sort of mistake which can be isolated and then at once corrected, leaving the original system intact. In Sir Karl's sense, on the other hand, a mistake infects an entire system and can be corrected only by replacing the system as a whole. No locutions and no similarities can disguise these fundamental differences, nor can it hide the fact that before infection set in the system had the full integrity of what we now call sound knowledge.

Quite possibly Sir Karl's sense of 'mistake' can be salvaged, but a successful salvage operation must deprive it of certain still current implications. Like the term 'testing', 'mistake' has been borrowed from normal science, where its use is reasonably clear, and applied to revolutionary episodes, where its application is at best problematic. That transfer creates, or at least reinforces, the prevalent impression that whole theories can be judged by the same sort of criteria that one employs when judging a theory's individual research applications. The discovery of applicable criteria then becomes a primary desideratum for many people. That Sir

Karl should be among them is strange, for the search runs counter to the most original and fruitful thrust in his philosophy of science. But I can understand his methodological writings since the *Logik der Forschung* in no other way. I shall now suggest that he has, despite explicit disclaimers, consistently sought evaluation procedures which can be applied to theories with the apodictic assurance characteristic of the techniques by which one identifies mistakes in arithmetic, logic, or measurement. I fear that he is pursuing a will-o'-the-wisp born from the same conjunction of normal and extraordinary science which made tests seem so fundamental a feature of the sciences.

III

In his *Logik der Forschung*, Sir Karl underlined the asymmetry of a generalization and its negation in their relation to empirical evidence. A scientific theory cannot be shown to apply successfully to all its possible instances, but it can be shown to be unsuccessful in particular applications. Emphasis upon that logical truism and its implications seems to me a forward step from which there must be no retreat. The same asymmetry plays a fundamental role in my *Structure of Scientific Revolutions*, where a theory's failure to provide rules that identify solvable puzzles is viewed as the source of professional crises which often result in the theory's being replaced. My point is very close to Sir Karl's, and I may well have taken it from what I had heard of his work.

But Sir Karl describes as 'falsification' or 'refutation' what happens when a theory fails in an attempted application, and these are the first of a series of related locutions that again strike me as extremely odd. Both 'falsification' and 'refutation' are antonyms of 'proof'. They are drawn principally from logic and from formal mathematics; the chains of argument to which they apply end with a 'Q.E.D.'; invoking these terms implies the ability to compel assent from any member of the relevant professional community. No member of this audience, however, still needs to be told that, where a whole theory or often even a scientific law is at stake, arguments are seldom so apodictic. All experiments can be challenged, either as to their relevance or their accuracy. All theories can be modified by a variety of *ad hoc* adjustments without ceasing to be, in their main lines, the same theories. It is important, furthermore, that this should be so, for it is often by challenging observations or adjusting theories that scientific knowledge grows. Challenges and adjustments are a standard part of normal research in empirical science, and adjustments, at least, play a dominant role in informal mathematics as well. Dr Lakatos's brilliant analysis of the permissible rejoinders to mathematical refutations provides the most telling arguments I know against a naive falsificationist position.[1]

Sir Karl is not, of course, a naive falsificationist. He knows all that has just been said and has emphasized it from the beginning of his career. Very early in his *Logic of Scientific Discovery*, for example, he writes: 'In point of fact, no conclusive disproof of a theory can ever be produced; for it is always possible to say that the experimental results are not reliable or that the discrepancies which are asserted to exist between the experimental results and the theory are only apparent and that they will disappear with the advance of our understanding.'[2] Statements like these display one more parallel between Sir Karl's view of science and my own, but what we make of them could scarcely be more different. For my view they are fundamental, both as evidence and as source. For Sir Karl's, in contrast, they are an essential qualification which threatens the integrity of his basic position. Having barred conclusive disproof, he has provided no substitute for it, and the relation he does employ remains that of logical falsification. Though he is not a naive falsificationist, Sir Karl may, I suggest, legitimately be treated as one.

If his concern were exclusively with demarcation, the problems posed by the unavailability of conclusive disproofs would be less severe and perhaps eliminable. Demarcation might, that is, be achieved by an exclusively syntactic criterion.[3] Sir Karl's view would then be, and perhaps is, that a theory is scientific if and only if *observation statements*—particularly the negations of singular existential statements—can be logically deduced from it, perhaps in conjunction with stated background knowledge. The difficulties (to which I shall shortly turn) in deciding whether the outcome of a particular laboratory operation justifies asserting a particular observation statement would then be irrelevant. Perhaps, though the basis for doing so is less apparent, the equally grave difficulties in deciding whether an observation statement deduced from an approximate (e.g. mathematically manageable) version of the theory should be considered consequences of the theory itself could be eliminated in the same way. Problems like these would belong not to the syntactics but to the pragmatics or semantics of the language in which the theory was cast, and they would therefore have no role in determining its status as a science. To be scientific a theory need be falsifiable only by an observation statement not by actual observation. The relation between statements, unlike that between

[1] Lakatos [1963–4]. [2] Popper [1959], p. 50.

[3] Though my point is somewhat different, I owe my recognition of the need to confront this issue to C. G. Hempel's strictures on those who misinterpret Sir Karl by attributing to him a belief in absolute rather than relative falsification. See his [1965], p. 45. I am also indebted to Professor Hempel for a close and perceptive critique of this paper in draft.

a statement and an observation, could be the conclusive disproof familiar from logic and mathematics.

For reasons suggested above (p. 9, footnote 1) and elaborated immediately below, I doubt that scientific theories can without decisive change be cast in a form which permits the purely syntactic judgements which this version of Sir Karl's criterion requires. But even if they could, these reconstructed theories would provide a basis only for his demarcation criterion, not for the logic of knowledge so closely associated with it. The latter has, however, been Sir Karl's most persistent concern, and his notion of it is quite precise. 'The logic of knowledge . . . ,' he writes, 'consists solely in investigating the methods employed in those systematic tests to which every new idea must be subjected if it is to be seriously entertained.'[1] From this investigation, he continues, result methodological rules or conventions like the following: 'Once a hypothesis has been proposed and tested, and has proved its mettle, it may not be allowed to drop out without "good reason". A "good reason" may be, for instance . . . the falsification of one of the consequences of the hypothesis.'[2]

Rules like these, and with them the entire logical enterprise described above, are no longer simply syntactic in their import. They require that both the epistemological investigator and the research scientist be able to relate sentences derived from a theory not to other sentences but to actual observations and experiments. This is the context in which Sir Karl's term 'falsification' must function, and Sir Karl is entirely silent about how it can do so. What is falsification if it is not conclusive disproof? Under what circumstances does the *logic* of knowledge require a scientist to abandon a previously accepted theory when confronted, not with statements about experiments, but with experiments themselves? Pending clarification of these questions, I am not clear that what Sir Karl has given us is a logic of knowledge at all. In my conclusion I shall suggest that, though equally valuable, it is something else entirely. Rather than a logic, Sir Karl has provided an ideology; rather than methodological rules, he has supplied procedural maxims.

That conclusion must, however, be postponed until after a last deeper look at the source of the difficulties with Sir Karl's notion of falsification. It presupposes, as I have already suggested, that a theory is cast, or can without distortion be recast, in a form which permits scientists to classify each conceivable event as either a confirming instance, a falsifying instance, or irrelevant to the theory. That is obviously required if a general law is to be falsifiable: to test the generalization $(x)\ \phi\ (x)$ by applying it to the constant a, we must be able to tell whether or not a lies within the

[1] Popper [1959], p. 31.

[2] Popper [1959], pp. 53 f.

range of the variable x and whether or not ϕ (a). The same presupposition is even more apparent in Sir Karl's recently elaborated measure of verisimilitude. It requires that we first produce the class of all logical consequences of the theory and then choose from among these, with the aid of background knowledge, the classes of all true and of all false consequences.[1] At least, we must do this if the criterion of verisimilitude is to result in a *method* of theory choice. None of these tasks can, however, be accomplished unless the theory is fully articulated logically and unless the terms through which it attaches to nature are sufficiently defined to determine their applicability in each possible case. In practice, however, no scientific theory satisfies these rigorous demands, and many people have argued that a theory would cease to be useful in research if it did so.[2] I have myself elsewhere introduced the term 'paradigm' to underscore the dependence of scientific research upon concrete examples that bridge what would otherwise be gaps in the specification of the content and application of scientific theories. The relevant arguments cannot be repeated here. But a brief example, though it will temporarily alter my mode of discourse, may be even more useful.

My example takes the form of a constructed epitome of some elementary scientific knowledge. That knowledge concerns swans, and to isolate its presently relevant characteristics I shall ask three questions about it: (*a*) How much can one know about swans without introducing explicit generalizations like 'All swans are white'? (*b*) Under what circumstances and with what consequences are such generalizations worth adding to what was known without them? (*c*) Under what circumstances are generalizations rejected once they have been made? In raising these questions my object is to suggest that, though logic is a powerful and ultimately an essential tool of scientific enquiry, one can have sound knowledge in forms to which logic can scarcely be applied. Simultaneously, I shall suggest that logical articulation is not a value for its own sake, but is to be undertaken only when and to the extent that circumstances demand it.

Imagine that you have been shown and can remember ten birds which have authoritatively been identified as swans; that you have a similar acquaintance with ducks, geese, pigeons, doves, gulls, etc.; and that you are informed that each of these types constitutes a natural family. A natural family you already know as an observed cluster of like objects,

[1] Popper [1963], pp. 233–5. Notice also, at the foot of the last of these pages, that Sir Karl's comparison of the relative verisimilitude of two theories depends upon there being 'no revolutionary changes in our background knowledge', an assumption which he nowhere argues and which is hard to reconcile with his conception of scientific change by revolutions.

[2] Braithwaite [1953], pp. 50–87, especially p. 76, and my [1962], pp. 97–101.

sufficiently important and sufficiently discrete to command a generic name. More precisely, though here I introduce more simplification than the concept requires, a natural family is a class whose members resemble each other more closely than they resemble the members of other natural families.[1] The experience of generations has to date confirmed that all observed objects fall into one or another natural family. It has, that is, shown that the entire population of the world can always be divided (though not once and for all) into perceptually discontinuous categories. In the perceptual spaces between these categories there are believed to be no objects at all.

What you have learned about swans from exposure to paradigms is very much like what children first learn about dogs and cats, tables and chairs, mothers and fathers. Its precise scope and content are, of course, impossible to specify, but it is sound knowledge nonetheless. Derived from observation, it can be infirmed by further observation, and it meanwhile provides a basis for rational action. Seeing a bird much like the swans you already know, you may reasonably presume that it will require the same food as the others and will breed with them. Provided swans are a natural family, no bird which closely resembles them on sight should display radically different characteristics on closer acquaintance. Of course you may have been misinformed about the natural integrity of the swan family. But that can be discovered from experience, for example, by the discovery of a number of animals (note that more than one is required) whose characteristics bridge the gap between swans and, say, geese by barely perceptible intervals.[2] Until that does occur, however, you will know a great deal about swans though you will not be altogether sure what you know or what a swan is.

Suppose now that all the swans you have actually observed are white. Should you embrace the generalization, 'All swans are white'? Doing so will change what you know very little; that change will be of use only in the unlikely event that you meet a non-white bird which otherwise resembles a swan; by making the change you increase the risk that the swan

[1] Note that the resemblance between members of a natural family is here a learned relationship and one which can be unlearned. Contemplate the old saw, 'To an occidental, all chinamen look alike'. That example also highlights the most drastic of the simplifications introduced at this point. A fuller discussion would have to allow for hierarchies of natural families with resemblance relations between families at the higher levels.

[2] This experience would not necessitate the abandonment of either the category 'swans' or the category 'geese', but it would necessitate the introduction of an *arbitrary* boundary between them. The families 'swans' and 'geese' would no longer be natural families, and you could conclude nothing about the character of a new swan-like bird that was not also true of geese. Empty perceptual space is essential if family membership is to have cognitive content.

family will prove not to be a natural family after all. Under those circumstances you are likely to refrain from generalizing unless there are special reasons for doing so. Perhaps, for example, you must describe swans to men who cannot be directly exposed to paradigms. Without superhuman caution both on your part and on that of your readers, your description will acquire the force of a generalization; this is often the problem of the taxonomist. Or perhaps you have discovered some grey birds that look otherwise like swans but eat different food and have an unfortunate disposition. You may then generalize to avoid a behavioural mistake. Or you may have a more theoretical reason for thinking the generalization worthwhile. For example, you may have observed that the members of other natural families share colouration. Specifying this fact in a form which permits the application of powerful logical techniques to what you know may enable you to learn more about the animal colour in general or about animal breeding.

Now, having made the generalization, what will you do if you encounter a black bird that looks otherwise like a swan? Almost the same things, I suggest, as if you had not previously committed yourself to the generalization at all. You will examine the bird with care, externally and perhaps internally as well, to find other characteristics that distinguish this specimen from your paradigms. That examination will be particularly long and thorough if you have theoretical reasons for believing that colour characterizes natural families or if you are deeply ego involved with the generalization. Very likely the examination will disclose other differentiae, and you will announce the discovery of a new natural family. Or you may fail to find such differentiae and may then announce that a black swan has been found. Observation cannot, however, force you to that falsifying conclusion, and you would occasionally be the loser if it could do so. Theoretical considerations may suggest that colour alone is sufficient to demarcate a natural family: the bird is not a swan because it is black. Or you may simply postpone the issue pending the discovery and examination of other specimens. Only if you have previously committed yourself to a full definition of 'swan', one which will specify its applicability to every conceivable object, can you be logically *forced* to rescind your generalization.[1] And why should you have offered such a definition? It could serve no cognitive function and would

[1] Further evidence for the unnaturalness of any such definition is provided by the following question. Should 'whiteness' be included as a defining characteristic of swans? If so, the generalization 'All swans are white' is immune to experience. But if 'whiteness' is excluded from the definition, then some other characteristic must be included for which 'whiteness' might have substituted. Decisions about which characteristics are to be parts of a definition and which are to be available for the statement of general laws are often arbitrary and, in practice, are seldom made. Knowledge is not usually articulated in that way.

expose you to tremendous risks.[1] Risks, of course, are often worth taking, but to say more than one knows solely for the sake of risk is foolhardy.

I suggest that scientific knowledge, though logically more articulate and far more complex, is of this sort. The books and teachers from whom it is acquired present concrete examples together with a multitude of theoretical generalizations. Both are essential carriers of knowledge, and it is therefore Pickwickian to seek a methodological criterion that supposes the scientist can specify in advance whether each imaginable instance fits or would falsify his theory. The criteria at his disposal, explicit and implicit, are sufficient to answer that question only for the cases that clearly do fit or that are clearly irrelevant. These are the cases he expects, the ones for which his knowledge was designed. Confronted with the unexpected, he must always do more research in order further to articulate his theory in the area that has just become problematic. He may then reject it in favour of another and for good reason. But no exclusively logical criteria can entirely dictate the conclusion he must draw.

IV

Almost everything said so far rings changes on a single theme. The criteria with which scientists determine the validity of an articulation or an application of existing theory are not by themselves sufficient to determine the choice between competing theories. Sir Karl has erred by transferring selected characteristics of everyday research to the occasional revolutionary episodes in which scientific advance is most obvious and by thereafter ignoring the everyday enterprise entirely. In particular, he has sought to solve the problem of theory choice during revolutions by logical criteria that are applicable in full only when a theory can already be presupposed. That is the largest part of my thesis in this paper, and it could be the entire thesis if I were content to leave altogether open the questions that have been raised. How do the scientists make the choice between competing theories? How are we to understand the way in which science does progress?

Let me at once be clear that having opened that Pandora's box, I shall close it quickly. There is too much about these questions that I do not understand and must not pretend to. But I believe I see the directions in which answers to them must be sought, and I shall conclude with an attempt briefly to mark the trail. Near its end we shall once more encounter a set of Sir Karl's characteristic locutions.

[1] This incompleteness of definitions is often called 'open texture' or 'vagueness of meaning', but those phrases seem decisively askew. Perhaps the definitions are incomplete, but nothing is wrong with the meanings. That is the way meanings behave!

I must first ask what it is that still requires explanation. Not that scientists discover the truth about nature, nor that they approach ever closer to the truth. Unless, as one of my critics suggests,[1] we simply define the approach to truth as the result of what scientists do, we cannot recognize progress towards that goal. Rather we must explain why science—our surest example of sound knowledge—progresses as it does, and we must first find out how, in fact, it does progress.

Surprisingly little is yet known about the answer to that descriptive question. A vast amount of thoughtful empirical investigation is still required. With the passage of time, scientific theories taken as a group are obviously more and more articulated. In the process, they are matched to nature at an increasing number of points and with increasing precision. Or again, the number of subject matters to which the puzzle-solving approach can be applied clearly grows with time. There is a continuing proliferation of scientific specialities, partly by an extension of the boundaries of science and partly by the subdivision of existing fields.

Those generalizations are, however, only a beginning. We know, for example, almost nothing about what a group of scientists will sacrifice in order to achieve the gains that a new theory invariably offers. My own impression, though it is no more than that, is that a scientific community will seldom or never embrace a new theory unless it solves all or almost all the quantitative, numerical puzzles that have been treated by its predecessor.[2] They will, on the other hand, occasionally sacrifice explanatory power, however reluctantly, sometimes leaving previously resolved questions open and sometimes declaring them altogether unscientific.[3] Turning to another area, we know little about historical changes in the unity of the sciences. Despite occasional spectacular successes, communication across the boundaries between scientific specialties becomes worse and worse. Does the number of incompatible viewpoints employed by the increasing number of communities of specialists grow with time? Unity of the sciences is clearly a value for scientists, but for what will they give it up? Or again, though the bulk of scientific knowledge clearly increases with time, what are we to say about ignorance? The problems solved during the last thirty years did not exist as open questions a century ago. In any age, the scientific knowledge already at hand virtually exhausts what there is to know, leaving visible puzzles only at the horizon of existing knowledge. Is it not possible, or perhaps even likely, that contemporary scientists know less of what there is to know about their world than the scientists of the eighteenth century knew of theirs? Scientific theories, it must be remembered, attach

[1] Hawkins [1963].

[2] Cf. Kuhn [1958].

[3] Cf. Kuhn [1962], pp. 102–8.

to nature only here and there. Are the interstices between those points of attachment perhaps now larger and more numerous than ever before?

Until we can answer more questions like these, we shall not know quite what scientific progress is and cannot therefore quite hope to explain it. On the other hand, answers to those questions will very nearly provide the explanation sought. The two come almost together. Already it should be clear that the explanation must, in the final analysis, be psychological or sociological. It must, that is, be a description of a value system, an ideology, together with an analysis of the institutions through which that system is transmitted and enforced. Knowing what scientists value, we may hope to understand what problems they will undertake and what choices they will make in particular circumstances of conflict. I doubt that there is another sort of answer to be found.

What form that answer will take is, of course, another matter. At this point, too, my sense that I control my subject matter ends. But again, some sample generalizations will illustrate the sorts of answers which must be sought. For a scientist, the solution of a difficult conceptual or instrumental puzzle is a principal goal. His success in that endeavour is rewarded through recognition by other members of his professional group and by them alone. The practical merit of his solution is at best a secondary value, and the approval of men outside the specialist group is a negative value or none at all. These values, which do much to dictate the form of normal science, are also significant at times when a choice must be made between theories. A man trained as a puzzle-solver will wish to preserve as many as possible of the prior puzzle-solutions obtained by his group, and he will also wish to maximize the number of puzzles that can be solved. But even these values frequently conflict, and there are others which make the problem of choice still more difficult. It is just in this connection that a study of what scientists will give up would be most significant. Simplicity, precision, and congruence with the theories used in other specialties are all significant value for the scientists, but they do not all dictate the same choice nor will they all be applied in the same way. That being the case, it is also important that group unanimity be a paramount value, causing the group to minimize the occasions for conflict and to reunite quickly about a single set of rules for puzzle solving even at the price of subdividing the specialty or excluding a formerly productive member.[1]

I do not suggest that these are the right answers to the problem of scientific progress, but only that they are the types of answers that must be sought. Can I hope that Sir Karl will join me in this view of the task still to be done? For some time I have assumed he would not, as a set of phrases

[1] Cf. my [1962], pp. 161–9.

that recurs in his work seems to bar the position to him. Again and again he has rejected 'the psychology of knowledge' or the 'subjective' and insisted that his concern was instead with the 'objective' or 'the logic of knowledge'.[1] The title of his most fundamental contribution to our field is *The* Logic *of Scientific Discovery*, and it is there that he most positively asserts that his concern is with the logical spurs to knowledge rather than with the psychological drives of individuals. Until very recently I have supposed that this view of the problem must bar the sort of solution I have advocated.

But now I am less certain, for there is another aspect of Sir Karl's work, not quite compatible with what precedes. When he rejects 'the psychology of knowledge', Sir Karl's explicit concern is only to deny the methodological relevance of an *individual's* source of inspiration or of an individual's sense of certainty. With that much I cannot disagree. It is, however, a long step from the rejection of the psychological idiosyncrasies of an individual to the rejection of the common elements induced by nurture and training in the psychological make-up of the licensed membership of a *scientific group*. One need not be dismissed with the other. And this, too, Sir Karl seems sometimes to recognize. Though he insists he is writing about the logic of knowledge, an essential role in his methodology is played by passages which I can only read as attempts to inculcate moral imperatives in the membership of the scientific group.

'Assume', Sir Karl writes, 'that we have deliberately made it our task to live in this unknown world of ours; to adjust ourselves to it as well as we can; . . . . and to explain it, *if* possible (we need not assume that it is) and as far as possible, with help of laws and explanatory theories. *If we have made this our task, then there is no more rational procedure than the method of . . . conjecture and refutation:* of boldly proposing theories; of trying our best to show that these are erroneous; and of accepting them tentatively if our critical efforts are unsuccessful.'[2] We shall not, I suggest, understand the success of science without understanding the full force of rhetorically induced and professionally shared imperatives like these. Institutionalized and articulated further (and also somewhat differently) such maxims and values may explain the outcome of choices that could not have been dictated by logic and experiment alone. The fact that passages like these occupy a prominent place in Sir Karl's writing is therefore further evidence of the resemblance of our views. That he does not, I think, ever see them for the social-psychological imperatives that they are is further evidence of the gestalt switch that still divides us deeply.

[1] Popper [1959], pp. 22 and 31 f., 46; and [1963], p. 52.

[2] Popper [1963], p. 51. Italics in original.

REFERENCES

Braithwaite [1953]: *Scientific Explanation*, 1953.
Guerlac [1961]: *Lavoisier—The Crucial Year*, 1961.
Hafner and Presswood [1965]: 'Strong Interference and Weak Interactions', *Science*, 149, pp. 503–10.
Hawkins [1963]: Review of Kuhn's 'Structure of Scientific Revolutions', *American Journal of Physics*, **31**.
Hempel [1965]: *Aspects of Scientific Explanation*, 1965.
Lakatos [1963–4]: 'Proofs and Refutations', *The British Journal for the Philosophy of Science*, **14**, pp. 1–25, 120–39, 221–43, 296–342.
Kuhn [1958]: 'The Role of Measurement in the Development of Physical Science', *Isis*, **49**, pp. 161–93.
Kuhn [1962]: *The Structure of Scientific Revolutions*, 1962.
Popper [1935]: *Logik der Forschung*, 1935.
Popper [1945]: *The Open Society and its Enemies*, 2 vols, 1945.
Popper [1957]: *The Poverty of Historicism*, 1957.
Popper [1959]: *Logic of Scientific Discovery*, 1959.
Popper [1963]: *Conjectures and Refutations*, 1963.
Stahlman [1956]: 'Astrology in Colonial America: An Extended Query', *William and Mary Quarterly*, **13**, pp. 551–63.
Thorndike [1923–58]: *A History of Magic and Experimental Science*, 8 vols, 1923–58.
Thorndike [1955]: 'The True Place of Astrology in the History of Science', *Isis*, **46**, pp. 273–8.

Against 'Normal Science'

JOHN WATKINS
London School of Economics

I

A few weeks ago I was asked to reply to Professor Kuhn this afternoon. Feyerabend and Lakatos were to have given the other papers; but the first could not come and the second found that, in arranging this colloquium, he had brought into existence a many-headed monster attending to whose multiplying demands would keep him busy approximately twenty-four hours a day.

This unexpected invitation made me very happy. Kuhn enjoys a unique position in the English speaking world as a philosophically-minded historian and historically-minded philosopher of science. I felt that it would be a privilege and a pleasure to reply to his paper.

For Kuhn, however, the programme change was not so agreeable. He had expected that Feyerabend and Lakatos would write independent papers so that his own would not need to be ready until this afternoon. Now he found that I was to reply to his paper, which rather suggested that I should see it beforehand. He responded herioically, rushing bits of his paper across the Atlantic as they left his typewriter. During much of last week I felt like a reader of a cliff-hanging serial, eagerly awaiting the next instalment. Thus my own paper has been written in a rush; and this has, I fear, aggravated my tendency to wave aside details and qualifications in trying to come to grips with someone's ideas.

In the turmoil of the last few days I have had one great stand-by. Kuhn's book, *The Structure of Scientific Revolutions*, is a famous book and one with which I am tolerably well acquainted. I was privileged to read it in manuscript in 1961 and to discuss it with its author. In 1963 it was discussed at length at Sir Karl Popper's seminar, where Mr Hattiangadi gave a paper on it (which he afterwards expanded into a very interesting dissertation). Later, I shall quote something which Popper said then; and I expect that my paper will contain some unconscious borrowings from our seminar discussions.

So my paper will be as much about Kuhn's book as about the paper he has just read. Fortunately, this is appropriate, since in his paper Kuhn has adopted a Sukarno-like policy of *confrontation* between the view of science propounded in his book, and Popper's view of science. I am glad

that he has done this. I remember suggesting to him in 1961 that he should bring out and discuss in his book the clash between his view of the scientific community as an essentially closed society, intermittently shaken by collective nervous breakdowns followed by restored mental unison, and Popper's view that the scientific community ought to be, and to a considerable degree actually is, an open society in which no theory, however dominant and successful, no 'paradigm' to use Kuhn's term, is ever sacred. Kuhn did not follow this suggestion at the time, but he has surely made the *amende honorable* this afternoon.

Yet two things leave me a little discontented with the way in which he has arranged the confrontation. For one thing, as presented by him, it is by no means as dramatic as it might be. Near the beginning he says: 'On almost all the occasions when we turn explicitly to the same problems, Sir Karl's view of science and my own are very nearly identical'.[1] My aim will be to bring out the larger *conflicts* between these two views. At this stage I will just cite one remark in Kuhn's paper which, as it were, incapsulates the main conflict in a sentence: 'it is precisely the abandonment of critical discourse that marks the transition to a science.'[2]

The second source of my discontent is different. A Sukarno-style confrontation involves, not only a major ideological clash, but also a good deal of local skirmishing. I hope Kuhn will forgive me if I confine most of my counter-skirmishing to a footnote.[3] In my text I shall concentrate upon his idea—it is an original and challenging idea—of Normal Science. There will be a certain conscious unfairness, or at least one-sidedness, in my dis-

[1] *This volume*, p. 1.

[2] *This volume*, p. 6.

[3] Kuhn's method is to pick out a few 'characteristic locutions', and to erect on these some construction at which he can nag away. But his constructions sometimes bear a rather faint resemblance to what was said in the books from which the locutions were picked. (Kuhn himself sometimes admits that a construction of his does not quite fit. Thus on p. 14 he writes: 'Though he is not a naive falsificationist, Sir Karl may, I suggest, legitimately be treated as one.') For instance, Kuhn ponders with much head-shaking the 'locution' that 'we can learn from our mistakes'. He seems unable to allow that Popper was using the word 'mistake' in a cheerfully guilt-free sense with no suggestion of personal failure, rule-transgression, etc. The physicist J. E. Wheeler was using the word in a Popperian spirit when he wrote: 'Our whole problem is to make the mistakes as fast as possible' (Wheeler [1956], p. 360).

Since Kuhn's main target was Popper's demarcation criterion, and since Popper has stated this pretty sharply, one might have expected that here, at least, Kuhn would have given chapter and verse. But no, he prefers once more to moot a construction of his own: 'Demarcation might . . . be achieved by an exclusively syntactic criterion. Sir Karl's view would then be, and perhaps is, that a theory is scientific if and only if observation statements —particularly the negations of singular existential statements—can be logically deduced from it . . .' (p. 14). If one consults Popper's [1934], section 21, one finds that this is full of mistakes (in Kuhn's sense).

cussion of this idea. I believe that it is of considerable sociological importance. A sociologist investigating the scientific profession as he might investigate, say, the medical profession, might do well to use it as his ideal type. But I shall consider it from a methodological point of view, and methodology, as I understand it, is concerned with science at its best, or with science as it should be conducted, rather than with hack science.

My programme will be this. I shall begin, in section II, by confronting Kuhn's account of Normal Science with the sort of appraisal which Popper would make of a scientific situation which lived up to—or down to—Kuhn's idea of Normal Science. Then, in section III, I shall ask why Kuhn should claim that Normal Science, as opposed to what he calls Extraordinary Science, constitutes the essence of science. Lastly, in section IV, I shall ask whether Normal Science could be as Kuhn describes it and yet give rise to Extraordinary Science. My answer will be, 'No'; and I will suggest that this answer happily rebuts Kuhn's view of scientific normalcy as a closed society of closed minds.

II

In considering Kuhn's idea of Normal Science from a Popperian point of view, it is natural that I should concentrate upon what Kuhn says about *testing* within Normal Science. Tests, he says, are being conducted all the time, but 'these tests are of a peculiar sort, for in the final analysis it is the individual scientist rather than current theory which is tested'.[1] His idea is this. So-called 'testing' in Normal Science is *not* testing of theories. Rather, it is part of puzzle-solving activity. Normal Science is governed by some paradigm (or dominant theory). The paradigm is trusted implicitly; but it will not fit experimental findings quite perfectly. There will always be apparent discrepancies or anomalies. Normal Research largely consists of resolving these anomalies by making suitable adjustments which leave the paradigm intact. The paradigm is taken as guaranteeing the existence of a solution to every puzzle generated by apparent discrepancies between it and observations. Hence, although the 'tests' carried out within Normal Science may *look* like tests of the prevailing theory if viewed through Popperian spectacles, they are really tests of something else, namely the experimenter's puzzle-solving skill. If the outcome of such a 'test' is negative, it does not hit the theory but backfires on the experimenter. *His* prestige may be lowered by the failure of his attempt to solve a puzzle; but the prestige of the paradigm within whose framework he makes the attempt is so high that it will scarcely be affected by any such little local difficulties.

[1] *This volume*, p. 5.

According to Kuhn it is only at a time of what he calls Extraordinary Science, when the prevailing theory itself is under attack, that something like genuine testing of theories may occur. *Then* a negative outcome of a test may be regarded, not as the personal failure of the experimenter, but as a failure of the theory. In Kuhn's words, 'A failure that had previously been personal may then come to seem the failure of a theory under test'.[1]

For Kuhn, Normal Science is, as the name suggests, the normal condition of science; Extraordinary Science is an abnormal condition; and within Normal Science, to repeat, the genuine testing of prevailing theories is rendered, in some rather mysterious psychological-cum-sociological way, impossible. (One can now see how Kuhn could be startled by a remark which he at the same time regards as 'virtually a cliché',[2] namely, Popper's remark that scientists put forwards statements and test them step by step. For Kuhn it *is* virtually a cliché to say that scientists normally engage in a lot of testing: they test their solutions to anomaly-generated puzzles; and it is, for him, startlingly incorrect to say that it is normal for scientists to test *theories*.)

That it is desirable that a theory should be defended with a certain dogmatism, so that it is not knocked out too quickly before its resources have been explored, Popper has never denied; but such dogmatism is healthy only so long as there are other people around who are not inhibited from criticizing and testing a tenaciously defended theory. If *everyone* were under some mysterious compulsion to preserve the current theories of science against awkward results, then those theories would, according to Popper, lose their scientific status and degenerate into something like metaphysical doctrines.

Thus we have the following clash: the condition which Kuhn regards as the normal and proper condition of science is a condition which, if it actually obtained, Popper would regard as *un*scientific, a state of affairs in which critical science had contracted into defensive metaphysics. Popper has suggested that the motto of science should be: *Revolution in permanence!* For Kuhn, it seems, a more appropriate maxim would be: *Not nostrums but normalcy!*

In his paper today Kuhn spoke of Popper's emphasis on the asymmetry between the falsifiability and the non-verifiability of scientific generalizations as 'a forward step from which there must be no retreat'.[3] He added that the 'same asymmetry plays a fundamental role in my *Structure of Scientific Revolutions* . . . I may well have taken it from what I had heard of his work.' But Kuhn's memory seems to have played a trick on him

[1] *This volume*, p. 7.
[2] *This volume*, p. 4.
[3] *This volume*, p. 13.

here: in his book he had referred explicitly to Popper's thesis that there is no verification and that falsification is what matters,[1] and he did so in order to *dismiss* that thesis as unrealistic, on the ground that in Normal Science there is no falsification of theories, while in Extraordinary Science the evidence which is taken as falsifying the paradigm which is being ushered out will *also* be taken as *verifying* the new paradigm which is being ushered in.[2]

In his *Structure of Scientific Revolutions* Kuhn did not advance any demarcation-criterion for science; he only set aside Popper's falsifiability-criterion. Now he advances an alternative criterion of his own:

> Finally, and this is for now my main point, a careful look at the scientific enterprise suggests that it is normal science, in which Sir Karl's sort of testing does not occur, rather than extraordinary science which most nearly distinguishes science from other enterprises. If a demarcation criterion exists (we must not, I think seek a sharp or decisive one), it may lie just in that part of science which Sir Karl ignores.[3]

That was cautiously worded. But on the next page Kuhn was bolder: 'of the two criteria, testing and puzzle-solving, the latter is at once the less equivocal and the more fundamental'.[4] And I will throw any remaining caution of Kuhn's to the winds and re-state his suggestion in an unguarded way: Normal Science (in which there is not really any testing of theories), is genuine science; Extraordinary Science (in which genuine testing of theories does occur) is so abnormal, so different from genuine science, that it can hardly be called science at all. Kuhn explains that it is because puzzle-solving is easily mistaken for testing that 'Sir Karl's line of demarcation and my own so frequently coincide'.[5] Well, the *lines* may coincide; but they divide the material in opposite ways. What is genuinely scientific for Kuhn is hardly science for Popper, and what is genuinely scientific for Popper is hardly science for Kuhn.

Kuhn advances the following consideration against Popper's criterion and in favour of his own: it has often happened in the history of science that a theory was replaced before it had failed a test but *not* 'before it had ceased adequately to support a puzzle-solving tradition'[6]; hence *testing* is not, after all, so very important: 'To rely on testing as the mark of a science is to miss what scientists mostly do and, with it, the most characteristic feature of their enterprise.'[7]

But first, what Popper relies on as the mark of a scientific theory is not that it has actually been tested but that it is test*able*, the more testable the

[1] Kuhn [1962], p. 145.

[2] 'But falsification, though it surely occurs, . . . might equally well be called verification since it consists in the triumph of a new paradigm over the old one' (Kuhn [1962], p. 146).

[3] *This volume*, p. 6.

[4] *This volume*, p. 7.

[5] *This volume*, p. 7.

[6] *This volume*, p. 10.

[7] *This volume*, p. 10.

better (other things being equal). So it is entirely in line with his philosophy of science that one scientific theory should be replaced by a more testable theory even though the previous theory has not yet failed a test.

Second, by contrast with the relatively sharp idea of testability, the notion of ceasing 'adequately to support a puzzle-solving tradition' is essentially vague; for since Kuhn insists that there are *always* anomalies and unsolved puzzles,[1] the difference between supporting, and failing to support, a puzzle-solving tradition is merely one of *degree*: there must be a critical level at which a tolerable turns into an intolerable amount of anomaly. Since we do not know what the critical level is, this is the sort of criterion that can be used only retrospectively: it entitles us to declare, *after* a paradigm-switch has occurred, that empirical pressure on the old paradigm *must* have become pretty intolerable. (This fits in well with Kuhn's idea that a reigning paradigm has such a sway over men's minds that only strong empirical pressure can dislodge it.)

But the history of science contains important examples of an empirically *successful* dominant theory being superseded by an incompatible and more testable theory. Let me mention one such example. Before Newton, Kepler's laws constituted the dominant theory of the solar system. I take it that it is no longer necessary to argue that Newtonian theory is strictly incompatible with Kepler's original laws—if we speak of the latter being incorporated in, or subsumed under, the former, then we should add that it is significantly modified versions of those laws that follow from Newton's theory.[2] If Kuhn allows that Kepler's theory was a paradigm and that it was incompatible with the Newtonian paradigm, then he must, I think, allow that this was a case of paradigm-change. So the question arises: is it plausible to maintain that the Keplerian paradigm 'had ceased adequately to support a puzzle-solving tradition'?

Well, there was, prior to Newton, an unsolved puzzle connected with Kepler's laws. Newton himself mentions 'a perturbation of the orbit of Saturn in every conjunction of this planet with Jupiter, so sensible, that astronomers are puzzled with it.'[3] But since, for Kuhn, there are always unsolved puzzles, this can hardly amount to failure 'to support a puzzle-

[1] Kuhn [1962], p. 81.

[2] Over fifty years ago Pierre Duhem wrote: '*The principle of universal gravity, very far from being derivable by generalization and induction from the observational laws of Kepler, formally contradicts these laws. If Newton's theory is correct, Kepler's laws are necessarily false*' (Duhem [1914], p. 193 of the 1954 English translation). For a more detailed analysis of inconsistencies between Newtonian theory and Kepler's laws—inconsistencies which mean that the latter have first to be corrected in important ways before they can be explained by the former—see Popper [1957] and [1963], p. 62 n.

[3] Newton [1687], discussion to Book III, Prop. xiii. Professor J. Agassi drew my attention to this passage. (He discusses it in his [1963], p. 79, footnote 5.)

solving tradition'. Newton, at any rate, seems to have been far from regarding the Keplerian system as having *failed* in any way. In the Proposition to which the above-quoted remark is annexed, he stated Kepler's first two laws in an uncorrected form,[1] thereby helping to initiate the legend perpetuated by Halley who, in his review of the *Principia*, wrote, 'Here [in Book III] the verity of the Hypothesis of Kepler is demonstrated.'[2]

It seems that a dominant theory may come to be replaced, not because of growing empirical pressure (of which there may be little), but because a new and incompatible theory (inspired perhaps by a different metaphysical outlook) has been freely elaborated: a scientific crisis may have theoretical rather than empirical causes.[3] If that is so, there is more free thinking in science than Kuhn supposes. I will revert to this issue in the last section.

III

Later, I shall argue that Normal Science cannot have the character Kuhn ascribes to it, if it is to be capable of giving rise to Extraordinary (or Revolutionary) Science. But for the time being I shall suppose that the history of science does indeed display a Kuhnian pattern; that is, I shall suppose that a typical cycle consists of a longish period of Normal Science, which gives way to a short and hectic bout of Extraordinary Science, after which a new period of Normal Science sets in.

The question I now ask is, Why is Kuhn concerned to up-value Normal Science and down-value Extraordinary Science? This question is prompted by several considerations. First, Normal Science seems to me to be rather boring and unheroic compared with Extraordinary Science. Kuhn himself thinks it a mistake, but a rather natural mistake, to regard Normal Science as 'an intrinsically uninteresting enterprise',[4] and he agrees that Normal Science is comparatively unproductive of new ideas. More accurate determinations of physical constants—that is the sort of thing achieved by the 'mopping-up operations' which constitute Normal Science.[5] Second, Kuhn has re-iterated this afternoon that he, like Popper, rejects 'the view that science progresses by accretion'[6]; but if he were asked in what manner Normal Science progresses, he would, presumably, say that it does so in

[1] Newton [1687], Book III, Prop. xiii. As to Kepler's third law, see Book I, Prop. iv, cor. vi., and also Newton [1669].

[2] Halley [1687], p. 410.

[3] The nearest Kuhn approaches this is in his admission that a new paradigm may emerge, 'at least in embryo, before a crisis has developed *far*' (Kuhn [1962], p. 86, my italics). That it might emerge before a crisis has developed *at all*, and might itself *generate* a crisis, is excluded by his idea of paradigm-dominance within Normal Science.

[4] *This volume*, p. 6.

[5] Kuhn [1962], pp. 24 and 27.

[6] *This volume*, p. 1.

an orderly, undramatic, step by step manner, i.e. it progresses by accretion. Why has Kuhn, despite his concern 'with the *dynamic process* by which scientific knowledge is acquired',[1] come to identify science with its periods of theoretical stagnation? Third, why has the author of one excellent book on the Copernican revolution, and of another more famous book on scientific revolutions generally, taken a sort of philosophical dislike to scientific revolutions? Why is he so enamoured with plodding, uncritical, Normal Science?

One answer, though I suspect that it is not the main answer, is that he has been impressed by sheer quantitative considerations: there is *much more* Normal Science, measured in man-hours, than Extraordinary Science. Normal Science, Kuhn says, 'accounts for the overwhelming majority of the work done in basic science'.[2] The sort of scientific developments with which Popper is concerned are 'very rare'.[3]

From a sociological point of view it may be quite in order to discount something on the ground that is rare. But from a methodological point of view, something rare in science—a path-breaking new idea or a crucial experiment between two major theories—may be far more important than something going on all the time.

But I do not think that these quantitative considerations were decisive for Kuhn. I suspect that a very different sort of consideration was at work. As this matter is a little personal and delicate, and as my evidence is all drawn from Kuhn's book, I will not blurt out my conjecture straightaway, but will lead up to it gradually. I will start by considering how far Kuhn's demarcation-criterion succeeds in excluding certain intellectual disciplines that few of us would want to call scientific.

It is interesting that Kuhn himself should have mentioned, in this connection, that he does not 'want to join Sir Karl in labelling astrology a metaphysic rather than a science'.[4] One can see why: the careful drawing up of a horoscope, or of an astrological calendar, fits Kuhn's idea of Normal Research rather nicely. The work is done under the aegis of a

[1] *This volume*, p. 1, my italics.

[2] *This volume*, p. 4.

[3] *This volume*, p. 5.

[4] This quotation is from the original draft of Kuhn's paper. He now says that 'Sir Karl is *right* to exclude astrology from the sciences' (p. 10, my italics)—right, but for the wrong reasons: for there *were* predictive failures in astrology (though these could always be 'explained'); on the other hand, astrologers 'had no puzzles to solve and therefore no science to practise' (p. 9).

This new revelation of the subtlety of Kuhn's puzzle-concept leaves me boggling. I knew that a predictive failure may be regarded as a mere puzzling anomaly, and that it may later, when the framework changes, come to be regarded as a refutation. I had not appreciated that there can be predictive failures which are regarded neither as refutations nor as posing any puzzle.

stable body of doctrine which is not discredited, in the eyes of astrologers, by predictive failures.

More interesting, apropos Kuhn's possible reasons for depreciating revolutionary science, is another sort of case which seems to fit his idea of Normal Research all too well. Consider a theological scholar working on an apparent inconsistency between two Biblical passages. Theological doctrine assures him that the Bible, properly understood, contains no inconsistencies. His task is to provide a gloss that offers a convincing reconciliation of the two passages. Such work seems essentially analogous to 'normal' scientific research as depicted by Kuhn; and there are grounds for supposing that he would not repudiate the analogy. For *The Structure of Scientific Revolutions* contains many suggestions, some explicit, others implicit in the choice of language, of a significant parallelism between science, especially Normal Science, and theology. Kuhn writes of a scientific education as a 'process of professional initiation'[1] which 'prepares the student for membership in the particular scientific community'.[2] He says that 'it is a narrow and rigid education, probably more so than any other *except perhaps in orthodox theology*'.[3] He also says that a scientific education involves the re-writing, in text-books, of history backwards, and that this indicates 'one of the aspects of scientific work that most clearly distinguishes it from every other creative pursuit *except perhaps theology*'.[4] In other places the suggestion of a science-theology parallelism, though less explicit, is no less obvious. For example, he says that Normal Science 'often suppresses fundamental novelties because they are necessarily subversive of its basic commitments.'[5] And when Kuhn discusses the personal process of repudiating an old paradigm and embracing a new one, he describes it as a 'conversion experience',[6] adding that 'a decision of that kind can only be made on faith.'[7]

My suggestion is, then, that Kuhn sees the scientific community on the analogy of a religious community and sees science as the scientist's religion. If that is so, one can perhaps see why he elevates Normal Science above Extraordinary Science; for Extraordinary Science corresponds, on the religious side, to a period of crisis and schism, confusion and despair, to a spiritual catastrophe.

IV

Hitherto, I have been considering Kuhn's comparative evaluations of Normal and Extraordinary Science on the supposition that the history of

[1] Kuhn [1962], p. 47.
[2] *Op. cit.* p. 11.
[3] *Op. cit.*, p. 165, my italics.
[4] *Op. cit.* p. 135, my italics.
[5] *Op. cit.* p. 5.
[6] *Op. cit.* p. 150.
[7] *Op. cit.* p. 157

science does in fact display a Normal Science/Extraordinary Science/ Normal Science cycle. I shall now challenge this supposition.

One way of challenging it would be to point to historical counter-examples, that is, to long stretches of scientific history in which no clear paradigm emerged and during which the typical symptoms of Normal Science were absent. I remember Popper saying (in the course of our seminar discussion of Kuhn's book) that, although Newtonianism did turn into something like a paradigm in Kuhn's sense, no such paradigm emerged during the long history of the theory of *matter*[1]: here from the pre-Socratics to the present day there has been an unending *debate* between discontinuous and continuous concepts of matter, between various atomic theories on the one hand, and ether, wave and field theories on the other.

I wish to raise a different objection. My objection concerns the possibility of the emergence of a new paradigm at the end of a period of Normal Science. I shall not criticize the epidemiological account he gave in his book of how, after a new paradigm has infected a few carriers, the epidemic is liable to spread among the scientific community. In what follows I shall focus attention on the *very first* scientist to take up a new paradigm. My thesis will be that a new paradigm never could emerge from Normal Science as characterized by Kuhn.

I begin by recapitulating some Kuhnian theses concerning paradigm change.

(1) It is in the nature of a paradigm to enjoy a monopoly in its hold on a scientist's thinking. A paradigm brooks no rivals: it is built into Kuhn's concept of a paradigm that one scientist cannot, while under the sway of one paradigm, seriously entertain a rival paradigm. If he has started toying with a rival paradigm, then the old paradigm is already defunct for him. I call this the Paradigm-Monopoly thesis.

(2) There is little or no interregnum between the end of the old paradigm's reign over a scientist's mind, and the beginning of the new paradigm's reign. A scientist does not flounder around for any substantial length of time with no paradigm to guide him. He abandons one paradigm only to embrace a new one. (It is as if his cry were, *The Paradigm is dead. Long live the Paradigm.*) I call this the No-Interregnum thesis.

(3) A new paradigm will be incompatible with the paradigm it supersedes.[2] (Indeed, Kuhn goes further and claims that the new paradigm will be *incommensurable* with the old one.[3] I will discuss the relation between incompatability and incommensurability later.) I call Kuhn's thesis concerning the clash between old and new paradigms the Incompatibility

[1] A similar point has been made independently by Dudley Shapere: cf. his [1964], p. 387.

[2] Kuhn [1962], pp. 91 and 102.

[3] *Op cit.* pp. 4, 102, 111 and 147

thesis. (This thesis obviously re-inforces the Paradigm-Monopoly thesis.)

(4) From the conjunction of the above three theses it follows that a scientist's change-over from an old paradigm to a new one must be pretty swift and decisive. Kuhn emphatically endorses this implication. We have already noticed him referring to a paradigm-switch as a 'conversion'; and from other passages in his book it is clear that he holds that such conversions are quick. He says that a paradigm-switch is 'a relatively sudden and unstructured event like the gestalt switch',[1] and that 'the transition between competing paradigms cannot be made a step at a time . . . Like the gestalt switch, it must occur all at once (though not necessarily in an instant)'.[2] I call this the Gestalt-Switch thesis.

(5) I now consider the implications of the fore-going theses for the *invention* of a new paradigm. Kuhn's view allows that it may take quite a time for a paradigm, once invented, to gain general acceptance. The question now is: how long may it take the original inventor to put together the rudiments of the new paradigm? To put it another way: what sort of prehistory could his new paradigm have? The answer implied the Gestalt-Switch thesis appears to be: none at all. Before he switched over to it his thinking was along irreconcilably different lines (by the Paradigm-Monopoly and Incompatibility theses). His *switch* to the new paradigm must be regarded as the very same thing as his *invention* of the new paradigm. (I am assuming that it was invented inside the scientific community and not imported from extra-scientific sources.) And since the switch to it was 'relatively sudden' the invention of it must have been relatively sudden, too. Kuhn endorses this implication. In his book he wrote: 'The new paradigm, or a sufficient hint to permit later articulation, emerges all at once, sometimes in the middle of the night, in the mind of a man deeply immersed in crisis'.[3] And this afternoon he repeated that theories are 'invented in one piece'.[4] I call this, a shade maliciously, the Instant-Paradigm thesis. (Instant coffee takes more than an instant to make; but it is made 'all at once', unlike steak-and-kidney pie, which might be said to 'be made a step at a time'.)

We must remember that the new paradigm is immediately powerful enough to induce our scientist to turn against the well-articulated and unrefuted paradigm that has dominated his scientific thinking hitherto. This means, I take it, that the new paradigm cannot begin as just a few fragmentary ideas, but must at the outset be large and definite enough for its striking potentialities to be fairly apparent to its inventor.

[1] *Op cit.* p. 121.
[2] *Op. cit.* p. 149.
[3] *Op cit.* p. 89.
[4] *This volume*, p. 12.

If that is so, the Instant-Paradigm thesis seems to me to be barely credible on pyschological grounds. I do not know how much a single genius might achieve in the middle of the night, but I suspect that this thesis expects *too* much of him. In any case, there are, surely historical counter-examples to it. To mention one: the Inverse Square Law was an important component of Newtonian theory (which Kuhn regards as a paradigm of paradigms); and Pierre Duhem has traced the long evolution of the Inverse Square Law back through Hooke, Kepler, and Copernicus, to Aristotle's idea that bodies seek the centre of the earth.[1] I conclude that the Instant-Paradigm thesis must be rejected.

The Instant-Paradigm thesis followed from the Gestalt-Switch thesis when the latter was applied to the first man to switch over. And the Gestalt-Switch thesis followed from the conjunction of the Paradigm-Monopoly, No-Interregnum, and Incompatibility theses. Hence at least one of these three theses must be rejected if the Instant-Paradigm thesis is rejected. I will consider the Incompatability thesis first.

There seems to be a certain internal incoherence in Kuhn's version of this thesis. He says that what 'emerges from a scientific revolution is not only incompatible but often actually incommensurable with what has gone before'.[2] But could two *incommensurable* theories be logically incompatible with each other? If someone holds that, say, Biblical myths and scientific theories are incommensurable, belong to different universes of discourse, he presumably implies that the Genesis account of the Creation should *not* be regarded as logically incompatible with geology, Darwinism, etc.: they *are* compatible and can peacefully co-exist just because they are incommensurable. But if the Ptolemaic system is logically incompatible with the Copernican system, or Newtonian theory with Relativity theory, peaceful co-existence is not possible: they were *rival* alternatives; and it was possible to make a rational choice between them partly because it was possible to devise crucial experiments between them (stellar parallax, star-shift, etc.).

So let us disengage Kuhn's Incompatibility thesis from the alien idea of incommensurability. Thus purified, this historical thesis of Kuhn's is in happy accord with a methodological thesis of Popper's. For if the new theory is to be highly testable, as Popper's methodology demands, it

[1] Duhem, *op. cit.* chapter vii, section 2. Duhem himself gave this example in support of his emphatically negative answer 'Surely no' to the question: 'Is [a man's] mind powerful enough to create a physical theory all out of one piece?' (*op. cit.* chapter vii, section 2). Agassi has labelled Duhem's own view of the evolution of scientific ideas 'the continuity theory' (Agassi [1963], pp. 31 ff.). Agassi attacks the historiographical method sponsored by this view; he does not, of course, advance the counter-claim that theories *are* invented in one piece.

[2] Kuhn [1962], p. 102.

should yield (not only some remarkable predictions beyond the predictive scope of existing theories, but) some predictions which *conflict* with those of existing theories, preferably in areas where the existing theories have been well tested and have not, so far, been faulted. Popper says, in effect, that major theoretical advances in science *ought* to have a revolutionary character; and Kuhn says, in effect, that they *do* have a revolutionary character. Good. Let us agree that the Incompatibility thesis should stay.

Then the Paradigm-Monopoly thesis and/or the No-Interregnum thesis must go. But these really hang together. The second says that a scientist's professional thinking is *always* paradigm-dominated, the first says that it is, at any one time, dominated by *one* paradigm. Against this I have maintained that since it takes time—a matter of years rather than of hours—to develop a potential new paradigm to the point where it may challenge an entrenched paradigm, heretical thinking must have been going on for a long time before paradigm-change can occur. This means that it is not true that a reigning paradigm exercises such a monopolizing sway over scientists' minds that they are all unable to consider it critically, or to toy with (without necessarily embracing) alternatives to it. It means that the scientific community is not, after all, a closed society whose chief characteristic is 'the abandonment of critical discourse'.

REFERENCES

Agassi [1963]: *Towards an Historiography of Science*, 1963.

Duhem [1914]: *La théorie Physique: son Objet et sa Structure*, 1914.

Halley [1687]: Review of Newton's *Principia, Philosophical Transactions*, 1687. Reprinted in I. B. Cohen (*ed.*): *Isaac Newton's Papers and Letters on Natural Philosophy*, 1958, pp. 405–11.

Kuhn [1962]: *The Structure of Scientific Revolutions*, 1962.

Newton [1669]: Manuscript, reprinted in Turnbull (*ed.*): *The Correspondence of Isaac Newton*, **1**, pp. 297–303.

Newton [1687]: *Philosophiae Naturalis Principia Mathematica*, 1687.

Popper [1934]: *Logik der Forschung*, 1935.

Popper [1957]: 'The Aim of Science', *Ratio*, **1**, pp. 24–35.

Popper [1963]: *Conjectures and Refutations*, 1963.

Shapere [1964]: 'The Structure of Scientific Revolutions', *The Philosophical Review*, **73**, pp. 383–94.

Wheeler [1956]: 'A Septet of Sibyls: Aids in the Search for Truth', *The American Scientist*, **44**, pp. 360–77.

Does the Distinction between Normal and Revolutionary Science Hold Water?

STEPHEN TOULMIN
University of Michigan

Professor T. S. Kuhn's contribution to this Symposium can be looked at from two angles: either as a critique of Sir Karl Popper's approach towards the philosophy of science, in the light of its contrasts with Professor Kuhn's own views, or alternatively, as a further instalment in the development of Kuhn's analysis of the process of scientific change. My concern here is with the second of these two aspects. I shall draw attention to certain significant changes in the position Kuhn now appears to be occupying from those which he adopted, first in his original paper on 'The Function of Dogma in Scientific Research' read at Worcester College, Oxford, in 1961,[1] and subsequently in his book *The Structure of Scientific Revolutions* published in 1962. And in the light of changes, I shall suggest how we might see our way beyond Kuhn's theory of 'scientific revolution' to a more adequate theory of scientific change.

The great merit of Professor Kuhn's insistence on the 'revolutionary' character of some changes in scientific theory is that it has compelled many people to face for the first time the full profundity of the conceptual transformations which have, at times, marked the historical development of scientific ideas. Yet from the beginning it was clear to many onlookers that Kuhn's original statement of his position was, in at least two respects, only provisional. Some of us have been waiting with interest to see in what direction his own intellectual development took him next. In the first place, although his choice of the word 'dogma' served well enough in the title of a thought-provoking paper at the Worcester College meeting, only a little closer examination was required to reveal the fact that its very effectiveness sprang from a certain built-in rhetorical exaggeration or play upon words. (To say 'all normal science rests on a foundation of dogma' was like saying 'we are all mad really'; which can make a point on a particular occasion, but . . .)

The nature of this play upon words becomes evident if we contrast the application of Kuhn's analysis to Newton's *Principia*, regarded as the founding document of classical mechanics, with its application to Newton's *Opticks*, which was so influential in eighteenth-century physics. Taking

[1] Printed in Crombie (ed), [1963], pp. 347–69.

Principia first, we can state a worthwhile philosophical point as follows: that the intellectual function of an established conceptual scheme is to determine the patterns of theory, the meaningful questions, the legitimate interpretations, etc., within which theoretical speculation is bounded for as long as that particular conceptual scheme retains intellectual authority within the natural science concerned. This (I repeat) is a *philosophical* point, which indicates something of what is involved in saying that scientific procedures, in the theoretical as well as the practical area, are 'methodical', and marked by plain good sense. However, this particular point does nothing at all to establish that *dogma* has any part to play in scientific theory. On the contrary, it was wholly reasonable—and undogmatic—for physicists between 1700 and 1880 to accept Newton's dynamics as their provisional starting-point. And it is always open to scientists to *challenge* the intellectual authority of the fundamental scheme of concepts within which they are provisionally working—the permanent right to challenge this authority being one of the things which (as Sir Karl Popper has always insisted) marks off an intellectual procedure as being 'scientific' at all. Incidentally, this first, philosophical point was stated rather more clearly and unambiguously, some twenty-five years ago, by R. G. Collingwood in his *Essay on Metaphysics*.[1] The *intellectual* function of Kuhn's 'paradigms' is precisely that of Collingwood's 'absolute presuppositions'.

Alternatively, if we take Newton's *Opticks* as our example, we may make a sociological point, as follows: that there is a tendency on the part of secondary workers in science to see only part of the intellectual picture in the subject with which they are concerned, and to restrict the choice of hypotheses by which they interpret their data, out of deference to the supposed example set them by a primary worker, whom they take as their master and whose *magisterial* authority they bow to. This is a *sociological* rather than a philosophical point: in this case, one can indeed speak of 'dogma' playing a part in the development of scientific ideas. But the very beginning of wisdom in any attempt to understand the nature of intellectual development in science must, surely, be to distinguish between the intellectual authority of an established conceptual scheme and the magisterial authority of a dominant individual. And it is only when secondary workers insist on retaining, say, a corpuscularian theory of light out of respect for the authority of Newton, even after legitimate alternatives have been put forward with as much experimental support, that the word 'dogma' has any relevance to science.

Kuhn, in moving on from his Oxford paper to the 1962 book, withdrew

[1] Collingwood [1940], esp. chapters iv–vi. Collingwood's argument is discussed, in parallel with Kuhn's, in my [1966].

his insistence on the term 'dogma', but attempted to retain a central distinction between 'normal science' and 'scientific revolutions'. Throughout the book he regarded the idea of 'revolutions' as having some power to illuminate and explain certain phases in scientific change. In this respect, too, his analysis was at best only provisional. As we know from political history, the term 'revolution' may serve as a useful descriptive *label*, but it has long since worn out its value as an explanatory *concept*. At one time, historians faced with political changes of a peculiarly drastic variety were quite ready to say, '... and then there was a revolution', and leave it at that: the implication was that, in the case of such drastic changes, no explanation could be given of the rational kind we rightly demand in the case of normal political developments. But in due course they were compelled to recognize that political change never in fact involves such an absolute and outright breach of continuity. Whether one considers the French Revolution, the American Revolution or the Russian Revolution, in each case the continuities in political and administrative structure and practice are quite as important as the changes. (Consider, for instance, the American legal system, the Russian practice of escorting tourists, and the French code of inheritance: the effect of political revolution was to change each of these only marginally, and the corresponding states of affairs in each country before and after the revolution in question were much more similar than the pre-revolutionary or post-revolutionary conditions in the different countries.) So, in the political sphere, statements about the occurrence of 'revolutions' are only preliminary to questions about the political mechanisms involved in revolutionary change. As the explanatory level, the difference between normal and revolutionary change in the political sphere turned out after all to be only one of degree.

The position Professor Kuhn adopted in his book has always appeared to me to demand similar qualifications. According to that argument, the differences between the kinds of change taking place during 'normal' and 'revolutionary' phases of scientific development are, at the intellectual level, absolute. As a result, the account he gave went too far by implying the existence of discontinuities in scientific theory far more profound and far less explicable than any which ever in fact occur. In his new paper, he appears to be withdrawing somewhat from that original, exposed position, to a less extreme one; yet the effect of doing so (I shall argue) is to demolish entirely his original distinction between 'normal' and 'revolutionary' phases. That is evidently not his intention, but the consequence is (in my view) inescapable.

Let me explain why I say this with the help of an analogy, taken from the history of palaeontology during the years between 1825 and 1860.

During those years, one of the two most influential palaeontological systems was built around the theory of 'castastrophes', put forward first by George Cuvier in France, and extensively developed by Louis Agassiz at Harvard. This theory emphasized the sheer discontinuities to be found in the geological and palaeontological record. It had the considerable merit of challenging the bland assumption (which formed a basic methodological axiom for the followers of James Hutton, including Charles Lyell in his early years) that all the agencies involved in geological and palaeontological change—both inorganic and organic—had been of exactly the same kinds, and had acted in exactly the same ways, at every phase in the earth's history. However, Cuvier went on from his original, quite authentic observation of geological and palaeontological discontinuities to insist that these discontinuities were evidence of 'super-natural' events—that is changes too sudden and violent to be explained in terms of natural physical and chemical processes. The discontinuities were, as he put it, evidence of 'catastrophes', and these (like the political historians' original 'revolutions') were something intellectually unbridgeable. When a geologist said, '... and then there was a catastrophe', this implied that for the change in question no rational explanation was possible, in terms of natural geological mechanisms such as accounted for the deposition of normal sedimentary stratas, for example. This theoretical interpretation of the geological and palaeontological discontinuities went too far. In some respects, it was true, the discontinuities evidenced in the earth's crust were quite as sharp as Cuvier said; but as investigation proceeded it turned out that they were neither universal in their extent nor beyond all hope of reasonable explanation.

How was this opposition between the uniformitarian theory and the theory of catastrophes resolved? That is the significant point for our purpose here. In due course, two kinds of things happened. On the one hand, uniformitarian geologists and palaeontologists of Lyell's generation were compelled, bit by bit, to acknowledge that *some* of the changes which formed the subject-matter of their inquiries had in fact taken place more drastically than they had hitherto supposed. Charles Darwin, for instance, observed on the coast of Chile the effects of recent earthquakes which had altered the relative location of different geological strata by as much as 20 feet in a single tremor, and this discovery convinced Lyell that past earthquakes might, after all, have been more severe than he had previously supposed. On the uniformitarian side, accordingly, ideas became progressively more 'catastrophic'. Meanwhile in the catastrophist camp ideas developed in the opposite direction. Louis Agassiz, in particular, found his studies compelling him to multiply the number of catastrophes invoked to explain the actual geological evidence, and to diminish their size. As a result, the original 'drastic and inexplicable' catastrophes, eventually became so many, and so minor, that they began to evince uniformities, so turning into geological and palaeontological phenomena in their own right. As such, the claim that they were not open to mechanistic or naturalistic explanation ceased any longer to be plausible, and the need—even in their case—to give some kind of an account of the mechanisms involved became unanswerable. In a word, the original 'catastrophes' became uniform and law-governed just like any other geological and palaeontological phenomena. What the catastrophist palaeontologists did not immediately appreciate was that this apparently innocent change within the structure of their theory destroyed their original criterion for distinguishing between 'normal' (or natural) and 'catastrophic' (or supernatural) changes in the earth's crust, and that the very distinction between the 'normal and the 'catastrophic' had thereby collapsed.

Let me now apply this analogy. As I read Professor Kuhn's present account of his position, he has moved away from the original 'normal'/'revolutionary' dichotomy, in the same direction that Agassiz moved away from Cuvier's original theory. Once again it was both worthwhile and important, at the outset, to insist that the development of scientific ideas involves, at times, changes so drastic that they introduce profound conceptual incongruities between the ideas accepted by successive generations of scientists. No theory of scientific growth and development would be adequate which did not recognize, and do justice to these intellectual discontinuities. In Kuhn's earlier accounts—in the 1962 book as much as in the 1961 paper—he depicted these 'revolutionary' discontinuities as being *absolute*. They created a situation in which there was, inevitably, complete incomprehension at the theoretical level between supporters of the older and newer systems of scientific thought; for example, between a supporter of the older, Newtonian dynamics and a supporter of the new Einsteinian dynamics. This incomprehension was inescapable because, when it came to organizing their experience, the two men shared no common language, no common viewpoint, nor even a common *gestalt*. In consequence, neither the Newtonian language nor the Einsteinian language would suffice to explain either point of view to supporters of the other. The occurrence of a 'scientific revolution' (it appeared) threw attempts at communication so completely out of joint that incomprehension was *guaranteed*.

Yet there was always an element of rhetorical exaggeration in this statement of the matter, just as much as in Kuhn's earlier use of the work 'dogma'. After all, the professional careers of numerous physicists spanned the years from 1890 to 1930, and these men lived through the change from the Newtonian to the Einsteinian system of thought. If the complete breakdown in scientific communication which Kuhn treats as the essential characteristic of a scientific revolution had in fact been manifested during this period, one should be able to document it from the experience of the men in question. What do we find? If the conceptual change involved in the transition was as deep as Kuhn claims, these physicists at any rate appeared curiously unaware of the fact. On the contrary, many of them were able to say, after the event, *why* they had changed their own personal position from a classical to a relativistic one—and when I say 'why' I mean 'for what reasons . . .'. Taking Kuhn at his word, however, such a change of position could have come about only as a result of a 'conversion'—the sort of mind-change which a man would have to describe by saying, 'I can no longer see Nature as I did before . . .'—or alternatively as the outcome of 'causes' rather than 'reasons'—'Einstein was so very persuasive . . .', or 'I found myself changing without knowing why . . .', or 'It was as much as my job was worth . . .'.

Accordingly, one may concede that the development of scientific thought does involve important conceptual discontinuities, and that the conceptual systems which displace one another within a scientific tradition may often be based on quite different, and even incongruous principles and axioms; but we must beware of going all the way with Kuhn's original 'revolutionary' hypothesis. For the displacement of one system of concepts by another is itself something that happens for perfectly good reasons, even though these particular 'reasons' cannot themselves be formalized into still broader concepts, or still more general axioms. For what is presupposed by both parties in such a debate—both those who cling on to the older view, and those who put forward a new one—is not a common body of principles and axioms: rather it is a common set of 'selection procedures' and 'selection rules', and these are not so much 'scientific principles' as 'principles constitutive of science'. (They, too, may change in the course of history, as Imre Lakatos has demonstrated in the case of the criteria of mathematical proof, but they do so more slowly than the theories which they are used to judge.)

Suppose, then, one concedes to Kuhn that 'conceptual incongruities' between the ideas of successive generations of scientists do introduce real discontinuities into the development of scientific thought. If this is the essence of his insight, then we shall have to go along with him down the next leg of his argument, corresponding to the 'modified catastrophism' of Agassiz. For whereas on Kuhn's original account, scientific revolutions were something that tended to happen in a given branch of science only once every two hundred years or so, the 'conceptual incongruities' with which he is now preoccupied are liable to turn up very much more frequently. On a small enough scale, indeed, they are very frequent indeed; and perhaps every new generation of scientists having any original ideas or 'slant' of its own finds itself, at certain points and in certain respects, at cross-purposes with the immediately previous generation. One may question, indeed, whether any natural science having a serious theoretical component *ever* develops by a process of 'accretion' alone.

In that case, however, the occurrence of a 'scientific revolution' no longer amounts to a dramatic interruption in the 'normal' continuous consolidation of science: instead, it becomes a mere 'unit of variation' within that very process of scientific change. As in palaeontology, the *hyper-rational* aspect of the discontinuities vanishes, and—in the process—the very basis for distinguishing between 'normal' and 'revolutionary' change in science which was the very heart and core of Kuhn's theory, collapses. For the 'absoluteness' of the transition involved in a scientific revolution provided the original criterion for recognizing that one had occurred at all. And, once we acknowledge that *no* conceptual change in science is ever absolute, we are left only with a sequence of greater and lesser conceptual modifications differing from one another in degree. The distinctive element in Kuhn's theory is thus destroyed, and we are left looking beyond it for a new sort of theory of scientific change. This theory will have to go beyond both Kuhn's concept of 'revolutions' and the naive uniformitarian views which he renounced, just as Darwin's evolutionary reinterpretation of palaeontology went beyond both the catastrophism of Cuvier and the uniformitarianism of Lyell.

Like Professor Kuhn, I believe that this new theory—when we have it—will have to be based in part on the results of new empirical studies of the actual development and growth of science; and that, as a result, it will tend to bring the logic of science closer together with its sociology and psychology. Yet it will remain as important as ever (as Sir Karl Popper emphasizes) to avoid *identifying* the logical criteria for appraising new scientific hypotheses with generalizations about the actual practice of scientists, taken either individually or collectively as professional groups.

What form should such a theory take? Once more, the experience of other historical disciplines may give us a hint. For again and again the fruitful direction for escaping the deadlock between revolutionary and uniformitarian views of historical change has been the same: by way of scrutinizing more closely the mechanisms involved, in particular, the mechanisms of variation and perpetuation. (Compare, for instance, Charles Darwin's *Origin of Species* with Crane Brinton's *Anatomy of Revolution*.)

Let me pursue this hint a little way, at the price of anticipating an argument to be set out at length elsewhere.[1]

Suppose we stop thinking of Kuhn's small-scale 'micro-revolutions' as units of effective *change* in scientific theory, and treat them instead as units of *variation*. We will then be faced with a picture of science in which the theories currently accepted at each stage serve as starting-points for a large number of suggested variants; but in which only a small fraction of these variants in fact survive and become established within the body of ideas passed on to the next generation. The single question, 'How do revolutions occur in science?' thus has to be reformulated, and gives rise to two distinct groups of questions. On the one hand we must ask, 'What factors determine the number and nature of theoretical variants proposed for consideration in a particular science during a given period?'—the counterpart, in biological evolution, to the genetical question about the origin of mutant forms. On the other hand we must ask, 'What factors and considerations determine which intellectual variants win acceptance, to become established in the body of ideas which serves as the starting-point for the next round of variations?'—the counterpart to biological questions about selection.

As in other historical disciplines, accordingly, the problem of historical change can usefully be restated as a problem of variation-and-selective-perpetuation. The advantages of such a restatement cannot be fully expounded here, but one thing at any rate is worth indicating. It not only helps us to locate the ambiguity which drives the debate between Kuhn and Popper into cross-purposes—the ambiguity between the philosophy of science, which is concerned with the question what consideration *should properly* determine the selection between new variants, and the psychology or sociology of science, which is concerned with the considerations that *in fact* settle the matter. It can also, I believe, help us to resolve some of the old perplexities about the relationship between external and internal factors in the development of an intellectual tradition. If scientific change is treated as a special case of a more general phenomenon of 'conceptual evolution', we can distinguish at least three different aspects of this evolution. The actual bulk, or *quantity*, of innovation going on in a given field at any time can be distinguished from the *direction* in which this innovation is predominantly tending; and both of these can be distinguished, in turn, from the *selection criteria* determining which variants are perpetuated within the tradition.

Once these distinctions are clearly made, it will be desirable to consider separately how far each aspect of scientific change is responsive either to

[1] See my [1966] for a brief analysis. The full exposition will be given in a forthcoming book on conceptual evolution and the problem of 'human understanding'.

internal or to external factors, and it will become naive to suppose that there need be any conflict between the two kinds of account. As a hint: the volume of innovation going on in any science presumably depends to a great extent on the opportunities provided in that social context for doing original work on the science in question—hence, the rate of innovation will be substantially responsive to factors external to science. On the other hand, the selection-criteria for appraising conceptual innovations in science will be very largely a professional and so an internal matter: many scientists, indeed, would expect them to be entirely an internal, professional matter—though this may, in practice, be no more than an unrealizable ideal. Finally, the direction of innovation in a particular science depends on a complex mixture of factors, internal and external: the sources of novel hypotheses are highly varied, and subject to influences and analogies remote from the detailed problems in hand.

The fuller ramifications of an 'evolutionary' theory of scientific change (as contrasted with Kuhn's 'catastrophism') must be held over for another occasion. For the moment, let me end with two questions, which will help to pin-point the transitional character of Kuhn's present position. (1) How extensive do the conceptual incongruities between the ideas of one scientific generation and those of the next have to be, if the transition between them is to constitute a 'scientific revolution' on Kuhn's present account? (I assume that none was ever, in fact, extensive enough to satisfy his original criterion; so we now need a new criterion to replace it.) (2) If *any* conceptual shift between the theories of successive generations capable of provoking incomprehension between them is to be accepted as a 'revolution', then can we not demand a general account of the role of *all* such conceptual shifts within the development of scientific thought? Are we not entitled, in a phrase, to treat these 'micro-revolutions' as the counterparts of the 'micro-catastrophes' of Agassiz and the later catastrophist geologists? And, if that is the case, are we not in fact entirely outgrowing the original implications of the term 'revolution'? Students of political history have by now outgrown any naive reliance on the idea of 'revolutions'. If I am right, and the 'micro-revolutions' of Kuhn's present position are the units of all scientific innovation, then the idea of 'scientific revolution' will have to follow that of 'political revolutions' out of the category of explanatory concepts and into that of mere descriptive labels.

REFERENCES

Collingwood [1940] *An Essay on Metaphysics*, 1940.

Crombie (*ed.*) [1963]: *Scientific Change*, 1963.

Toulmin [1966]: 'Conceptual Revolutions in Science', in Cohen-Wartofsky (*eds.*): *Boston Studies in the Philosophy of Science*, **3**, 1967, pp. 331–47.

Normal Science, Scientific Revolutions and the History of Science

L. PEARCE WILLIAMS
Cornell University

I should like to comment very briefly on the Kuhn-Popper disagreement over the *essential* nature of science and the genesis of scientific revolutions. If I understand Sir Karl Popper correctly, science is basically and constantly *potentially* on the verge of revolution. A refutation, at least if it is big enough, constitutes such a revolution. Professor Kuhn argues, on the other hand, that *most* of the time devoted to the pursuit of science is what he calls 'normal' science—that is, problem solving, or working out chains of argument implicit in previous work. Thus, for Kuhn, a scientific revolution is a long time a-building and occurs only rarely because *most* people are *not* trying to refute current theories. Both sides have presented their positions in considerable detail but there seems to me to be a very important gap in both theories. It is, simply, how do we know what science is all about? The question may sound startlingly naive, but I shall now attempt to justify it.

There are, essentially, two respectable scholarly ways to go about answering the question. One is sociological; the scientific community may be treated like any other community and subjected to sociological analysis. Note that this 'may' be done, but that it has not yet been done. To put it another way, most scientific activity *may* be directed toward refutation or toward 'problem solving', but we don't know whether it is or not. I may just interject here that I am not impressed with Miss Masterman's observation that the paradigm is eagerly grasped by researchers in such fields as computer science and the social sciences. After all, the figure of the drowning man and the straw is a familiar one. I do not believe that Dr Kuhn intended to restrict his analysis to embryo sciences and I am interested in what practitioners of mature sciences think they are doing. To repeat, we simply do not have this information. The difficulties in the way of compiling it are enormous. Do we want simply a quantitative sample? Is what *most* scientists do really relevant to what science, in the long run, is? Do we weight the opinion of, say, Peter Debye equally with that of a man who accurately measures nuclear cross-sections? I am no sociologist, but I should think that approaching the problem through sociology would be to run a course filled with thorns.

Yet it should be noted that both Kuhn and Popper base their systems on (in Kuhn's case) what scientists *do* (with no hard evidence that they do do science this way) or (in Popper's case) on what they *ought* to do (with very few examples to persuade us that this is right). Both Kuhn and Popper *really* base their views of the structure of science on the history of science and the main point of my remarks here is that the history of science cannot bear such a load at this time. We simply do not know enough to permit a philosophical structure to be erected on a historical foundation. For example, there could be no better illustration of 'normal' science than the experimental researches in electricity of Michael Faraday in the 1830s. Beginning with the 'accidental' discovery of electromagnetic induction in 1831, each new step seemed to follow clearly from the previous one. Here was puzzle-solving with a vengeance. This is the traditional view of Faraday, master experimentalist who, if one reads Tyndall or even Thompson, never had a theoretical *idea* in his life. Yet, the minute one moves behind the published papers to the *Diary* and the manuscript notes and letters, a strange Faraday emerges. From 1821 on he was *testing* fundamental hypotheses on the nature of matter and force. How many 'normal' scientists (as defined by their published papers) are really revolutionaries at heart? Hopefully, some day the history of science will be able to answer this, but as of now, no one can say.

Before I give too much comfort to the followers of Popper, I should like to raise before them the spectre of the history of spectroscopy between 1870 and 1900. I think it fair to describe this period as one of mapping, in which the spectra of the elements were described with every increasing precision. There is precious little 'refutation' going on here, yet it would be hard to deny Ångstrom the title of scientist. Nor should it be forgotten that one of the most successful 'problem solvers' in the history of science was Max Planck who was also the most reluctant revolutionary of all time.

As a historian, then, I must view both Popper and Kuhn with a somewhat jaundiced eye. Both have raised issues of fundamental importance; both have provided deep insights into the nature of science; but neither has amassed sufficient hard evidence to lead me to believe that the essence of the scientific quest has been captured. I shall continue to use both as guides to my researches, always keeping in mind Lord Bolingbroke's remark that 'history is philosophy teaching by example.' We need a lot more examples.

Normal Science and its Dangers

KARL POPPER
London School of Economics

Professor Kuhn's criticism of my views about science is the most interesting one I have so far come across. There are, admittedly, some points, more or less important, where he misunderstands me or misinterprets me. For example, Kuhn quotes with disapproval a passage from the beginning of the first chapter of my book, *The Logic of Scientific Discovery*. Now I should like to quote a passage overlooked by Kuhn, from the Preface to the First Edition. (In the first edition the passage stood immediately before the passage quoted by Kuhn; later I inserted the Preface to the English Edition between these two passages.) While the brief passage quoted by Kuhn may, out of context, sound as if I had been quite unaware of the fact, stressed by Kuhn, that scientists necessarily develop their ideas within a definite theoretical framework, its immediate predecessor of 1934 almost sounds like an anticipation of this central point of Kuhn's.

After two mottos taken from Schlick and from Kant, my book begins with the following words: 'A scientist engaged in a piece of research, say in physics, can attack his problem straight away. He can go at once to the heart of the matter: that is, to the heart of an organized structure. For a structure of scientific doctrines is already in existence; and with it, a generally accepted problem-situation. This is why he may leave it to others to fit his contribution into the framework of scientific knowledge.' I then go on to say that the philosopher finds himself in a different position.

Now it seems pretty clear that the passage quoted describes the 'normal' situation of a scientist in a way very similar to Kuhn: there is an edifice, an organized structure of science which provides the scientist with a generally accepted problem-situation into which his own work can be fitted. This seems very similar to one of Kuhn's main points: that 'normal' science, as he calls it, or the 'normal' work of a scientist, presupposes an organized structure of assumptions, or a theory, or a research programme, needed by the community of scientists in order to discuss their work rationally.

The fact that Kuhn overlooked this point of agreement and that he fastened on what came immediately after, and what he thought was a point of disagreement, seems to me significant. It shows that one never reads or understands a book except with definite expectations in one's mind. This indeed may be regarded as one of the consequences of my thesis that *we approach everything in the light of a preconceived theory*. So also a book. As a consequence one is liable to pick out these things which one either likes or dislikes or which one wants for other reasons to find in the book; and so did Kuhn when reading my book.

Yet in spite of such minor points, Kuhn understands me very well—better, I think, than most critics of mine I know of; and his two main criticisms are very important.

The first of these criticisms is, briefly, that I have completely overlooked what Kuhn calls 'normal' science, and that I have been exclusively engaged in describing what Kuhn calls 'extraordinary research', or 'extraordinary science'.

I think that the distinction between these two kinds of enterprise is perhaps not quite as sharp as Kuhn makes it; yet I am very ready to admit that I have at best been only dimly aware of this distinction; and further, that the distinction points out something that is of great importance.

This being so it is a minor matter, comparatively, whether or not Kuhn's terms 'normal' science and 'extraordinary science' are somewhat question begging, and (in Kuhn's sense) 'ideological'. I think that they are all this; but this does not diminish my feelings of indebtedness to Kuhn for pointing out the distinction, and for thus opening my eyes to a host of problems which previously I had not seen quite clearly.

'Normal' science, in Kuhn's sense, exists. It is the activity of the non-revolutionary, or more precisely, the not-too-critical professional: of the science student who accepts the ruling dogma of the day; who does not wish to challenge it; and who accepts a new revolutionary theory only if almost everybody else is ready to accept it—if it becomes fashionable by a kind of bandwagon effect. To resist a new fashion needs perhaps as much courage as was needed to bring it about.

You may say, perhaps, that in so describing Kuhn's 'normal' science, I am implicitly and surreptitiously criticizing him. I shall therefore state again that what Kuhn has described does exist, and that it must be taken into account by historians of science. That it is a phenomenon which I dislike (because I regard it as a danger to science) while he apparently does not dislike it (because he regards it as 'normal') is another question; admittedly, a very important one.

In my view the 'normal' scientist, as Kuhn describes him, is a person one ought to be sorry for. (According to Kuhn's views about the history of science, many great scientists must have been 'normal'; yet since I do not feel sorry for them, I do not think that Kuhn's views can be quite right.) The 'normal' scientist, in my view, has been taught badly. I believe, and so do many others, that all teaching on the University level (and if possible below) should be training and encouragement in critical thinking. The 'normal' scientist, as described by Kuhn, has been badly taught. He has been taught in a dogmatic spirit: he is a victim of indoctrination. He has learned a technique which can be applied without asking for the reason why (especially in quantum mechanics). As a consequence, he has become what may be called an *applied scientist*, in contradistinction to what I should call a *pure scientist*. He is, as Kuhn puts it, content to solve 'puzzles'.[1] The choice of this term seems to indicate that Kuhn wishes to stress that it is not a really fundamental problem which the 'normal' scientist is prepared to tackle: it is, rather, a routine problem, a problem of applying what one has learned: Kuhn describes it as a problem in which a dominant theory (which he calls a 'paradigm') is applied. The success of the 'normal' scientist consists, entirely, in showing that the ruling theory can be properly and satisfactorily applied in order to reach a solution of the puzzle in question.

Kuhn's description of the 'normal' scientist vividly reminds me of a conversation I had with my late friend, Philip Frank, in 1933 or thereabouts. Frank at that time bitterly complained about the uncritical approach to science of the majority of his Engineering students. They merely wanted to 'know the facts'. Theories or hypotheses which were not 'generally accepted' but problematic, were unwanted: they made the students uneasy. These students wanted to know only those things, those facts, which they might apply with a good conscience, and without heart-searching.

I admit that this kind of attitude exists; and it exists not only among engineers, but among people trained as scientists. I can only say that I see a very great danger in it and in the possibility of its becoming normal (just as I see a great danger in the increase of specialization, which also is an undeniable historical fact): a danger to science and, indeed, to our civilization. And this shows why I regard Kuhn's emphasis on the existence of this kind of science as so important.

I believe, however, that Kuhn is mistaken when he suggests that what he calls 'normal' science is normal.

Of course, I should not dream of quarrelling about a term. But I wish to suggest that few, if any, scientists who are recorded by the history

[1] I do not know whether Kuhn's use of the term 'puzzle' has anything to do with Wittgenstein's use. Wittgenstein, of course, used it in connection with his thesis that there are *no genuine problems* in philosophy—only puzzles, that is to say, pseudo-problems connected with the improper use of language. However this may be, the use of the term 'puzzle' instead of 'problem' is certainly indicative of a wish to show that the problems so described are not very serious or very deep.

of science were 'normal' scientists in Kuhn's sense. In other words, I disagree with Kuhn both about some historical facts, and about what is characteristic for science.

Take as an example Charles Darwin *before* the publication of *The Origin of Species*. Even after this publication he was what might be described as a 'reluctant revolutionary', to use Professor Pearce Williams's beautiful description of Max Planck; before it he was hardly a revolutionary at all. There is nothing like a conscious revolutionary attitude in his description of *The Voyage of the Beagle*. But it is brim full of problems; of genuine, new and fundamental problems, and of ingenious conjectures—conjectures which often compete with each other—about possible solutions.

There can be hardly a less revolutionary science than descriptive botany. Yet the descriptive botanist is constantly faced with genuine and interesting problems: problems of distribution, problems of characteristic locations, problems of species or sub-species differentiation, problems like those of symbiosis, characteristic enemies, characteristic diseases, resistant strains, more or less fertile strains, and so on. Many of these descriptive problems force upon the botanist an experimental approach; and this leads on to plant physiology and thus to a theoretical and experimental (rather than purely 'descriptive') science. The various stages of these transitions merge almost imperceptibly, and genuine problems rather than 'puzzles' arise at every stage.

But perhaps Kuhn calls a 'puzzle' what I should call a 'problem'; and surely, we do not want to quarrel about words. So let me say something more general about Kuhn's typology of scientists.

Between Kuhn's 'normal scientist' and his 'extraordinary scientist' there are, I assert, many gradations; and there must be. Take Boltzmann: there are few greater scientists. But his greatness can hardly be said to consist in his having staged a major revolution for he was, to a considerable extent, a follower of Maxwell. But he was as far from a 'normal scientist' as anybody could be: he was a valiant fighter who resisted the ruling fashion of his day—a fashion which, incidentally, ruled only on the continent and had few adherents, at that time, in England.

I believe that Kuhn's idea of a typology of scientists and of scientific periods is important, but that it needs qualification. His schema of 'normal' periods, dominated by *one* ruling theory (a 'paradigm' in Kuhn's terminology) and followed by exceptional revolutions, seems to fit astronomy fairly well. But it does not fit, for example, the evolution of the theory of matter; or of the biological sciences since, say, Darwin and Pasteur. In connection with the problem of matter, more especially, we have had at least three dominant theories competing since antiquity: the continuity theories, the atomic theories, and those theories which tried to combine the two. In addition, we had for a time Mach's version of Berkeley—the theory that 'matter' was a metaphysical rather than a scientific concept: that there was no such thing as a physical theory of the structure of matter; and that the phenomenological theory of heat should become *the one paradigm* of all physical theories. (I am using here the word 'paradigm' in a sense slightly different from Kuhn's usage: to indicate not a *dominant theory*, but rather a *research programme*—a mode of explanation which is considered so satisfactory by some scientists that they demand its general acceptance.)

Although I find Kuhn's discovery of what he calls 'normal' science most important, I do not agree that the history of science supports his doctrine (essential for his theory of rational communication) that 'normally' we have *one* dominant theory—a 'paradigm'—in each scientific domain, and that the history of a science consists in a sequence of dominant theories, with intervening revolutionary periods of 'extraordinary' science; periods which he describes as if communication between scientists had broken down, owing to the absence of a dominant theory.

This picture of the history of science clashes with the facts as I see them. For there was, ever since antiquity, constant and fruitful discussion between the competing dominant theories of matter.

Now in his present paper, Kuhn seems to propose the thesis that the logic of science has little interest and no explanatory power for the historian of science.

It seems to me that coming from Kuhn this thesis is almost as paradoxical as the thesis 'I do not use hypotheses' was when it was pronounced in Newton's *Optics*. For as Newton used hypotheses, so Kuhn uses logic—not merely in order to argue, but precisely in the same sense in which I speak of the *Logic of Discovery*. He uses, however, a logic of discovery which in some points differs radically from mine: Kuhn's logic is the logic of *historical relativism*.

Let me first mention some points of agreement. I believe that science is essentially critical; that it consists of bold conjectures, controlled by criticism, and that it may, therefore, be described as revolutionary. But I have always stressed the need for some dogmatism: the dogmatic scientist has an important role to play. If we give in to criticism too easily, we shall never find out where the real power of our theories lies.

But this kind of dogmatism is not what Kuhn wants. He believes in the domination of a ruling dogma over considerable periods; and he does not believe that the method of science is, normally, that of bold conjectures and criticism.

What are his main arguments? They are not psychological or historical—they are logical: Kuhn suggests that the rationality of science presupposes the acceptance of a common framework. He suggests that rationality *depends* upon something like a common language and a common set of assumptions. He suggests that rational discussion, and rational criticism, is only possible if we have agreed on fundamentals.

This is a widely accepted and indeed a fashionable thesis: the thesis of *relativism*. And it is a *logical* thesis.

I regard the thesis as mistaken. I admit, of course, that it is much easier to discuss puzzles within an accepted common framework, and to be swept along by the tide of a new ruling fashion into a new framework, than to discuss fundamentals—that is, the very framework of our assumptions. But the relativistic thesis that the framework *cannot* be critically discussed is a thesis which *can* be critically discussed and which does not stand up to criticism.

I have dubbed this thesis *The Myth of the Framework*, and I have discussed it on various occasions. I regard it as a logical and philosophical mistake. (I remember that Kuhn does not like my usage of the word 'mistake'; but this dislike is merely part of his relativism.)

I should like just to indicate briefly why I am not a relativist:[1] I do believe in 'absolute' or 'objective' truth, in Tarski's sense (although I am, of course, not an 'absolutist' in the sense of thinking that I, or anybody else, has the truth in his pocket). I do not doubt that this is one of the points on which we are most deeply divided; and it is a logical point.

I do admit that at any moment we are prisoners caught in the framework of our theories; our expectations; our past experiences; our language. But we are prisoners in a Pickwickian sense: if we try, we can break out of our framework at any time. Admittedly, we shall find ourselves again in a framework, but it will be a better and roomier one; and we can at any moment break out of it again.

The central point is that a critical discussion and a comparison of the various frameworks is always possible. It is just a dogma—a dangerous dogma—that the different frameworks are like mutually untranslatable languages. The fact is that even totally different languages (like English and Hopi, or Chinese) are not untranslatable, and that there are many Hopis or Chinese who have learnt to master English very well.

The Myth of the Framework is, in our time, the central bulwark of irrationalism. My counter-thesis is that it simply exaggerates a difficulty

[1] See, for example, Chapter 10 of my *Conjectures and Refutations*, and the first *Addendum* to the 4th (1962) and later editions of volume ii of my *Open Society*.

into an impossibility. The difficulty of discussion between people brought up in different frameworks is to be admitted. But nothing is more fruitful than such a discussion; than the culture clash which has stimulated some of the greatest intellectual revolutions.

I admit that an intellectual revolution often looks like a religious conversion. A new insight may strike us like a flash of lightning. But this does not mean that we cannot evaluate, critically and rationally, our former views, in the light of new ones.

It would thus be simply false to say that the transition from Newton's theory of gravity to Einstein's is an irrational leap, and that the two are not rationally comparable. On the contrary, there are many points of contact (such as the role of Poisson's equation) and points of comparison: it follows from Einstein's theory that Newton's theory is an excellent approximation (except for planets or comets moving on elliptic orbits with considerable eccentricities).

Thus in science, as distinct from theology, a critical comparison of the competing theories, of the competing frameworks, is always possible. And the denial of this possibility is a mistake. In science (and only in science) can we say that we have made genuine progress: that we know more than we did before.

Thus the difference between Kuhn and myself goes back, fundamentally, to logic. And so does Kuhn's whole theory. To his proposal: 'Psychology rather than Logic of Discovery' we can answer: all your own arguments go back to the thesis that the scientist is *logically forced* to accept a framework, since no rational discussion is possible between frameworks. This is a logical thesis—even though it is mistaken.

Indeed, as I have explained elsewhere, 'scientific knowledge' may be regarded as subjectless.[1] It may be regarded as a system of theories on which we work as do masons on a cathedral. The aim is to find theories which, in the light of critical discussion, get nearer to the truth. Thus the aim is the increase of the truth-content of our theories (which, as I have shown,[2] can be achieved only by increasing their content).

I cannot conclude without pointing out that to me the idea of turning for enlightenment concerning the aims of science, and its possible progress, to sociology or to psychology (or, as Pearce Williams recommends, to the history of science) is surprising and disappointing.

In fact, compared with physics, sociology and psychology are riddled

[1] See now my lecture 'Epistemology Without a Knowing Subject' in *Proceedings of the Third International Congress for Logic, Methodology and Philosophy of Science*, Amsterdam 1967.

[2] See my paper 'A Theorem on Truth-Content' in the Feigl Festschrift *Mind, Matter, and Method*, edited by P. K. Feyerabend and Grover Maxwell, 1966.

with fashions, and with uncontrolled dogmas. The suggestion that we can find anything here like 'objective, pure description' is clearly mistaken. Besides, how can the regress to these often spurious sciences help us in this particular difficulty? Is it not sociological (or psychological, or historical) *science* to which you want to appeal in order to decide what amounts to the question 'What is *science*?' or 'What is, in fact, normal in science?' For clearly you do not want to appeal to the sociological (or psychological or historical) lunatic fringe? And whom do you want to consult: the 'normal' sociologist (or psychologist, or historian) or the 'extraordinary' one?

This is why I regard the idea of turning to sociology or psychology as surprising. I regard it as disappointing because it shows that all I have said before against sociologistic and psychologistic tendencies and ways, especially in history, was in vain.

No, this is not the way, as mere logic can show; and thus the answer to Kuhn's question 'Logic of Discovery or Psychology of Research?' is that while the Logic of Discovery has little to learn from the Psychology of Research, the latter has much to learn from the former.

The Nature of a Paradigm[1]

MARGARET MASTERMAN
Cambridge Language Research Unit

1. *The initial difficulty: Kuhn's multiple definitions of a paradigm.*
2. *The originality of Kuhn's sociological notion of a paradigm: the paradigm is something which can function when the theory is not there.*
3. *The philosophic consequence of Kuhn's insistence on the centrality of normal science: philosophically speaking, a paradigm is an artefact which can be used as a puzzle-solving device; not a metaphysical world-view.*
4. *A paradigm has got to be a concrete 'picture' used analogically; because it has got to be a 'way of seeing'.*
5. *Conclusion: preview of the logical characteristics of a paradigm.*

The purpose of this paper is to elucidate T. S. Kuhn's conception of a paradigm; and it is written on the assumption that T. S. Kuhn is one of the outstanding philosophers of science of our time.

It is curious that, up to now, no attempt has been made to elucidate this notion of paradigm, which is central to Kuhn's whole view of science as set out in his [1962].[2] Perhaps this is because this book is at once scientifically perspicuous and philosophically obscure. It is being widely read, and increasingly appreciated, by actual research workers in the sciences, so that it must be (to a certain extent) scientifically perspicuous. On the other hand, it is being given widely diverse interpretations by philosophers, which gives some reason to think that it is philosophically obscure. The reason for this double reaction, in my view, derives from the fact that Kuhn has really looked at actual science, in several fields, instead of confining his field of reading to that of the history and philosophy of science, i.e. to one field. Insofar, therefore, as his material is recognizable and familiar to actual scientists, they find his thinking about it easy to understand. In so far as this same material is strange and unfamiliar to

[1] This paper is a later version of an earlier paper which I had been asked to read when there was to have been a panel discussion of T. S. Kuhn's work in this Colloquium; and which I was prevented from writing by getting severe infective hepatitis. This new version is therefore dedicated to the doctors, nurses and staff of Block 8, Norwich Hospital, who allowed a Kuhn subject-index to be made on a hospital bed.

It has been tailored in shape to conform, as closely as possible, to the convalescent contribution which I actually made from the floor at the Symposium.

[2] The view presented in this paper is based on Kuhn's [1962], not on the rest of his published work. All page-numbers given in the text refer to Kuhn's [1962].

philosophers of science, they find any thinking that is based on it opaque. Kuhn's form of thinking, however, is not in fact opaque, but complex, since, philosophically speaking, it reflects the complexity of its material. In an analogous way Lakatos, in *Proofs and Refutations*[1] has introduced a new complexity and realism into our conception of mathematics, because he has taken a close look at what mathematicians really do when they refine and change each other's devices and ideas. As philosophers, therefore, we ought to progress beyond the new 'point of realism' about science these two have established, not regress from it. And as scientists, we ought to examine closely the work of both these two detailed thinkers, since, even if only as a general guide, it might be of use actually within science.

The present paper is written more from a scientific point of view than a philosophic one; though it should be said at once that I work not in the physical, but in the computer sciences. That being so, far from querying the existence of Kuhn's 'normal science', I am going to assume it. There is no need to keep on invoking history here. That there is normal science—and that it is exactly as Kuhn says it is—is the outstanding, the crashingly obvious fact which confronts and hits any philosophers of science who set out, in a practical or technological manner, to do any actual scientific research. It is because Kuhn—at last—has noticed this central fact about all real science (basic research, applied, technological, are all alike here), namely that it is normally a habit-governed, puzzle-solving activity, not a fundamentally upheaving or falsifying activity (not, in other words, a *philosophical* activity), that actual scientists are now, increasingly reading Kuhn instead of Popper: to such an extent, indeed, that, in new scientific fields particularly, 'paradigm' and not 'hypothesis' is now the 'O.K. word'. It is thus scientifically urgent, as well as philosophically important, to try to find out what a Kuhnian paradigm is.

Since my overall viewpoint is scientific, this paper also assumes that science as it is actually done—i.e. science roughly as Kuhn describes it—is also science as it ought to be done. For if there is not some self-correcting mechanism which operates within science itself, then there is no hope that, scientifically speaking, things ever will be set right when they go wrong. For the one thing working scientists are not going to do is to change their ways of thinking, in doing science, *ex more philosophico*, because they have Popper and Feyerabend pontificating at them like eighteenth-century divines; particularly as both Popper and Feyerabend normally pontificate at even more than eighteenth-century length.[2]

[1] Lakatos [1963–4].

[2] Feyerabend [1962], p. 60. (This more-than-prophetic outburst includes within itself a meta-outburst against contemporary linguistic Oxford philosophy.) See also, more briefly, Watkins in the present symposium.

This preface is, I fear, a shade aggressive; compression of material and indignation with what I shall call in the paper 'philosophy-of-science-aetherialism' have caused this. In any case, in view especially of some of the more extreme phrases used by Watkins,[1] a little pro-Kuhn aggressiveness injected into this symposium will not do any harm.

I. THE INITIAL DIFFICULTY: KUHN'S MULTIPLE DEFINITIONS OF A PARADIGM

Two vital difficulties arise for those who take Kuhn's 'new image of science'[2] seriously. On the first, which is his conception of verification in experience (or the absence of it), I do not agree with him and on this it seems to me that the philosophical empiricist world has indeed a case against him. But on the second, which is his conception of a paradigm, he has a case against them. For not only is Kuhn's paradigm, in my view, a fundamental idea and a new one in the philosophy of science, and therefore one which deserves examination, but also, although Kuhn's whole general view of the nature of scientific revolutions depends on it, those who attack him have never taken the trouble to find out what it is. Instead, they assume without question either that a paradigm is a 'basic theory' or that it is a 'general metaphysical viewpoint'; whereas I think it is in fact quite easy to show that, in its primary sense, it cannot be either of these.

Kuhn, of course, with that quasi-poetic style of his, makes paradigm-elucidation genuinely difficult for the superficial reader. On my counting, he uses 'paradigm' in not less than twenty-one different senses in his [1962], possibly more, not less. Thus he describes a paradigm:

(1) as a universally recognized scientific achievement (p. *x*): '[Paradigms] I take to be universally recognized scientific achievements that for a time provide model problems and solutions to a community of practitioners'.

(2) As a myth (p. 2): 'Historians confront growing difficulties in distinguishing the 'scientific' component of past observation and belief from what their predecessors had readily labelled 'error' and 'superstition'. The more carefully they study, say, Aristotelian dynamics, phlogistic chemistry, or caloric thermodynamics, the more

[1] For example, in the comparison between Kuhn's view of 'the scientific community as an essentially closed society, intermittently shaken by collective nervous breakdowns followed by restored mental unison', and Popper's (noble) view of it as an open society; see Watkins, *this volume*, p. 26, footnote 2 and pp. 29–30. The latter contains a really very gross distortion of Kuhn's real view—a distortion repeated on pp. 31–32, and in the whole tone of the passage, accusing Kuhn of 'seeing science as the scientist's religion'; and in that of his discussion of what he calls 'The Instant-Paradigm Thesis'. It is only fair to say that Watkins also apologizes twice for the unnecessary violence of his style; once when he correctly accuses himself of 'a certain conscious unfairness'; and once when he confesses to speaking 'a trifle maliciously'. But that a serious philosopher of his calibre should consider himself justified in being concurrently thus superficial and inaccurate in criticism, and thus violent in style—this is not only matter for comment, but also for surprise.

[2] Kuhn [1962], pp. 1 and 3.

certain they feel that those once current views of nature were, as a whole, neither less scientific nor more the product of human idiosyncrasy than those current today. If these out-of-date beliefs are to be called myths, then myths can be produced by the same sorts of methods and held for the same sorts of reasons that now lead to scientific knowledge. If, on the other hand, they are to be called science, then science has included bodies of belief quite incompatible with the ones we hold today.'

(3) As a 'philosophy', or constellation of questions (pp. 4–5): '[No] scientific group could practise its trade without some set of received beliefs. Nor does it make less consequential the particular constellation to which the group, at a given time, is in fact committed. Effective research scarcely begins before a scientific community thinks it has acquired firm answers to questions like the following: What are the fundamental entities of which the universe is composed? How do these interact with each other and with the senses? What questions may legitimately be asked about such entities and what techniques employed in seeking solutions?'.

(4) As a textbook, or classic work (p. 10): ' "Normal science" means research firmly based upon one or more past scientific achievements, achievements that some particular scientific community acknowledges for a time as supplying the foundation for its further practice. Today such achievements are recounted, though seldom in their original form, by science textbooks, elementary and advanced. These textbooks expound the body of accepted theory, illustrate many or all of its successful applications, and compare these applications with exemplary observations and experiments. Before such books became popular early in the nineteenth century (and until even more recently in the newly matured sciences), many of the famous classics of science fulfilled a similar function. Aristotle's *Physica*, Ptolemy's *Almagest*, Newton's *Principia* and *Opticks*, Franklin's *Electricity*, Lavoisier's *Chemistry*, and Lyell's *Geology*—these and many other works served for a time implicitly to define the legitimate problems and methods of a research field for succeeding generations of practitioners. They were able to do so because they shared two essential characteristics. Their achievement was sufficiently unprecedented to attract an enduring group of adherents away from competing modes of scientific activity. Simultaneously, it was sufficiently open-ended to leave all sorts of problems for the redefined group of practitioners to resolve. Achievements that share these two characteristics I shall henceforth refer to as "paradigms".'

(5) As a whole tradition, and in some sense, as a model (pp. 10–11): '. . . some accepted examples of actual scientific practice—examples which include law, theory, application, and instrumentation together—provide models from which spring particular coherent traditions of scientific research. These are the traditions which the historian describes under such rubrics as "Ptolemaic astronomy" (or "Copernican"), "Aristotelian dynamics" (or "Newtonian"), "corpuscular optics" (or "wave optics"), and so on. The study of paradigms, including many that are far more specialised than those named illustratively above, is what mainly prepares the student for membership in the particular scientific community with which he will later practise.'

(6) As a scientific achievement (p. 11): 'Because in this essay the concept of a paradigm will often substitute for a variety of familiar notions, more will need to be said about the reasons for its introduction. Why is the concrete scientific achievement, as a locus of professional commitment, prior to the various concepts, laws, theories, and points of view that may be abstracted from it? In what sense is the shared paradigm a fundamental unit for the student of scientific development, a unit that

cannot be fully reduced to logically atomic components which might function in its stead?'

(7) As an analogy (p. 14): 'One early group of theories, following seventeenth-century practice, regarded attraction and frictional generation as the fundamental electrical phenomena. This group tended to treat repulsion as a secondary effect due to some sort of mechanical rebounding and also to postpone for as long as possible both discussion and systematic research on Gray's newly discovered effect, electrical conduction. Other "electricians" (the term is their own) took attraction and repulsion to be equally elementary manifestations of electricity and modified their theories and research accordingly. (Actually, this group is remarkably small—even Franklin's theory never quite accounted for the mutual repulsion of two negatively charged bodies.) But they had as much difficulty as the first group in accounting simultaneously for any but the simplest conduction effects. Those effects, however, provided the starting point for still a third group, one which tended to speak of electricity as a "fluid" that could run through conductors rather than as an "effluvium" that emanated from non-conductors.'

(8) As a successful metaphysical speculation (pp. 17–18): '. . . in the early stages of the development of any science different men confronting the same range of phenomena, but not usually all the same particular phenomena, describe and interpret them in different ways. What is surprising, and perhaps also unique in its degree to the fields we call science, is that such initial divergences should ever largely disappear . . . To be accepted as a paradigm, a theory must seem better than its competitors, but it need not, and in fact never does, explain all the facts with which it can be confronted.'

(9) As an accepted device in common law (p. 23): 'In its established usage, a paradigm is an accepted model or pattern, and that aspect of its meaning has enabled me, lacking a better word, to appropriate "paradigm" here. But it will shortly be clear that the sense of "model" and "pattern" that permits the appropriation is not quite the one usual in defining "paradigm". In grammar, for example, "*amo, amas, amat*" is a paradigm because it displays the pattern to be used in conjugating a large number of other Latin verbs, e.g. in producing "*laudo, laudas, laudat*". In this standard application, the paradigm functions by permitting the replication of examples any one of which could in principle serve to replace it. In a science, on the other hand, a paradigm is rarely an object for replication. Instead, like an accepted judicial decision in the common law, it is an object for further articulation and speculation under new or more stringent conditions.'

(10) As a source of tools (p. 37): '. . . the conceptual and instrumental tools the paradigm supplies.'

(11) As a standard illustration (p. 43): 'Close historical investigation of a given speciality at a given time discloses a set of recurrent and quasi-standard illustrations of various theories in their conceptual, observational, and instrumental applications. These are the community's paradigms, revealed in its textbooks, lectures, and laboratory exercises. By studying them and by practising with them, the members of the corresponding community learn their trade. The historian, of course, will discover in addition a penumbral area occupied by achievements whose status is still in doubt, but the core of solved problems and techniques will usually be clear. Despite occasional ambiguities, the paradigms of a mature scientific community can be determined with relative ease.'

(12) As a device, or type of instrumentation (pp. 59–60): '. . . they denied previously

paradigmatic types of instrumentation their right to that title. In short, consciously or not, the decision to employ a particular piece of apparatus and to use it in a particular way carries an assumption that only certain sorts of circumstances will arise. There are instrumental as well as theoretical expectations, and they have often played a decisive role in scientific development. One such expectation is, for example, part of the story of oxygen's belated discovery. Using a standard test for "the goodness of air", both Priestley and Lavoisier mixed two volumes of their gas with one volume of nitric oxide, shook the mixture over water, and measured the volume of the gaseous residue. The previous experience from which this standard procedure had evolved assured them that with atmospheric air the residue would be one volume and that for any other gas (or for polluted air) it would be greater. In the oxygen experiments both found a residue close to one volume and identified the gas accordingly. Only much later and in part through an accident did Priestley renounce the standard procedure and try mixing nitric oxide with his gas in other proportions. He then found that with quadruple the volume of nitric oxide there was almost no residue at all. His commitment to the original test procedure—a procedure sanctioned by much previous experience—had been simultaneously a commitment to the non-existence of gases that could behave as oxygen did. Illustrations of this sort could be multiplied by reference, for example, to the belated identification of uranium fission. One reason why that nuclear reaction proved especially difficult to recognise was that men who knew what to expect when bombarding uranium chose chemical tests aimed mainly at elements from the upper end of the periodic table. Ought we to conclude from the frequency with which such instrumental commitments prove misleading that science should abandon standard tests and standard instruments? That would result in an inconceivable method of research. Paradigm procedures and applications are as necessary to science as paradigm laws and theories. . . .'

(13) As an anomalous pack of cards.[1]

(14) As a machine-tool factory (p. 76): 'So long as the tools a paradigm supplies continue to prove capable of solving the problems it defines, science moves fastest and penetrates most deeply through confident employment of those tools. The reason is clear. As in manufacture so in science—retooling is an extravagance to be reserved for the occasion that demands it.'

(15) As a gestalt figure which can be seen two ways (p. 85): '. . . the marks on paper that were first seen as a bird are now seen as an antelope, or vice versa. That parallel can be misleading. Scientists do not see something *as* something else; instead, they simply see it. We have already examined some of the problems created by saying that Priestley saw oxygen as dephlogisticated air. In addition, the scientist does not preserve the gestalt subject's freedom to switch back and forth between ways of seeing. Nevertheless, the switch of gestalt, particularly because it is today so familiar, is a useful elementary prototype for what occurs in full-scale paradigm shift.'

(16) As a set of political institutions (p. 92): '. . . it is crisis alone that attenuates the role of political institutions as we have already seen it attenuate the role of paradigms.'

(17) As a 'standard' applied to quasi-metaphysics (p. 102): 'And as the problems change, so, often, does the standard that distinguishes a real scientific solution from a mere metaphysical speculation, word game, or mathematical play.'

[1] Cf. Kuhn's discussion of the Bruner-Postman experiment, *op. cit.* pp. 62–3.

(18) As an organizing principle which can govern perception itself (p. 112): 'Surveying the rich experimental literature from which these examples are drawn makes one suspect that something like a paradigm is prerequisite to perception itself.'
(19) As a general epistemological viewpoint (p. 120): '. . . philosophical paradigm initiated by Descartes and developed at the same time as Newtonian dynamics.'
(20) As a new way of seeing (p. 121): 'Scientists . . . often speak of the "scales falling from the eyes" or of the "lightning flash" that "inundates" a previously obscure puzzle, enabling its components to be seen in a new way. . . .'
(21) As something which defines a broad sweep of reality (p. 128): 'Paradigms determine large areas of experience at the same time.'

It is evident that not all these senses of 'paradigm' are inconsistent with one another: some may even be elucidations of others. Nevertheless, given the diversity, it is obviously reasonable to ask: 'Is there anything in common between all these senses? Is there, philosophically speaking, anything definite or general about the notion of a paradigm which Kuhn is trying to make clear? Or is he just a historian-poet describing different happenings which have occurred in the course of the history of science, and referring to them all by using the same word "paradigm"?'

Preliminary attempts to answer this query by textual criticism make clear that Kuhn's twenty-one senses of 'paradigm' fall into three main groups. For when he equates 'paradigm' with a set of beliefs (p. 4), with a myth (p. 2), with a successful metaphysical speculation (p. 17), with a standard (p. 102), with a new way of seeing (pp. 117–21), with an organizing principle governing perception itself (p. 120), with a map (p. 108), and with something which determines a large area of reality (p. 128), it is clearly a metaphysical notion or entity, rather than a scientific one, which he has in his mind. I shall therefore call paradigms of this philosophical sort *metaphysical paradigms*, or *metaparadigms*; and these are the only kind of paradigm to which, to my knowledge, Kuhn's philosophical critics have referred. Kuhn's second main sense of 'paradigm', however, which is given by another group of uses, is a sociological sense. Thus he defines 'paradigm' as a universally recognized scientific achievement (p. *x*), as a concrete scientific achievement (pp. 10–11), as like a set of political institutions (p. 91), and as like also to an accepted judicial decision (p. 23). I shall call paradigms of this sociological sort *sociological paradigms*. Finally, Kuhn uses 'paradigm' in a more concrete way still, as an actual textbook or classic work (p. 10), as supplying tools (pp. 37 and 76), as actual instrumentation (pp. 59 and 60); more linguistically, as a grammatical paradigm (p. 23), illustratively, as an analogy (e.g. on p. 14); and more psychologically, as a gestalt-figure and as an anomalous pack of cards (pp. 63 and 85). I shall call paradigms of this last sort *artefact paradigms* or *construct paradigms*.

From now on I shall assume (though with some apology to scholars) that textual criticism of Kuhn gives us, in the end, only metaphysical paradigms, sociological paradigms and construct paradigms; and I will discuss the sociological sense of 'paradigm' first.

2. THE ORIGINALITY OF KUHN'S SOCIOLOGICAL NOTION OF A PARADIGM: THE PARADIGM IS SOMETHING WHICH CAN FUNCTION WHEN THE THEORY IS NOT THERE

Seen sociologically (as opposed to being seen philosophically) a paradigm is a set of scientific habits. By following these, successful problem-solving can go on: thus they may be intellectual, verbal, behavioural, mechanical, technological; any or all of these; it depends on the type of problem which is being solved. The only explicit definition of a paradigm, in fact, which Kuhn ever gives is in terms of these habits, though he lumps them all together under the name of a concrete scientific achievement. 'Normal science', he says (p. 10), means 'research based upon one or more past scientific achievements that some particular community acknowledges for a time as supplying the foundation for its further practice'. These achievements are (i) 'sufficiently unprecedented to attract an enduring group of adherents away from competing modes of scientific activity', and (ii) 'sufficiently open-ended to leave all sorts of problems for the redefined group of practitioners to solve. Achievements that share these two characteristics I shall henceforward refer to as *paradigms*'. Thus, by assigning the central place, in real science, to a concrete achievement rather than to an abstract theory, Kuhn, alone among philosophers of science, puts himself in a position to dispel the worry which so besets the working scientist confronted for the first time with professional philosophy-of-science, 'How can I be using a theory which isn't there?'

Kuhn himself has no doubt, moreover, that his paradigms, thus sociologically defined, are prior to theory. (This is part of the reason why he wants a new word, other than 'theory' to describe them.) For 'why', he asks himself (p. 11) is the paradigm, or scientific achievement, 'as a locus of professional commitment, prior to the various concepts, laws, theories and points of view that may be abstracted from it?' Unfortunately (and typically), having posed this highly germane question, Kuhn gives himself no answer, and the reader is left to work out the answer for himself, if he can. But at least it is made clear that, for Kuhn, something sociologically describable, and above all, concrete, already exists in actual science, at the early stages, when the theory is not there.

It is worth remarking also that, whatever synonym-patterns Kuhn may get trapped into establishing in the heat of his arguments, he never, in fact, equates 'paradigm', in any of its main senses, with 'scientific theory'. For his metaparadigm is something far wider than, and ideologically prior to, theory: i.e. a whole *Weltanschauung*. His sociological paradigm, as we have seen, is also prior to theory, and other than theory, since it is something concrete and observable: i.e. a set of habits. And his construct-paradigm is less than a theory, since it can be something as little theoretic as a single piece of apparatus: i.e. anything which can cause actual puzzle-solving to occur.

Thus the widely-held popular views that Kuhn is not really saying anything new: or that in so far as he is a philosopher at all, his views are essentially the same as Feyerabend's; or that he must be trying to say the same things as Popper (since Popper first said everything that is true about the philosophy of science), but that he does not say them very efficiently or with the right kind of emphasis; all these judgements can be shown, from actual examination of Kuhn's text, to be false.[1] It is, in fact, the very differences between Kuhn's 'new image' of science (or, as I shall from now on call it, the 'paradigm view' of science) and all other philosophies of science which are known to me, which is causing Kuhn's book to be so widely read, and which is prompting me to write the present paper.

I will therefore try to say, in the next section, what I think it is in the paradigm view which, by successfully establishing the characteristic scientificness of science, successfully combats the aetherial philosophicness of the Popperian 'falsifiable metaphysics' view. After that I will try to say something about the kind of effect that Kuhn's paradigm view has on the older and tighter 'hypothetico-deductive view'; for the paradigm view surprisingly seems to me to be much nearer to the second of these views than to the first. In conclusion I will hint at what I think are going to turn out to be the distinctive and revolutionary logical characteristics of Kuhn's paradigm, once it has been stripped of its sociological environment and looked at generally and philosophically. I shall derive all these logical characteristics from the paradigm's basic property, which I shall call concreteness or 'crudeness'.

Before starting all this, and to round off this section, I will try to sketch in, in an impressionistic manner, how I think Kuhn's view of science differs from Feyerabend's, since Feyerabend is both the philosopher of science who, so far, is nearest to Kuhn, and also the one who has given Kuhn's work most study.[2] The main difference, I think, is that, owing to his general sociological bias, Kuhn's interests are much more inclusive

[1] I could document all of these; but I won't.

[2] Feyerabend [1962], p. 32. What is given here is a very cavalier account of Feyerabend's paper, for which I ought to apologize, since I have given a positive and summary impression of what is in fact a series of negative results.

than Feyerabend's. Kuhn is interested in both the rise and fall of science, in the whole process of human beings trying to achieve a scientific explanation. Feyerabend is interested only in the fall; all his analyses are about that sense of explanation in which it is thought to be synonymous with reduction; Feyerabend, for instance, presupposes at least one fully articulated theory already to exist. But Kuhn does not presuppose anything; not even, initially, his paradigms.[1] He researches into the real history, and broods; he reads scientific teaching textbooks, and wonders. An investigation into the originality of Kuhn, then, is also an investigation into the crude forms and early stages of a science.

And this is, above all, what makes his work attractive to scientists in new fields; pre-eminently, of course, to scientists in the social sciences, and in experimental psychology. One of the reasons why professional philosophy of science at present looks aethereal to actual research scientists, is that modern philosophers of science, taken as a group, have worked backwards. First we had the hypothetico-deductive view, the datum of which is the single, apparently all-inclusive, self-consistent, fully articulated, complete and tight interpreted deductive system—that ideal which no science reaches, but to which, if Kuhn is right, every teaching textbook, in an advanced hard science, tries to approximate.[2] Subsequent to this, we have had Feyerabend's newer conception (following on Popper's) of the stage which comes before: that is, of two much younger, much less completely finished-off theories which compete to cover what (though only in a Pickwickian sense) can be called 'the same field'. No modern philosopher of science has, as yet, gone back earlier; to the stages when there are either no theories at all, as I am about to say in the next section, or far too many theories (if the word 'theory' is used metaphysically or colloquially) and no clear field. In view of the current proliferation of alleged new sciences, however, it is overdue, if the philosophy of science is to become, as it should, a scientifically useful guide to actual research workers, that some informed philosophic move backwards should now be made.

Kuhn, in my view, has made this move; or tried to.

3. THE PHILOSOPHIC CONSEQUENCE OF KUHN'S INSISTENCE ON THE CENTRALITY OF NORMAL SCIENCE: PHILOSOPHICALLY SPEAKING, A PARADIGM IS AN ARTEFACT WHICH CAN BE USED AS A PUZZLE-SOLVING DEVICE; NOT A METAPHYSICAL WORLD-VIEW

It might be said, by those who are impressed by the analytic primacy which Kuhn gives to sociology as opposed to philosophy, as giving the

[1] Before he took up his present intellectual position, Kuhn's development ranged over a number of fields and went through at least six stages (see his [1962], preface pp. vii–x).

[2] Kuhn [1962], p. 1; pp. 1–2; p. 10; pp. 135 ff.; p. xi; and see also section IV, *below*.

main clues to the foundations of real science, 'Why do you flog this notion of "paradigm" any further? It's just Kuhn's name for a set of habits, that's all. These exist, granted; but the fact is of no philosophical importance.'

This is not correct, even about Kuhn. Besides his sociological paradigms (sense 2), he has metaphysical paradigms (sense 1), and also artefact paradigms, or construct paradigms (sense 3). It is easy to show that he has at least these. But quite apart from what Kuhn, taken now as a philosopher, has actually said about paradigms, there is a deeper and more immediate reason for not being satisfied with a purely sociological sense of 'paradigm', which is that any definition of this is bound to be circular. For, to establish the paradigm's (temporal) priority to theory in scientific action, we have to define it, sociologically, as an *already known* concrete scientific achievement, an *already established* set of habits. But how does the scientist himself, in a new science, first find out that what he is following is going to become a concrete scientific achievement, unless he already knows that he is following a paradigm? There is clearly a circularity here: first we define a paradigm as an already finished achievement; and then, from another point of view, describe the achievement as building up round some already existent paradigm.

It could be argued, of course, that if we seriously undertook the detailed sociology, obtained by observation, of fresh contemporary new sciences, instead of confining ourselves to the detailed history, obtained through hindsight, of stale past sciences, this circularity, for practical purposes, could be broken down; since if they existed, we could then detect paradigms in the process of being formed. But even then, how would we know that it was *paradigms* which we were looking for, as opposed to other things, unless we already knew, non-sociologically, what a paradigm was? The primary sense of 'paradigm', clearly, has got to be a philosophic one; and the paradigm has got to exist prior to the theory. This once established, the man who says, 'What, in actual fact, is this "paradigm", this entity?', can then indeed be answered by being told to go and look at what is happening in a new scientific field. For in a new science, not only is the formal theory almost sure to be missing; but also a very great deal of high-powered scientific activity is aimed at the right choice of the moment when it will be worth the labour to construct it. The alternative is 'just going on as we are now'; that is, with some trick, or embryonic technique, or picture, *and an insight that this is applicable in this field.* And it is this trick, plus this insight, which together constitute the paradigm. The explicit metaphysics (what the scientist himself calls 'the philosophy' or 'the gas'), the fuller mathematicizing innovation, the more developed experimental procedures —all those things which, taken together, will later become 'the concrete established scientific achievement'—nearly always come long after the initial practical trick-which-works-sufficiently-for-the-choice-of-it-to-embody-a-potential-insight, that is, after the first tryout of the paradigm. In fact, and in genuine and live science, the very effort to establish a 'concrete scientific achievement' has to justify itself. For the resultant theory (and/or the more exact and expensive technique) to be acceptable, it must enable results to be obtained which could not be obtained otherwise. No good scientist wants to establish such an achievement just to figure later in books on the philosophy of science. Still less does he want theoretically to clean up his subject at the cost of removing from the hitherto used colloquial description of the facts any possible analysis of the real centres of difficulty. Thus the real problem, in getting a philosophy of new science, is to describe philosophically the original trick, or device, on which the sociological paradigm (i.e. the set of habits) is itself founded.

With all this in mind, it is enlightening to turn again comparatively to Kuhn's first and third senses of 'paradigm'. As has been seen, if we ask what a Kuhnian paradigm *is*, Kuhn's habit of multiple definition poses a problem. If we ask, however, what a paradigm *does*, it becomes clear at once (assuming always the existence of normal science) that the construct sense of 'paradigm', and not the metaphysical sense or metaparadigm, is the fundamental one. *For only with an artefact can you solve puzzles.* And though, having initially asserted (p. 36) that he is going to use 'puzzle' in the literal, standard, dictionary sense, Kuhn later weakens and talks (p. 42) about 'the metaphor that relates normal science to puzzle-solving', yet, in general, he has a steady, literal and very concrete idea of what he means by the puzzle-solving activity of normal science. A normal-scientific puzzle always has a solution (p. 36) which is guaranteed by the paradigm, but which it takes ingenuity and resourcefulness to find. Typically (p. 35), the solution is known beforehand, as with any other puzzle, but the step-by-step route to it is not. The normal scientist is a puzzle-solving addict (p. 37); it is in this puzzle-solving —not just vague 'problem-solving', but *puzzle*-solving—that normal science prototypically consists. And a puzzle is always an artefact. It is all very well to say that the paradigm 'supplies tools' (pp. 37 and 76) or, vaguely, that it makes problem-solving possible. It remains true that for any puzzle which is really a puzzle to be solved by using a paradigm, this paradigm must be a construct, an artefact, a system, a tool; together with the manual of instructions for using it successfully and a method of interpretation of what it does.

However, if it is true that it is Kuhn's construct-paradigm, and not either of his two other main senses of 'paradigm', which provides the philosophical clue to what paradigms in a new science really are, by pinpointing the trick or device which starts off a new science; if all this, then why is it that all philosophers of science other than myself have thought it evident that by 'paradigm' Kuhn meant a metaphysical world-view, that his primary sense of 'paradigm' was sense 1, not sense 3? The immediate explanation of this is easy. They did not take Kuhn's account of normal science seriously. However, it might still be thought that by saying all this I intend to repudiate all that philosophers of science are currently saying about science emerging out of metaphysics (the 'falsifiable metaphysics' view); or that I am ignoring what Kuhn himself says about preparadigm science[1]; or that I am laying down the law in a Marxist manner about the motivation for all new science being technological. This is not so. It is obvious that one of the roots of scientific achievement is metaphysical, as Popper, Kuhn himself and many others have said. But the current philosophic bias has gone so much towards examining what is conceptual, in thinking about the nature of any science, that philosophers have all but forgotten to allow for what is practical. Thus Kuhn has not seen the relevance, in discussing the verification problem, of final technological application[2]; and Popper has not seen the relevance, in discussing the emergence of science out of metaphysics and philosophy, of the technical trick which starts off each new science. Though he must have heard the old saw to the effect that science is a marriage between metaphysics and technology, Popper never asks himself how the copulation occurs; consequently, the fatal weakness of the Popperian view of science is that the Popperians can provide no answer to the question, 'If a scientific system is essentially a metaphysical system which is falsifiable, how can the metaphysics itself be used as a model, and subjected to test?'.

This brings me to my promised comparison of Kuhn and Popper; or, more exactly, to a comparison between the paradigm view of new science, and the Popperian view. For the gross *lacuna* which I assert to be in the Popperian view—namely, that Popper cannot account for how any new research line suddenly starts up—this is not due, as is sometimes alleged by cynics, to the fact that Popperian philosophers of science are incapable of understanding technology, or that technologists are incapable of thinking Popperianwise about the philosophy of science. Neither of these assertions is true, and both are irrelevant. What has caused the trouble, in my judgement, is excessive reliance upon Newton. Newtonian mechanics, just because it has lasted so long, is in the unique position, among scientific

[1] And indeed I am being cavalier about what Kuhn says about preparadigm science; just as I was earlier cavalier about Feyerabend. See, however, the discussion of it at the end of this section.

[2] Kuhn [1962], pp. xii, 19, 69 and 166–7; Kuhn thinks technology is outside the sphere of the philosophy of science.

theories, of being able to be regarded either as quasi-metaphysics, or as the very prototype of deductive theory, or (now) as technology, according to how you choose to look at it. Moreover, reliance upon Newtonian mechanics, as being always there to be ambiguously pointed at in any crisis as to what science is, is slavish. If all the philosophers of science who derive from Kant had not been able to equate science with Newtonian mechanics, where would the philosophy of science be? Popper himself, indeed, in *Conjectures and Refutations*, sees one great difficulty in making this equation; but whereas Popper thinks that the trouble lies (for us, but not for Kant) in the fact that we must now consider Newton's theory 'as a hypothesis whose truth is problematic' since 'Einstein has shown that it is possible, using basic principles very far removed from those of Newton, to do justice to the entire range of the data of experience',[1] in fact the trouble with Newtonian mechanics is that it works so completely that it has now become part of technology, namely the technology of sending up space satellites. On Kuhnian principles, therefore, and I think also on Popper's, it is no longer part of the philosophy of science.

Self-deprived of Newton, Popper thenceforward makes a very much poorer show of giving a realistic account of creative thinking in science. 'We invent our myths and our theories and we try them out', he says[2]—to which the answer is: 'How?' 'When?' 'Where?' Theories 'are seen to be the free creations of our minds', he continues, 'the result of an almost poetic intuition'[3]—to which the short answer is: 'Who so sees them?' 'We do not try to prove them . . . but . . . to *refute* them'[4]—to which the only answer is: 'In fact, do we?' At the earliest opportunity, moreover, Popper leaves discussion of scientific theories altogether to turn to philosophic theories, in order to analyse, brilliantly, whether these are not also, in a more direct way, refutable. He then, bar a hairsbreadth, equates these with scientific theories[5]; and one suspects that—apart from Newton—it is these, and not science as it really is, which he has had at the back of his mind all the time.

It is this virtual equation (bar Newton) of scientific thinking with speculative philosophical thinking which, more than any one other thing, currently gives rise to what I described at the beginning as 'philosophy-of science aetherialism'. By contrast with this abstractness, Kuhn, by insisting on the sociological importance of the actual set of habits which, in fact, characterize any new science, and which are prior to any formulation of theory, has succeeded in establishing, as central to his philosophy, the essential concreteness which is characteristic of science; i.e. in remaking

[1] Popper [1963], p. 191. [2] Popper [1963], p 192. [3] *Loc. cit.*
[4] *Loc. cit.* [5] Popper [1963], pp. 199–200.

the distinction which the scientist himself in his talk makes between the actual 'picture', or the 'model', and the 'philosophy'. This 'model' (the operation of which I have described earlier as the trick, or device, which starts off any new science or research line) becomes for Kuhn his construct-paradigm (paradigm sense 3), the use of which enables the puzzle-solving of normal science to be performed. And it is this identification in its turn—i.e. the fact that it is the construct-sense of paradigm and not the metaphysical sense of paradigm which has to be Kuhn's primary sense of paradigm—which enables him then to make a new interrelation between model-using and metaphysics. For instead of asking 'How is it that a metaphysical system can be used as a model?'—i.e. instead of asking the question which I said earlier that the Popperians could not answer—Kuhn can now ask: 'How is it that a puzzle-solving construct (i.e. a paradigm, sense 3) can be used metaphysically? How, in fact, can a construct-paradigm become a "way of seeing"?'

Consideration of this question forces us sharply back from the Popperian impression of science in general to a more sophisticated re-evaluation of the hypothetico-deductive conception of the exact function of a scientific theory. For, after all, an hypothetico-deductive system—if it can be constructed at all—is, by its nature, a problem-solving artefact. Before we go on to this, though, there is one confusion to clear up; which is what Kuhn himself says about the nature of new, or first stage, or preparadigm, science. For I said earlier that a consideration of the originality of Kuhn was also an investigation into the crude origins, and early stages, of any science; and I confirmed this by advancing reasons for thinking—as well as by showing that Kuhn thinks—that paradigms already exist when the theory is not there. But this immediately prompts the further question: 'What does Kuhn think exists, then, even before the paradigm?'

This is one of the points on which I disagree with Kuhn, in that his general view of the preparadigm science seems to me both confused and incompletely analysed. As I see it, he fails to distinguish from one another three relevant states of affairs, which I will call respectively *non-paradigm science*, *multiple-paradigm science*, and *dual-paradigm science*. *Non-paradigm science* is the state of affairs right at the beginning of the process of thinking about any aspect of the world, i.e. at the stage when there is no paradigm. Of this state Kuhn says (p. 15) that in it only the easily accessible facts are collected, and these in a casual manner, unless some more recondite facts have been made available by technology; that this is because, at this stage, all facts seem equally relevant; and that different but overlapping sets of facts are interpreted in differing metaphysical or quasi-fanciful ways. He further says (p. 11) that 'there can be a sort of scientific research without paradigms . . .', but that it is non-esoteric; and (pp. 13, 100 and 163) that in such research 'though the field's practitioners were scientists, the net result of their activity was something less than science'. He further notes (p. 20) that in such situations the book (as opposed to the article) possesses 'the same relation to professional achievement that it still retains in other creative fields'; that every individual scientist starts over again from the beginning (p. 13); that there are a number of competing schools directing their publications primarily against one another (p. 25); that there is continual philosophic discussion over fundamentals (p. 159); and no progress (pp. 159 and 163). In short, non-paradigm science is barely distinguishable, if at all, from 'the philosophy of' the relevant subject, and is covered by Popperian analysis.

This pre-scientific and philosophic state of affairs sharply contrasts, however, with *multiple-paradigm science*, with that state of affairs in which, far from there being no paradigm, there are on the contrary too many. (This is the present overall situation in the psychological, social and information sciences.) Here, within the sub-field defined by each paradigmatic technique, technology can sometimes become quite advanced, and normal research puzzle-solving can progress. But each sub-field as defined by its technique is so obviously more trivial and narrow than the field as defined by intuition, and also the various operational definitions given by the techniques are so grossly discordant with one another, that discussion on fundamentals remains, and long-run progress (as opposed to local progress) fails to occur. This state of affairs is brought to an end when someone invents a deeper, though cruder (p. 23) paradigm, which gives a more central insight into the nature of the field, though restricting it and making research into it more rigid, esotoric, precise (pp. 18 and 37). This (p. 16) either by causing rival, more shallow paradigms to collapse, or alternatively, by attaching them somehow or other to itself, triumphs over the rest, so that advanced scientific work can set in, with only one total paradigm. Thus multiple-paradigm science is full science, on Kuhn's own criteria; with the proviso that these criteria have to be applied by treating each sub-field as a separate field.

During the period of crisis, however, just before a scientific revolution, Kuhn says (pp. 84 and 86) that many of the characteristics of pre-paradigm science again set in, 'except that the locus of difference [between the competing schools] is both smaller and more defined'. During this period there are always two competing paradigms struggling for the mastery (pp. 75 and 91); I have therefore described it as *dual-paradigm science.*

The reason that Kuhn fails sufficiently to distinguish non-paradigm science from multiple-paradigm science, and therefore sufficiently to connect multiple-paradigm science with dual-paradigm science, is due partly to a confusion; having said that there can be a sort of scientific research without paradigms, he then adds, 'or at least without any so unequivocal and binding as the ones named above' (p. 11), as though these two states of affairs were identical. It is also partly due to the insufficient place within science which he gives to technology, which exists plentifully and sometimes excessively in multiple-paradigm science, but only non-relevantly, if at all, in non-paradigm science.

As opposed to this complicated and confused preparadigm survey of Kuhn's (and taking his notion of 'normal science' seriously) I have simplified the position by saying squarely that when 'normal science' sets in, *anywhere*, there you have science, and where it does not set in, there you have philosophy or something else, not science, and that it is always some construct-using, puzzle-solving trick which starts off normal science. This assertion exposes me to attack in two ways. Firstly I can be attacked for not being able to distinguish a single new research line from a total new science (see, for instance, the passage earlier where I equated the two with one another), and therefore in the terminology just given above, of not being able to distinguish multiple-paradigm science from mature single paradigm science. This attack is correct. In my view the two can only be distinguished from one another later, by hindsight, when a total new science with one vast paradigm, is ultimately seen to have been created through the convergence of a number of paradigm-guided research lines which mutually threw light on one another. The second attack which can be made on me is that if I distinguish 'science' from 'philosophy' *only* because within science there always somewhere occurs normal science, what about the converse case where 'normal science' prematurely sets in in some unjustified manner, by a set of fashion-following scientists starting to imitate one another without proper pre-examination of the paradigm (i.e. without the alleged insight that a certain paradigm is relevant to a particular field being a genuine insight)? My reply to this is: 'What indeed?' Do we not see premature 'normal science' (which is also called 'phoney science' and 'pseudo-science' by soured critics) setting in all round us in a nightmarish manner, in the newer sciences, especially where computers can be grandiosely used to give a spurious impression of genuine scientific efficiency? But the fact that new science can be exceedingly bad does not cause it to cease to be bad *science* (as opposed to bad philosophy, bad painting or other bad what-have-you). In the end phoney scientific normal-research lines collapse, or fail to yield any results, or topple, or evaporate—or so one hopes; and so in the past (e.g. in the case of astrology which, as Watkins says, was in some ways exceedingly 'normal') it has finally proved.

Having done what can be done to establish non-sociologically a Kuhnian paradigm as a genuinely insightful puzzle-solving trick or device, let us now both examine further the nature of the device, and also, if possible, the nature of the insight.

4. A PARADIGM HAS GOT TO BE A CONCRETE 'PICTURE' USED ANALOGICALLY; BECAUSE IT HAS GOT TO BE A 'WAY OF SEEING'

If a paradigm were only to be an interpretable construct or artefact the use of which had become an established social institution, it might be hard to distinguish Kuhn's paradigm view of science from some sociologically sophisticated hypothetico-deductive view; all the more so, as I think it can be shown that Kuhn's paradigm-view of science has a little more in common with the hypothetico-deductive view than a superficial reading of his book would imply. For in spite of his apparently vague and poetic style, both he and the hypothetico-deductivists are struggling to say something about the development of science which is exact. What distinguishes the two views from one another is that a puzzle-solving paradigm, unlike a puzzle-solving hypothetico-deductive system, has also got to be a concrete 'way of seeing'.

With hypothetico-deductive system-making in mind, let us see what Kuhn does say. He repeatedly compares the switch from one scientific paradigm to another to the operation of 're-seeing' an ambiguous gestalt-figure[1] or to being subjected to a gestalt-psychological experiment.[2] Note therefore that each of these is a completely specifiable artefact which has been specially constructed to be itself a 'way of seeing'; in fact, to be two alternative ways of seeing. When, however, we compare the paradigm itself to a gestalt-figure, the comparison becomes trivial; because if, to do so, we now ask ourselves what a gestalt-figure is like when it represents only one way of seeing, we get the trivial answer that it is a perfectly ordinary picture of a simple concrete object. Moreover, the gestalt-figure comparison fails also in yet another respect, in that an ambiguous gestalt-figure, unlike a paradigm, cannot be extended or developed, since any extra detail which is added will be bound to bias it either towards one of its interpretations, or towards the other.[3]

What Kuhn must be feeling his way to, in talking about an artefact which is also a 'way of seeing', is an assertion, not about the nature of his artefact, but about its use: namely, that being a picture of one thing, it is

[1] Kuhn [1962], pp. 85, 110, 113, 116, 119, 121, 125 and 149.
[2] *Ibid.* pp. 62, 64, 111, 112 and 125.
[3] This can be seen particularly clearly from examination of the perceptually ambiguous gestalt-figures in Gregory [1966].

used to represent another—for example, the geometrical model made of wire and beads, though it is primarily a glorification of a well-known kind of child's toy, is used in science to represent a protein molecule.

It is, in fact, actual artefacts used analogically which Kuhn is after, as have been many other philosophers of science from Norman Campbell to Hesse. But Kuhn's artefact, unlike Hesse's,[1] cannot be a simple four-point analogy or a material analogy, because it has got to be an organized puzzle-solving gestalt which is itself a 'picture' of something, *A*, if it is then to be applied, non-obviously, to provide a new 'way of seeing' something else, *B*.

Unlike Kuhn's two-way gestalt-figure, however, his paradigm does not itself have to be ambiguous as well as non-obvious in its application; it can, therefore, with caution, be developed, like other analogies. The question sharply arises, though: *how* is it to be developed? And is there any real sense in which an analogy, as contrasted with a model or a mathematical system, can be an artefact?

Before, in conclusion, we discuss this, more must be said of how Kuhn's paradigm is to be distinguished from a scientific hypothetico-deductive theory by the fact that it is a 'way of seeing'. To say that it is a concrete constructed 'picture' or device used analogically is not enough. For, it could be replied, a mathematical system itself, and even when uninterpreted, is, notoriously, a very abstract 'way of seeing'. For the man using it can always be asked, particularly in a new science, for example, 'Why are you using *that* mathematical system, and no other?', or, 'Are you sure that this mathematical picture which you are building up is giving you the kind of space which you will later want when your experimental evidence has been organized more sharply?' Moreover, according to the hypothetico-deductive view, the mathematics used in science are not uninterpreted. They are coloured—'faintly tinted', would be a better description, for the colouring-mechanism is never sufficiently made clear—by the still more highly coloured concrete truths which form the lower, more particular parts of the system. On this view, concreteness and interpretation are supposed to seep back and up, somehow, from the more concrete lower parts to the more abstract and aethereal higher parts; thus making of the whole hypothetico-deductive edifice an artefact which could indeed be held to be a 'way of seeing', *par excellence*.

Kuhn's paradigm's 'way of seeing', however, really is different from this—and not only because, as asserted earlier, his paradigm already exists when the theory is not there. It is different because his paradigm is a concrete 'picture' of something, *A*, which is used analogically to describe a concrete something else, *B*. (That is, the trick which, as I said earlier,

[1] Hesse [1963], pp. 70–3.

starts off every new science, is that a known construct, an artefact, becomes a 'research vehicle', and at the same moment, if successful, it becomes a paradigm, by being used to apply to new material, and in a non-obvious way.) It thus has two kinds of concreteness, not one: the concreteness which it brought with it through being a 'picture' of *A*, and *the second concreteness which it has now acquired, through becoming applied to B*. This second kind of concreteness is the kind which the hypothetico-deductive view of science tries to account for; but the first, on the hypothetico-deductive view, is not accounted for at all. If, however, we complicate the hypothetico-deductive view by saying, as Campbell, in effect, does,[1] but Hesse, I think, does not,[2] that there is always an analogy or a concrete model at the heart of any mathematics used in science, and that this model is not merely something attached afterwards, to be used heuristically or as a mechanical aid; if we say further, as indeed Campbell more than once does say, that it is this analogy which guides and restricts the theory's articulation, excising and removing, by the need to preserve it, the otherwise excessive possibilities of abstract development inherent in all mathematics, then the first kind of concreteness (call this A-concreteness) is accounted for as well as the second kind (call this B-concreteness). For A-concreteness now becomes the concreteness which the analogy brings with it to the mathematics from the time before it was an analogy, when it was only a 'picture' of A; whereas B-concreteness is what seeps back into the mathematics from the field of application, B. The abstract entities in the resulting theory can then be doubly interpreted—as indeed in a new science they have to be—firstly A-wise, in terms of the generating analogy, and secondly B-wise (that is, operationally, and, as the theory develops, increasingly) in terms of data taken from the field to which the theory is being applied.

That there are, quite obviously and in fact, A-components as well as B-components in scientific theories, will be seen at once as soon as philosophers of science start looking around them at fresh science instead of looking only backwards at stale science, or alternatively, and in a self-obfuscating attempt to be up-to-date, goggling only from afar at the increasing variegation of chaos in theoretic quantum mechanics. The most striking example I know of the distinction is given by the Genetic Code. Here the

[1] Campbell [1920]; see especially, pp. 129–30.

[2] Hesse's mind is split on the question as to whether analogy is at the heart of theory, as Campbell says, or only an aid to it. In her [1963] she argues brilliantly, in effect, for a Campbellian view; but in her [1964] she says only that 'the deductive model of scientific explanation should be modified and supplemented by a view of theoretical explanation as metaphoric redescription of the domain of the explanandum' (p. 1), thus still putting the mathematical cart before the metaphorical horse.

initial A-concreteness is given by a 'picture' of language, which has now been extended to include not only 'letters' and 'words', but also 'sentences' and 'punctuation'; whereas the operational B-reinterpretation in terms of operational procedures is biochemical.

I will take it from now on that I have established that there are two operational components, the A-component and the B-component, even in an idealized scientific theory; and that, whereas the hypothetico-deductive view only allows for the second, Kuhn's paradigm-view stresses the first. Both have to be distinguished, in behaviour, from their common mathematical clothing: further considerations which assist in making this distinction are given in conclusion below. Enough has been said, however, to give the case for saying that, within the current scope of the philosophy of science, the primary enterprise, in discovering the philosophical nature of a Kuhnian paradigm, now becomes that of prising out the A-component of a developed theory, the paradigm, from its also B-interpretable mathematical envelope.[1]

5. CONCLUSION: PREVIEW OF THE LOGICAL CHARACTERISTICS OF A PARADIGM

If a paradigm has got to have the property of concreteness, or 'crudeness', this means that it must either be, literally, a model; or, literally, a picture; or, literally, an analogy-drawing sequence of word-uses in natural language; or, some combination of these.

In any of these cases, I wish to say that a paradigm draws a 'crude analogy'; and further to define a crude analogy as an analogy which has the following logical characteristics:

(*a*) a crude analogy is finite in extensibility
(*b*) it is incomparable with any other crude analogy
(*c*) it is extensible only by an inferential process of 'replication', which can be examined by using the computer-programming technique of 'inexact matching', but not by the normal methods of examining inference.

The problem of saying something philosophical and yet exact about such a paradigm (which now becomes that of saying something general and exact about the nature and methods of operation of a concrete artefact, constructed of pigments, or wire, or language) cannot be attacked within the confines of this paper; all the more so as it is, I think, the same problem

[1] It is worth remarking that, on this view, the domain of the philosophical paradigm, or crude paradigm, is narrower by far than that of the sociological paradigm, or total paradigm, seen historically and by hindsight. For this second includes within itself everything the operation of which could become a habit; including, ideally, the mathematical part and the B-experimentation of an hypothetico-deductive system.

which Black tries to attack when he tries to discover the nature of an archetype,[1] or when he asks himself how he is going to formalize the 'interaction view' of metaphor used in language.[2] In my view, the new 'way of seeing' produced by Black's metaphoric 'interaction' is an alternative form of that produced by Kuhn's gestalt-switch.

Here, I will only point out, in conclusion, that once the concreteness, or 'crudeness' of an initial paradigm is granted, then great simplification can be achieved in several areas of the philosophy of science. For instance, when Kuhn says that his paradigms are not directly comparable with one another, his word for this is 'incommensurable', and the context makes clear that he is thinking of advanced science. But if one tries to construct a general and exact notion of this incommensurability, as Feyerabend does, then it can be shown, I think, that the attempt leads to great philosophic difficulties, as well as producing a *reductio ad absurdum* of real science. And if we merely envisage a concrete paradigm which draws a crude analogy, then, notoriously, in so far as it is really crude, it will not be directly comparable with any other crude analogy. (How, for instance, can you compare, 'Man, the paragon of animals', to 'Man, that wolf'?) Note also that this agreed non-comparability depends on the crudeness. It does not hold when the paradigm in question has become embedded in mathematical form, except in so far as it is the A-component and not the B-component which is in question. For the A-component, being concrete, draws a crude analogy; whereas the B-component, being mathematical-cum-operational, draws, if any analogy, only a mathematical analogy; and analogies between pieces of mathematics render them not incommensurable, but conversely, comparable.

The property of crudeness allows a comparable simplification to be made of Kuhn's statements to the effect that a paradigm must be finite in extensibility. For in so far as the crude analogy drawn by a paradigm is not merely *like* that drawn by a speaker in natural language but *is* one, then it is notorious that it cannot be developed too far (all poets know this); whereas, by contrast, mathematical extensibility is always imagined as being capable of going on and on by accretion, indefinitely.

In this matter I have to confess that (inspired by Feyerabend) I also was not content with the simplification produced by the postulate of paradigm-crudeness, but tried to construct an abstract general notion of non-extensibility. I started with the logicians' traditional generalizing device of analogy—as expounded, say by Jevons[3]—and then tried to prove

[1] Black [1962], chapter xiii.

[2] *Ibid.* chapter iii.

[3] Jevons [1873]: see *Analogy* in the index; and also chapter ii, on the logic of terms, and especially pp. 25–7.

finiteness in extensibility by using the logic of terms. To do this requires saying that the kind of analogy we want, i.e. one which makes the application of a whole A-organization to a B-field, counts as an instance of the qualification by a complex 'adjective' of a general name, or 'noun'. If we could allow this, we could then say that the intension-extension law of the logic of terms would apply also to this case, in the form that when the meaning-in-extension of such an 'adjective-noun' is indefinitely increased by the addition of further 'adjectives', its meaning-in-extension is correspondingly decreased. Thus, no matter what threshold or zero limit of intelligible meaning one sets up, there comes a stage when the continually extending sequence goes over it; thereby exhibiting the phenomenon of 'death by a thousand qualifications'. But I do not think that the development of the paradigm-analogy which is made when a good new scientific research line is started, can count as a straight-forward extra term-qualification, since the whole point of it is that it causes new features of the field of application to be discovered which would never have been noticed without the help of the paradigmatic analogy, thus *increasing* the meaning-in-extension of the whole term-sequence by adding to that which it denotes, i.e. to the field.

So, my attempt to be abstract about paradigm-extensibility failed, and I found myself left with the inescapable property of crudeness, trying to explain *ignotum per ignotius*; i.e. of explaining a logically unknown entity, a paradigm, by means of an even more unknown logical property, crudeness. The heart of the problem is that of envisaging a crude analogy stated in ambiguous words as an artefact; pictures and wire models can be fitted in with comparative ease, after this first central problem has been faced. And faced it must be; because, if what the scientist working in a new science is actually doing is constructing and extending a crude analogy *by using speech*, with or without the help of mechanical apparatus or of mathematics, then the evident fact that natural language is continually used by scientists, and sometimes to draw crude analogies with, this skeleton, has got to come out of the philosophico-logical cupboard. This is especially so as an increasing number of papers in the literature now discuss 'semantics' or 'meanings' within science, and, through absence of explicit confrontation with the problem of word ambiguity, say some very extraordinary things about it indeed.[1]

[1] See not only Feyerabend's [1962] but also Brodbeck [1962] and Putnam [1962]; and the earlier Ryle-Toulmin-Scriven bibliography that they refer to.

Of these, Feyerabend's error seems to me to be philosophic: fulminating indiscriminately against linguistic philosophers, he fails to distinguish the *truisms* of natural language from the combinatorial *resources* of natural language. Brodbeck makes statements to the effect that physicists' colloquial conversation is elliptical, allusive and laconic, as well as context-bound, whereas their official reports are explicit, comprehensible, logically complete, and

Kuhn's own account of the limits and extensibility of a paradigm is both sketchy and faulty, for which fact he himself apologizes.[1] On the other hand, the way in which he describes a paradigm breaking down, by the emergence within it of anomaly which deepens into crisis, is at once illuminating and realistic, when applied to a new science. His essential point is that an anomaly is an untruth, or a should-be-soluble-but-is-insoluble problem, or a germane but unwelcome result, or a contradiction, or an absurdity, *which is thrown up by the paradigm itself being pushed too far*[2]; not just an incidental counter-argument to the theory, or an awkward fact, which Kuhn correctly characterizes as merely an 'irritant'.[3] Neither is it an extra-paradigmatic novelty,[4] nor a problem which used to exist within the field at an earlier stage, but which the developers of the paradigm have now suppressed and rendered invisible, because it is incompatible with the paradigm's 'basic commitment'.[5] The anomaly, to be a true anomaly, has

context-free; or at least when they are not, it is because they fail to approximate to an ideal Platonic Physicist's Report which they might have written but did not (pp. 237–8). She also makes unsophisticated remarks of the kind that the ordinary-language philosophers have correctly criticized, e.g. to the effect that, 'it is necessary that white horses are white', is a statement from normal English Prose, whereas, in fact, it is either an obvious example from a logic-book, or an untrue but rather splendid poetic remark which might refer to waves, pubs, favourite writers, and angelic transport, as well as to natural animals (p. 238). She further asserts that logicians' language is useful to the philosopher 'precisely because and only in so far as it is a reconstruction of a large part of the language that we speak' (*ibid.*). Putnam profoundly wrestles with Quine's 'All bachelors are unmarried'; but in the course of doing so asserts not only that 'bachelor' is context-free (thus forgetting bachelor's buttons, bachelors of arts, medieval knights' assistants—and Fodor and Katz); but also that it is not law-clustered (forgetting equally the effect on the use of the word 'male' of testerone experiments and chromosome-aberrancies of intersexes). Likewise—though in an exceedingly interesting article—he commits himself to the rash statement (p. 362) that there are no synonymies and analyticities underlying language (after all, Strawson might be right); and to the false statement that linguists know how to describe a natural language in terms of a set of rules (pp. 389–90).

Clearly, when exceptional thinkers make remarks like these, a new kind of insight is needed on the whole subject.

[1] Kuhn [1962], pp. 86 and 89. At only one point does Kuhn argue that paradigms *must* be non-extensible (on pp. 95–6); mostly he just retreats into history, and says that they are.

[2] Kuhn [1962], p. 65 (see also, pp. 5, 52 and 78).

[3] Kuhn [1962], pp. 78–9. Kuhn's actual phrase (on p. 78, line 12) is 'minor irritant'.

[4] See *above*, footnote 2; especially p. 5, on the suppression of fundamental novelty; and all other passages which would be listed under 'novelty' in a Kuhn index if there were a Kuhn index. See also, in the same imagined index, 'anomaly'.

[5] Kuhn [1962]: p. 5 again (for the notion of 'basic commitment'); p. 102, '. . . the reception of a new paradigm often necessitates a redefinition of the corresponding science. Some old problems may be relegated to another science or declared entirely "unscientific" '; p. 37, '. . . one of the things a scientific community acquires with a paradigm is a criterion for choosing problems that, while the paradigm is taken for granted, can be assumed to have solutions. To a great extent these are the only problems that the community will admit as scientific or encourage its members to undertake. Other problems, including many that

got to be produced from within the paradigm. So that, if the paradigm is to be conceived as a crude analogy, the anomaly, in its simplest and crudest form, will correspond to Hesse's neutral analogy which turns out to be negative analogy[1]; i.e. to a set of statements (or laws) developed from within the analogy itself, which should have been true, had the analogy held that far, but which, since the analogy does not hold that far, turn out to be false. In this simple situation, attempts will inevitably be made to adjust the analogy; in the more complex, mathematicized situation, attempts are made either to derestrict or complicate the mathematics, to produce variants of the theory, or to dig out the theory's fundamental assumptions, to try to make the analogy fit again.[2] Anomaly deepens into crisis when these attempts fail; when, for example, the complexity of the theory increases faster than its accuracy[3]; or, the area of trouble grows larger, and not smaller, until the very fundamentals of the paradigm are thrown in question[4]; or, some rank outsiders with a quite different viewpoint and rudimentary new technique succeed in solving with ease the main problem which was causing all the trouble, so that the whole present paradigm, together with all its commitments, derivations, and assumptions, is made to look dreamlike. Putting it more generally, it is not only the case that a fully extended paradigm, or theory, reaches a point where further extensions of it produce diminishing returns. The situation is worse. The paradigm itself goes bad on you, if it is stretched too far, producing conceptual inconsistency, absurdity, misexpectation, disorder, complexity and confusion, in exactly the same way as a crude analogy does, if pressed too far, say, in a poem, but quite unlike the way in which a system of pure mathematics does, when it yields undecidable formulae or contradictions, or fails to yield proofs; i.e. when an exact statement of what has gone wrong can still be made.

No philosopher of science before Kuhn had described this deterioration. All had blamed the gradual collapse of various scientific theories on the fact that they were eventually falsified in experience by, say, the emergence of new facts; i.e. on the non-cooperation, as it were, of nature. None had

---

had previously been standard, are rejected as metaphysical, as the concern of another discipline, or sometimes as just too problematic to be worth the time.' For examples of basic problems which later science rendered 'invisible', see pp. 103–7; for the general discussion of the 'invisibility', see the whole chapter on *Revolutions as Changes in World View.*

[1] Hesse [1963], pp. 9 ff.

[2] Kuhn [1962], p. 90: 'The proliferation of competing articulations, the willingness to try anything, the expression of explicit discontent, the recourse to philosophy and to debate over fundamentals, all these are symptoms of a transition from normal to extraordinary research.' See also the comparison of crisis-science with pre-paradigm science (p. 84).

[3] Kuhn [1962], pp. 68–70.

[4] *Ibid.* p. 65.

blamed it on the fact that theories, since they have to have concrete analogical paradigms at the heart of them to define their basic commitments, and since the effect of these paradigms is drastically to restrict their fields, collapse, when extended too far, by their own make-up; without any necessary accentuating irritation from nature at all.

And now, to end, we get to the heart of the matter: that of envisaging a crude analogy as an artefact. And the heart of considering this consists in asking the question: 'How does a crude paradigm extend itself?', or 'What (if anything) does Kuhn mean by "replication"?'

I will start with the second question, since it leads to the first. One sign that Kuhn takes seriously the notion that normal science consists of puzzle-solving (and therefore that a paradigm has got to be an artefact) is that he immediately asks himself (p. 38), 'If there is puzzle-solving, where are the rules?' He is then brought up short (pp. 42–6) by the fact that, three quarters of the time, there are no rules. Faced with his own inability to find any rules, Kuhn then takes two incompatible ways out. The first (pp. 42–4) is to assert tough-mindedly that there need not be any rules. The second, characteristically, is to say (pp. 38–9) that by 'rule' he did not really mean 'rule', but 'preconception', or 'established viewpoint'. This second suggestion, in puzzle-solving, just won't do, for rules either are rules or they are not; and that Kuhn knows this, really, is shown by the fact that thenceforward, and indeed throughout the book, he pursues his own first enterprise of trying to find out how paradigms operate independently of rules. His suggestions are the following. Maybe, he says, paradigms add new developments and parts to themselves by exploiting a 'network of overlapping and crisscross' Wittgensteinian 'family resemblances' (p. 45), each resemblance only holding with regard to some properties and between some of the parts. Or perhaps paradigms 'may relate by resemblance and by modelling to one or another part of the scientific corpus which the community in question already recognizes as among its established achievements . . .' (p. 45). Earlier (p. 23), in defining 'paradigm', he had talked about an exact grammatical replication-relation which, however, he said 'rarely holds between a paradigm and its exemplifications'; and later (pp. 32 ff.) he talks of the 'articulation' or 'reformulation' of a paradigm as a process which, when it occurs in a qualitative science, cannot be described in terms of normal mathematical inference. Of course, it may be that all these Kuhnian resemblance-relations do not form a *genus*: they may all essentially differ from one another; but again (see above, the discussion of the different senses of 'paradigm') if they do, Kuhn, philosophically speaking, is saying nothing definite at all. If, however, they do form a genus; and still more, if—as I shall from now on presuppose—they are all different ways of doing the same thing; then Kuhn is indeed saying something philosophically new.

Within normal science (says Kuhn, on this reading) paradigms are capable of expansion and development in two quite different ways. They develop, in the end, by mathematical or other rule-governed inference—which alone enables true puzzles to be solved. But they also develop, initially, by intuitive 'articulation' (or 'family resemblance', or, 'direct modelling', or 'replication', in an extended sense—any or all of these). This second process also is a form of inference in a wider sense of 'inference'—in that sense in which 'inference' is literally *any* kind of permission to pass from one unit or sequence of units or states of affairs to another unit or sequence of units or states of affairs—but it is intuitive; it does not go by rules.

And this brings us back to our first question, of how a crude paradigm extends itself. If the answer is, 'By intuitive inference', we then ask: 'What is this so-called intuitive inference, and is it really intuitive?' For if there is one operation more than another which is not intuitive, it is the entirely mechanizable operation of marking a replica, B', of an original, B. This replication cannot therefore be what Kuhn means. He much more means, that when B' is a replication of B, B' reproduces what, for some known purpose, P, are taken to be the *main features* of B. When a mathematical model, M, for instance, is 'hung on to' a crude paradigm, C, in the manner which we have been describing, M, for some P, reproduces the *main features* of C. It may be, as Max Black says,[1] in describing this model-original form of relation, that many of what superficially seem to be the main features of M—for instance, its scale—may be irrelevant to building up the replication between M and C; they are not included in the statement of the purpose P. But, as between M and C, there must be some corresponding main features; otherwise, we should not say that M is a model of C.

Now there are two forms of formal thinking which are relevant to the analysis of main-feature replication; both of these have emerged from the computer sciences. The first of these, on which there is now quite a literature,[2] is the mathematics of classification, or of 'clumps'; i.e. the formalization of the process of finding Wittgensteinian families. The second of these, on which there is almost no literature, apart from the general literature on mechanized pattern recognition,[3] is the set of

[1] Black [1962], pp. 219–23. As Black shows, the model-original form of relation tends to be, in reality, more complicated than I have here defined it.

[2] Parker-Rhodes and Needham [1960]; Parker-Rhodes [1961]; Needham [1961a] and [1961b]; Needham [1963]; Needham and Spärck-Jones [1964] and Needham [1965].

[3] See, for example, Barus [1962].

procedures for making a digital computer make an 'inexact match' between two formulae which are highly similar to one another, but not quite the same.

In both of these methods, the conglomerates of data in question have to be characterized by reference to a set of properties with regard to which an answer can always be given to the question, 'Has this conglomerate this property or not?' If it has, a *1* is written in its characteristics; if not, an *0*. At the end of the characterization, binary numbers of equal lengths will have been produced for all the conglomerates of data; and, for the case of all pieces of data which, according to the characterization, come out exactly the same, the binary numbers, of course, come out equal. But for the cases in which there is 'some similarity', as we say, but not complete likeness, two things can be done: (*a*) in the mathematics of clumps, a similarity-criterion can be formulated,[1] according to which all conglomerates scrutinized as similar will come out as being in the same family or clump; or (*b*) weight some properties of the data, or some combinations of properties of the data, as 'main features' of the data, in such a way that a unique answer can be given to the question, 'Which, of all the set of conglomerates of data, $D_1 \ldots D_n$, is "most similar in its main features" to another conglomerate of data, D', which comes from outside the set; i.e. which D "inexactly matches" with D'?' It is this last procedure which is so exceedingly difficult to reduce to programme-form (not that the programming of the mathematics of clumps itself is easy); in fact, it is so difficult that it is a well-known non-numerical data-programmer's horror.[2] Nevertheless, a strong *prima facie* case could be made for saying that this 'inexact matching', if and when it can be achieved, is the 'replication-relation' which we are looking for. It is not certain in what sense it is a relation: it is reflexive and symmetric, for instance, but not transitive (from the fact that A has its main features similar to those of B, and B to those of C, it by no means follows that A has its main features similar to those of C, unless each replication has an identical P). Thus replication-logic, in its crude state, cited in note is a one-step-at-a-time logic which never gets off the ground; a

[1] Various similarity-criteria are mentioned in the papers cited in note 2 to p. 85 *above*. The earliest to be formulated was that of Tanimoto [1958]. See also, Sneath and Sokal [1963].

[2] A vicious infinite regress is apt to set in of the following form:

(i) the tests for similarity of main features over a certain threshold cannot be applied until tests for mainness of feature have first been applied. A second calculus of mainness has thus to be created.

(ii) the tests for mainness of feature cannot be applied until they have first been ordered since they turn out not to be independent of each other. A third calculus giving the ordering of the criteria for testing mainness of feature thus has to be created.

(iii) These ordering-considerations themselves depend on bracketing-considerations . . . (etc.).

In other words, the process of progressive detection of complexity increases faster than the invention of means of dealing with it.

logic in which the whole putative effort is to see under what conditions, and with what weighting, and with what feedback of information to change the weighting, and at what cost to the richness and completeness of the characterization scheme, a limited amount of recursiveness within some particular sequential pattern of replications, can be established. One feature of the logic is always transitive, namely that of temporal succession; for if *A* occurs earlier in a replication-sequence than *B*, and *B* than *C*, then *A* occurs earlier than *C*; and this can be important if what is being studied is the gradual accentuation, through a sequence of sequences of replication, each feeding back output into some other, of some prechosen main feature.

It is not even certain that replication is, strictly speaking, a form of inference. I do not see, for instance, how any inference-theorem could be proved of it. In fact, when contrasted with normal simple deduction, replication, and controlling replications, is logically horrible. It is however what of all things, the human brain in its unconscious recognition-processes seems most easily to do; the artificial intelligence men have now thrown new light on it[1]; and it is (I think) how Kuhn's paradigm extends itself. Quite a few very simply replicating-systems have actually been made; within the field of information-retrieval, for instance, any retrieval algorithm which has a scale-of-relevance-procedure attached to it counts as a replication-system within the description which I have given, as does any search-procedure which distinguishes main features and which is built into a character-reader. But such procedures have not yet been thought of in general terms, so that no general analysis of the operation of main-feature recognition has as yet been made.

In view of the obvious difficulties of handling, even with a machine, such an entity as Kuhn's crude paradigm has turned out to be (that is, if I am right as to what it has turned out to be) and in view of the obvious scepticism which even the suggestion that we should take Kuhn's paradigm seriously and philosophically is bound to arouse, it is worth reminding ourselves, in a final paragraph, of what happens if we do *not* follow up Kuhn's thought any further; i.e. what happens if we drop his whole paradigm idea?

It may be difficult both to ascertain Kuhn's thought, and to develop it; but if we do not make the effort to do this, then it seems to me that we are left in a very disturbing position indeed. For, as historians, however much we may cavil at Kuhn's conclusions in detail, we are not going to be able to go back to where we were before Kuhn and his immediate predecessors began to get at us. Their protest against the unconscious dishonesty and the swings of bias with which the history of science has been done in

[1] See particularly, the notion of 'regeneration' in Good [1965].

scientific textbooks up to now cuts far too deep; and so does their outcry against the oversimple and distorted accumulative view of science which has resulted from reading the textbooks as though they were the real history. On the other hand, if no more adequate overall view of science results, in the end, from doing the history of science better, what is the point of doing this history at all—except perhaps as an esoteric hobby? The history of science, by its nature as part of the history of ideas, has got to be a discipline which helps actual scientists to get a deeper insight into the real nature of their own science. If it does not do this, it becomes trivial—the activity of making a pedagogic collection of, in themselves, minor facts. So, if we retreat from all further consideration of Kuhn's 'new image' of science, we run the risk of totally disconnecting the new-style realistic history of science from its old-style philosophy: a disaster.

And if we go forward, and if I am right in my analysis, we have got to re-examine what is true of analogy in the light of what Kuhn has shown to be true of paradigms.

REFERENCES

Barus [1962]: 'A Scheme for Recognizing Patterns for an Unspecified Class', in Fischer, Pollock, Raddack and Stevens (*eds.*): *Optical Character Recognition*, 1962.

Black [1962]: *Models and Metaphors*, 1962.

Brodbeck [1962]: 'Explanation, Prediction and "Imperfect Knowledge" ', in Feigl and Maxwell (*eds.*): *Minnesota Studies in the Philosophy of Science*, **3**, pp. 231–72.

Campbell [1920]: *Foundations of Science*, 1920.

Feyerabend [1962]: 'Explanation, Reduction and Empiricism', in Feigl and Maxwell (*eds.*): *Minnesota Studies in the Philosophy of Science*, **3**, pp. 28–97.

Good [1965]: *Speculations Concerning the First Ultra-Intelligent Machine*, 1965.

Gregory [1966]: *Eye and Brain*, 1966.

Hesse [1963]: *Models and Analogies in Science*, 1963.

Hesse [1964]: 'The Explanatory Function of Metaphor', in Bar-Hillel (*ed.*): *Logic, Methodology and Philosophy of Science*, 1966, pp. 249–59.

Jevons [1873]: *The Principles of Science*, 1873.

Kuhn [1962]: *The Structure of Scientific Revolutions*, 1962.

Lakatos [1963–64]: 'Proofs and Refutations', *The British Journal for the Philosophy of Science*, **14**, pp. 1–25, 120–39, 221–43 and 296–342.

Needham [1961*a*]: 'The Theory of Clumps, II', *Cambridge Language Research Unit Working Papers*, **139**.

Needham [1961*b*]: 'Research on Information Retrieval, Classification and Clumping, 1957–61', *Ph.D. Thesis, Cambridge* 1961.

Needham [1963]: 'A Method for Using Computers in Information Classification', in *Information Process 62: Proceedings of the International Federation for Information Processing Congress, Amsterdam*, 1962.

Needham and Sparck Jones [1964]: 'Keywords and Clumps', *Journal of Documentation*, **20**, no. 1.

Needham [1965]: 'Applications of the Theory of Clumps', *Mechanical Translation*, **8**, pp. 113–27.

Parker-Rhodes and Needham [1960]: 'The Theory of Clumps', *Cambridge Language Research Unit Working Papers*, **126**.

Parker-Rhodes [1961]: 'Contributions to the Theory of Clumps', *Cambridge Language Research Unit Working Papers*, **138**.
Popper [1963]: *Conjectures and Refutations*, 1963.
Putnam [1962]: 'The Analytic and the Synthetic', in Feigl and Maxwell (*eds.*): *Minnesota Studies in the Philosophy of Science*, **3**, pp. 358–97.
Sneath and Sokal [1963]: *Principles of Numerical Taxonomy*, 1963.
Tanimoto [1958]: 'An Elementary Mathematical Theory of Classification and Prediction' *I.B.M. Research*, 1958.

Falsification and the Methodology of Scientific Research Programmes[1]

IMRE LAKATOS
London School of Economics

1. SCIENCE: REASON OR RELIGION?

For centuries knowledge meant proven knowledge—proven either by the power of the intellect or by the evidence of the senses. Wisdom and intellectual integrity demanded that one must desist from unproven utterances and minimize, even in thought, the gap between speculation and established knowledge. The proving power of the intellect or the senses was

[1] This paper is a considerably improved version of my [1968*b*] and a crude version of my [1970]. Some parts of the former are here reproduced without change with the permission of the Editor of the *Proceedings of the Aristotelian Society*. In the preparation of the new version I received much help from Tad Beckman, Colin Howson, Clive Kilmister, Larry Laudan, Eliot Leader, Alan Musgrave, Michael Sukale, John Watkins and John Worrall.

questioned by the sceptics more than two thousand years ago; but they were browbeaten into confusion by the glory of Newtonian physics. Einstein's results again turned the tables and now very few philosophers or scientists still think that scientific knowledge is, or can be, proven knowledge. But few realize that with this the whole classical structure of intellectual values falls in ruins and has to be replaced: one cannot simply water down the ideal of proven truth—as some logical empiricists do—to the ideal of 'probable truth'[1] or—as some sociologists of knowledge do—to 'truth by [changing] consensus'.[2]

Popper's distinction lies primarily in his having grasped the full implications of the collapse of the best-corroborated scientific theory of all times: Newtonian mechanics and the Newtonian theory of gravitation. In his view virtue lies not in caution in avoiding errors, but in ruthlessness in eliminating them. Boldness in conjectures on the one hand and austerity in refutations on the other: this is Popper's recipe. Intellectual honesty does not consist in trying to entrench, or establish one's position by proving (or 'probabilifying') it—intellectual honesty consists rather in specifying precisely the conditions under which one is willing to give up one's position. Committed Marxists and Freudians refuse to specify such conditions: this is the hallmark of their intellectual dishonesty. *Belief* may be a regrettably unavoidable biological weakness to be kept under the control of criticism: but *commitment* is for Popper an outright crime.

Kuhn thinks otherwise. He too rejects the idea that science grows by accumulation of eternal truths.[3] He too takes his main inspiration from Einstein's overthrow of Newtonian physics. His main problem too is *scientific revolution*. But while according to Popper science is 'revolution in permanence', and criticism the heart of the scientific enterprise, according to Kuhn revolution is exceptional and, indeed, extra-scientific, and criticism is, in 'normal' times, anathema. Indeed for Kuhn the transition

[1] The main contemporary protagonist of the ideal of 'probable truth' is Rudolf Carnap. For the historical background and a criticism of this position, cf. Lakatos [1968*a*].

[2] The main contemporary protagonists of the ideal of 'truth by consensus' are Polanyi and Kuhn. For the historical background and a criticism of this position, cf. Musgrave [1969*a*], Musgrave [1969*b*] and Lakatos [1970].

[3] Indeed he introduces his [1962] by arguing against the 'development-by-accumulation' idea of scientific growth. But his intellectual debt is to Koyré rather than to Popper. Koyré showed that positivism gives bad guidance to the historian of science, for the history of physics can only be understood in the context of a succession of 'metaphysical' research programmes. Thus scientific changes are connected with vast cataclysmic metaphysical revolutions. Kuhn develops this message of Burtt and Koyré and the vast success of his book was partly due to his hard-hitting, direct criticism of justificationist historiography—which created a sensation among ordinary scientists and historians of science whom Burtt's, Koyré's (or Popper's) message has not yet reached. But, unfortunately, his message had some authoritarian and irrationalist overtones.

from criticism to commitment marks the point where progress—and 'normal' science—begins. For him the idea that on 'refutation' one can demand the rejection, the elimination of a theory, is 'naive' falsificationism. Criticism of the dominant theory and proposals of new theories are only allowed in the rare moments of 'crisis'. This last Kuhnian thesis has been widely criticized[1] and I shall not discuss it. My concern is rather that Kuhn, having recognized the failure both of justificationism and falsificationism in providing rational accounts of scientific growth, seems now to fall back on irrationalism.

For Popper scientific change is rational or at least rationally reconstructible and falls in the realm of the *logic of discovery*. For Kuhn scientific change—from one 'paradigm' to another—is a mystical conversion which is not and cannot be governed by rules of reason and which falls totally within the realm of the *(social) psychology of discovery*. Scientific change is a kind of religious change.

The clash between Popper and Kuhn is not about a mere technical point in epistemology. It concerns our central intellectual values, and has implications not only for theoretical physics but also for the underdeveloped social sciences and even for moral and political philosophy. If even in science there is no other way of judging a theory but by assessing the number, faith and vocal energy of its supporters, then this must be even more so in the social sciences: truth lies in power. Thus Kuhn's position would vindicate, no doubt, unintentionally, the basic political *credo* of contemporary religious maniacs ('student revolutionaries').

In this paper I shall first show that in Popper's logic of scientific discovery two different positions are conflated. Kuhn understands only one of these, 'naive falsificationism' (I prefer the term 'naive methodological falsificationism'); I think that his criticism of it is correct, and I shall even strengthen it. But Kuhn does not understand a more sophisticated position the rationality of which is not based on 'naive' falsificationism. I shall try to explain—and further strengthen—this stronger Popperian position which, I think, may escape Kuhn's strictures and present scientific revolutions as constituting rational progress rather than as religious conversions.

2. FALLIBILISM VERSUS FALSIFICATIONISM

(*a*) *Dogmatic* (*or naturalistic*) *falsificationism. The empirical basis.*

To see the conflicting theses more clearly, we have to reconstruct the problem situation as it was in philosophy of science after the breakdown of 'justificationism'.

[1] Cf. e.g. Watkins's and Feyerabend's contributions to this volume.

According to the 'justificationists' scientific knowledge consisted of proven propositions. Having recognized that strictly logical deductions enable us only to infer (transmit truth) but not to prove (establish truth), they disagreed about the nature of those propositions (axioms) whose truth can be proved by extra-logical means. *Classical intellectualists* (or 'rationalists' in the narrow sense of the term) admitted very varied—and powerful—sorts of extralogical 'proofs' by revelation, intellectual intuition, experience. These, with the help of logic, enabled them to prove every sort of scientific proposition. *Classical empiricists* accepted as axioms only a relatively small set of 'factual propositions' which expressed the 'hard facts'. Their truth-value was established by experience and they constituted the *empirical basis* of science. In order to prove scientific *theories* from nothing else but the narrow empirical basis, they needed a logic much more powerful than the deductive logic of the classical intellectualists: '*inductive logic*'. All justificationists, whether intellectualists or empiricists, agreed that a singular statement expressing a 'hard fact' may *disprove* a universal theory[1]; but few of them thought that a finite conjunction of factual propositions might be sufficient to *prove* 'inductively' a universal theory.[2]

Justificationism, that is, the identification of knowledge with proven knowledge, was the dominant tradition in rational thought throughout the ages. Scepticism did not deny justificationism: it only claimed that there was (and could be) no proven knowledge and *therefore* no knowledge whatsoever. For the sceptics 'knowledge' was nothing but animal belief. Thus justificationist scepticism ridiculed objective thought and opened the door to irrationalism, mysticism, superstition.

This situation explains the enormous effort invested by classical rationalists in trying to save the synthetical *a priori* principles of intellectualism and by classical empiricists in trying to save the certainty of an empirical basis and the validity of inductive inference. For all of them *scientific honesty demanded that one assert nothing that is unproven.* However, both were defeated: Kantians by non-Euclidean geometry and by non-Newtonian physics, and empiricists by the logical impossibility of establishing

[1] Justificationists repeatedly stressed this asymmetry between singular factual statements and universal theories. Cf. e.g. Popkin's discussion of Pascal in Popkin [1968], p. 14 and Kant's statement to the same effect as quoted in the new *motto* of the third 1969 German edition of Popper's *Logik der Forschung*. (Popper's choice of this time-honoured cornerstone of elementary logic as a *motto* of the new edition of his classic shows his main concern: to fight *probabilism*, in which this asymmetry becomes irrelevant; for probabilists theories may become almost as well established as factual propositions.)

[2] Indeed, even some of these few shifted, following Mill, the rather obviously insoluble problem of inductive proof (of universal from particular propositions) to the slightly less obviously insoluble problem of proving *particular* factual propositions from other *particular* factual propositions.

an empirical basis (as Kantians pointed out, facts cannot prove propositions) and of establishing an inductive logic (no logic can infallibly increase content). It turned out that *all theories are equally unprovable.*

Philosophers were slow to recognize this, for obvious reasons: classical justificationists feared that once they conceded that theoretical science is unprovable, they would have also to concede that it is sophistry and illusion, a dishonest fraud. The philosophical importance of *probabilism* (or '*neojustificationism*') lies in the denial that such a concession is necessary.

Probabilism was elaborated by a group of Cambridge philosophers who thought that although scientific theories are equally unprovable, they have different degrees of probability (in the sense of the calculus of probability) relative to the available empirical evidence.[1] *Scientific honesty then requires less than had been thought: it consists in uttering only highly probable theories; or even in merely specifying, for each scientific theory, the evidence, and the probability of the theory in the light of this evidence.*

Of course, replacing proof by probability was a major retreat for justificationist thought. But even this retreat turned out to be insufficient. It was soon shown, mainly by Popper's persistent efforts, that under very general conditions all theories have zero probability, whatever the evidence; *all theories are not only equally unprovable but also equally improbable.*[2]

Many philosophers still argue that the failure to obtain at least a probabilistic solution of the problem of induction means that we 'throw over almost everything that is regarded as knowledge by science and common sense.'[3] It is against this background that one must appreciate the dramatic change brought about by falsificationism in evaluating theories, and in general, in the standards of intellectual honesty. Falsificationism was, in a sense, a new and considerable retreat for rational thought. But since it was a retreat from utopian standards, it cleared away much hypocrisy and muddled thought, and thus, in fact, it represented an advance.

First I shall discuss a most important brand of falsificationism: dogmatic (or 'naturalistic')[4] falsificationism. Dogmatic falsificationism admits the fallibility of *all* scientific theories without qualification, but it retains a sort of infallible empirical basis. It is strictly empiricist without being inductivist: it denies that the certainty of the empirical basis can be transmitted

[1] The founding fathers of probabilism were intellectualists; Carnap's later efforts to build up an empiricist brand of probabilism failed. Cf. my [1968*a*], p. 367 and also p. 361, footnote 2.

[2] For a detailed discussion, cf. my [1968*a*], especially pp. 353 ff.

[3] Russell [1943], p. 683. For a discussion of Russell's justificationism, cf. my [1962], especially pp. 167 ff.

[4] For the explanation of this term, cf. *below*, p. 98, footnote 1.

to theories. *Thus dogmatic falsificationism is the weakest brand of justificationism.*

It is extremely important to stress that admitting [fortified] empirical counterevidence as a final arbiter against a theory does not make one a dogmatic falsificationist. Any Kantian or inductivist will agree to such arbitration. But both the Kantian and the inductivist, while bowing to a negative crucial experiment, will also specify conditions of how to establish, entrench one unrefuted theory more than another. Kantians held that Euclidean geometry and Newtonian mechanics were established with certainty; inductivists held they had probability 1. For the dogmatic falsificationist, however, empirical *counter*evidence is the *one and only* arbiter which may judge a theory.

The hallmark of dogmatic falsificationism is then the recognition that all theories are equally conjectural. Science cannot *prove* any theory. But although science cannot *prove*, it can *disprove*: it 'can perform with complete logical certainty [the act of] repudiation of what is false',[1] that is, there is an absolutely firm empirical basis of facts which can be used to disprove theories. Falsificationists provide new—very modest—standards of scientific honesty: they are willing to regard a proposition as 'scientific' not only if it is a proven factual proposition, but even if it is nothing more than a falsifiable one, that is, if there are factual propositions available at the time with which it may clash, or, in other words, if it has potential falsifiers.[2]

Scientific honesty then consists of specifying, in advance, an experiment such that if the result contradicts the theory, the theory has to be given up.[3] The falsificationist demands that once a proposition is disproved, there must be no prevarication: the proposition must be unconditionally rejected. To (non-tautologous) unfalsifiable propositions the dogmatic falsificationist gives short shrift: he brands them 'metaphysical' and denies them scientific standing.

Dogmatic falsificationists draw a sharp demarcation between the theoretician and the experimenter: the theoretician proposes, the experimenter—in the name of Nature—disposes. As Weyl put it: 'I wish to record my unbounded admiration for the work of the experimenter in his struggle to wrest interpretable facts from an unyielding Nature who knows so well how to meet our theories with a decisive *No*—or with an inaudible

[1] Medawar [1967], p. 144.

[2] This discussion already indicates the vital importance of a demarcation between provable factual and unprovable theoretical propositions for the dogmatic falsificationist.

[3] '*Criteria of refutation* have to be laid down beforehand: it must be agreed which observable situations, if actually observed, mean that the theory is refuted' (Popper [1963], p. 38, footnote 3).

Yes.'[1] Braithwaite gives a particularly lucid exposition of dogmatic falsificationism. He raises the problem of the objectivity of science: 'To what extent, then, should an established scientific deductive system be regarded as a free creation of the human mind, and to what extent should it be regarded as giving an objective account of the facts of nature?'. His answer is: 'The form of a statement of a scientific hypothesis and its use to express a general proposition, is a human device; what is due to Nature are the observable facts which refute or fail to refute the scientific hypothesis . . . [In science] we hand over to Nature the task of deciding whether any of the contingent lowest-level conclusions are false. This objective test of falsity it is which makes the deductive system, in whose construction we have very great freedom, a deductive system of scientific hypotheses. Man proposes a system of hypotheses: Nature disposes of its truth or falsity. Man invents a scientific system, and then discovers whether or not it accords with observed fact.'[2]

According to the logic of dogmatic falsificationism, science grows by repeated overthrow of theories with the help of hard facts. For instance, according to this view, Descartes's vortex theory of gravity was refuted—and eliminated—by the *fact* that planets moved in ellipses rather than in Cartesian circles; Newton's theory, however, explained successfully the then available facts, both those which had been explained by Descartes's theory and those which refuted it. Therefore Newton's theory replaced Descartes's theory. Analogously, as seen by falsificationists, Newton's theory was, in turn, refuted—proved false—by the anomalous perihelion of Mercury, while Einstein's explained that too. Thus science proceeds by bold speculations, which are never proved or even made probable, but some of which are later eliminated by hard, conclusive refutations and then replaced by still bolder, new and, at least at the start, unrefuted speculations.

Dogmatic falsificationism, however, is untenable. It rests on two false assumptions and on a too narrow criterion of demarcation between scientific and non-scientific.

The *first assumption* is that there is a natural, *psychological* borderline between theoretical or speculative propositions on the one hand and factual or observational (or basic) propositions on the other. (I shall call this—following Popper—the *naturalistic doctrine of observation.*)

The *second assumption* is that if a proposition satisfies the psychological

[1] Quoted in Popper [1934], section 85, with Popper's comment: 'I fully agree.'

[2] Braithwaite [1953], pp. 367-8. For the 'incorrigibility' of Braithwaite's observed facts, cf. his [1938]. While in the quoted passage Braithwaite gives a forceful answer to the problem of scientific objectivity, in another passage he points out that 'except for the straightforward generalizations of observable facts . . . complete refutation is no more possible than is complete proof' ([1953], p. 19). Also cf. *below*, p. 113, footnote 4.

criterion of being factual or observational (or basic) then it is true; one may say that it was *proved* from facts. (I shall call this the *doctrine of observational (or experimental) proof.*)[1]

These two assumptions secure for the dogmatic falsificationist's deadly disproofs an empirical basis from which proven falsehood can be carried by deductive logic to the theory under test.

These assumptions are complemented by a *demarcation criterion*: only those theories are 'scientific' which forbid certain observable states of affairs and therefore are factually disprovable. *Or, a theory is 'scientific' if it has an empirical basis.*[2]

But both assumptions are false. Psychology testifies against the first, logic against the second, and, finally, methodological judgment testifies against the demarcation criterion. I shall discuss them in turn.

(1) A first glance at a few characteristic examples already undermines the *first assumption.* Galileo claimed that he could 'observe' mountains on the moon and spots on the sun and that these 'observations' refuted the time-honoured theory that celestial bodies are faultless crystal balls. But his 'observations' were not 'observational' in the sense of being observed by the—unaided—senses: their reliability depended on the reliability of his telescope—and of the optical theory of the telescope—which was violently questioned by his contemporaries. It was not Galileo's—pure, untheoretical—*observations* that confronted Aristotelian *theory* but rather Galileo's 'observations' in the light of his optical theory that confronted the Aristotelians' 'observations' in the light of their theory of the heavens.[3] This leaves us with two inconsistent theories, *prima facia* on a par. Some empiricists may concede this point and agree that Galileo's 'observations' were not genuine observations; but they still hold that there is a 'natural demarcation' between statements impressed on an empty and passive mind directly by the senses—only these constitute genuine 'immediate knowledge'—and between statements which are suggested by impure, theory-impregnated sensations. Indeed, *all* brands of justificationist theories of knowledge which acknowledge the senses as a source (whether as *one*

[1] For these assumptions and their criticism, cf. Popper [1934], sections 4 and 10. It is because of this assumption that—following Popper—I call this brand of falsificationism 'naturalistic'. Popper's 'basic propositions' should not be confused with the basic propositions discussed in this section; cf. *below*, p. 106, footnote 4.

It is important to point out that these two assumptions are also shared by many justificationists who are not falsificationists: they may add to experimental proofs 'intuitive proofs'—as did Kant—or 'inductive proofs'—as did Mill. Our falsificationist accepts experimental proofs *only*.

[2] The empirical basis of a theory is the set of its potential falsifiers: the set of those observational propositions which may disprove it.

[3] Incidentally, Galileo also showed—with the help of his optics—that if the moon was a faultless crystal ball, it would be invisible (Galileo [1632]).

source or as *the* source) of knowledge are bound to contain a *psychology of observation*. Such psychologies specify the 'right', 'normal', 'healthy', 'unbiased', 'careful' or 'scientific' state of the senses—or rather the state of mind as a whole—in which they observe truth as it is. For instance, Aristotle—and the Stoics—thought that the right mind was the medically healthy mind. Modern thinkers recognized that there is more to the right mind than simple 'health'. Descartes's right mind is one steeled in the fire of sceptical doubt which leaves nothing but the final loneliness of the *cogito* in which the *ego* can then be re-established and God's guiding hand found to recognize truth. All schools of modern justificationism can be characterized by the particular *psychotherapy* by which they propose to prepare the mind to receive the grace of proven truth in the course of a mystical communion. In particular, for classical empiricists the right mind is a *tabula rasa*, emptied of all original content, freed from all prejudice of theory. But it transpires from the work of Kant and Popper—and from the work of psychologists influenced by them—that such empiricist psychotherapy can never succeed. For there are and can be no sensations unimpregnated by expectations and therefore *there is no natural (i.e. psychological) demarcation between observational and theoretical propositions.*[1]

(2) But even if there was such a natural demarcation, logic would still destroy the *second assumption* of dogmatic falsificationism. For the truth-value of the 'observational' propositions cannot be indubitably decided: *no factual proposition can ever be proved from an experiment*. Propositions can only be derived from other propositions, they cannot be derived from facts: one cannot prove statements from experiences—'no more than by thumping the table.'[2] This is one of the basic points of elementary logic, but one which is understood by relatively few people even today.[3]

If factual propositions are unprovable then they are fallible. If they are fallible then clashes between theories and factual propositions are not 'falsifications' but merely inconsistencies. Our imagination may play a greater role in the formulation of 'theories' than in the formulation of

[1] True, most psychologists who turned against the idea of justificationist sensationalism did so under the influence of pragmatist philosophers like William James who denied the possibility of any sort of objective knowledge. But, even so, Kant's influence through Oswald Külpe, Franz Brentano and Popper's influence through Egon Brunswick and Donald Campbell played a role in the shaping of modern psychology; and if psychology ever vanquishes psychologism, it will be due to an increased understanding of the Kant-Popper mainline of objectivist philosophy.

[2] Cf. Popper [1934], section 29.

[3] It seems that the first philosopher to emphasize this might have been Fries in 1837 (cf. Popper [1934], section 29, footnote 3). This is of course a special case of the general thesis that logical relations, like probability or consistency, refer to *propositions*. Thus, for instance, the proposition 'nature is consistent' is false (or, if you wish, meaningless), for nature is not a proposition (or a conjunction of propositions).

'factual propositions',[1] but they are both fallible. Thus *we cannot prove theories and we cannot disprove them either*.[2] The demarcation between the soft, unproven 'theories' and the hard, proven 'empirical basis' is non-existent: *all* propositions of science are theoretical and, incurably, fallible.[3]

(3) Finally, even if there were a natural demarcation between observation statements and theories, and even if the truth-value of observation statements could be indubitably established, dogmatic falsificationism would still be useless for eliminating the most important class of what are commonly regarded as scientific theories. For even if experiments *could* prove experimental reports, their disproving power would still be miserably restricted: *exactly the most admired scientific theories simply fail to forbid any observable state of affairs*.

To support this last contention, I shall first tell a characteristic story and then propose a general argument.

The story is about an imaginary case of planetary misbehaviour. A physicist of the pre-Einsteinian era takes Newton's mechanics and his law of gravitation (N), the accepted initial conditions, I, and calculates, with their help, the path of a newly discovered small planet, p. But the planet deviates from the calculated path. Does our Newtonian physicist consider that the deviation was forbidden by Newton's theory and therefore that, once established, it refutes the theory N? No. He suggests that there must be a hitherto unknown planet p' which perturbs the path of p. He calculates the mass, orbit, etc., of this hypothetical planet and then asks an experimental astronomer to test his hypothesis. The planet p' is so small that even the biggest available telescopes cannot possibly observe it: the experimental astronomer applies for a research grant to build yet a bigger one.[4] In three years' time the new telescope is ready. Were the unknown planet p' to be discovered, it would be hailed as a new

[1] Incidentally, even this is questionable. Cf. *below*, pp. 127 ff.

[2] As Popper put it: 'No conclusive disproof of a theory can ever be produced'; those who wait for an infallible disproof before eliminating a theory will have to wait for ever and 'will never benefit from experience' ([1934], section 9).

[3] Kant and his English follower, Whewell, both realized that all scientific propositions, whether *a priori* or *a posteriori*, are equally theoretical; but both held that they are equally provable. Kantians saw clearly that the propositions of science are theoretical in the sense that they are not written by sensations on the *tabula rasa* of an empty mind, nor deduced or induced from such propositions. A factual proposition is only a special kind of theoretical proposition. In this Popper sided with Kant against the empiricist version of dogmatism. But Popper went a step further: in his view the propositions of science are not only theoretical but they are all also *fallible*, conjectural for ever.

[4] If the tiny conjectural planet were out of the reach even of the biggest *possible* optical telescopes, he might try some quite novel instrument (like a radiotelescope) in order to enable him to 'observe it', that is, to ask Nature about it, even if only indirectly. (The new 'observational' theory may itself not be properly articulated, let alone severely tested, but he would care no more than Galileo did.)

victory of Newtonian science. But it is not. Does our scientist abandon Newton's theory and his idea of the perturbing planet? No. He suggests that a cloud of cosmic dust hides the planet from us. He calculates the location and properties of this cloud and asks for a research grant to send up a satellite to test his calculations. Were the satellite's instruments (possibly new ones, based on a little-tested theory) to record the existence of the conjectural cloud, the result would be hailed as an outstanding victory for Newtonian science. But the cloud is not found. Does our scientist abandon Newton's theory, together with the idea of the perturbing planet and the idea of the cloud which hides it? No. He suggests that there is some magnetic field in that region of the universe which disturbed the instruments of the satellite. A new satellite is sent up. Were the magnetic field to be found, Newtonians would celebrate a sensational victory. But it is not. Is this regarded as a refutation of Newtonian science? No. Either yet another ingenious auxiliary hypothesis is proposed or . . . the whole story is buried in the dusty volumes of periodicals and the story never mentioned again.[1]

This story strongly suggests that even a most respected scientific theory, like Newton's dynamics and theory of gravitation, may fail to forbid any observable state of affairs.[2] Indeed, *some scientific theories forbid an event occurring in some specified finite spatio-temporal region* (*or briefly, a 'singular event'*) *only on the condition that no other factor* (possibly hidden in some distant and unspecified spatio-temporal corner of the universe) *has any influence on it*. But then *such theories never alone contradict a 'basic' statement:* they contradict at most a conjunction of a basic statement describing a spatio-temporally singular event and of a universal non-existence statement saying that no other relevant cause is at work anywhere in the universe. And the dogmatic falsificationist cannot possibly claim that such universal non-existence statements belong to the empirical basis: that they can be observed and proved by experience.

Another way of putting this is to say that some scientific theories are normally interpreted as containing a *ceteris paribus* clause[3]: in such cases it is always a specific theory *together* with this clause which may be refuted. But such a refutation is inconsequential for the *specific* theory under test

[1] At least not until a new research programme supersedes Newton's programme which happens to explain this previously recalcitrant phenomenon. In this case, the phenomenon will be unearthed and enthroned as a 'crucial experiment'; cf. *below*, pp. 154 ff.

[2] Popper asks: 'What kind of clinical responses would refute to the satisfaction of the analyst not merely a particular diagnosis but psychoanalysis itself?' ([1963], p. 38, footnote 3.) But what kind of observation would refute to the satisfaction of the Newtonian not merely a particular version but Newtonian theory itself?

[3] [*Added in press:*] This '*ceteris paribus*' clause must not normally be interpreted as a separate premise. For a discussion, cf. *below*, p. 186.

because by replacing the *ceteris paribus* clause by a different one the *specific* theory can always be retained whatever the tests say.

If so, the 'inexorable' disproof procedure of dogmatic falsificationism breaks down in these cases *even if* there were a firmly established empirical basis to serve as a launching pad for the arrow of the *modus tollens*: the prime target remains hopelessly elusive.[1] And as it happens, it is exactly the most important, 'mature' theories in the history of science which are *prima facie* undisprovable in this way.[2] Moreover, by the standards of dogmatic falsificationism all probabilistic theories also come under this head: for no finite sample can ever *disprove* a universal probabilistic theory[3]; probabilistic theories, like theories with a *ceteris paribus* clause, have no empirical basis. But then the dogmatic falsificationist relegates the most important scientific theories *on his own admission* to metaphysics where rational discussion—consisting, by his standards, of proofs and disproofs—has no place, since a metaphysical theory is neither provable nor disprovable. The demarcation criterion of dogmatic falsificationism is thus still strongly antitheoretical.

(Moreover, *one can easily argue that ceteris paribus clauses are not exceptions, but the rule in science.* Science, after all, must be demarcated from a curiosity shop where funny local—or cosmic—oddities are collected and displayed. The assertion that 'all Britons died from lung cancer between 1950 and 1960' is logically possible, and might even have been true. But if it has been only an occurrence of an event with minute probability, it would have only curiosity value for the crankish fact-collector, it would have a macabre entertainment value, but no scientific value. A proposition might be said to be scientific only if it aims at expressing a causal connection: such connection between being a Briton and dying of lung cancer may not even be intended. Similarly, 'all swans are white', if true, would be a mere curiosity unless it asserted that swanness *causes* whiteness. But then a black swan would not refute this proposition, since it may only indicate *other causes* operating simultaneously. Thus 'all swans are white' is either an oddity and easily disprovable or a scientific proposition with a *ceteris paribus* clause and therefore undisprovable. *Tenacity of a theory against empirical evidence would then be an argument for rather than against regarding it as 'scientific'. 'Irrefutability' would become a hallmark of science.*[4])

[1] Incidentally, we might persuade the dogmatic falsificationist that his demarcation criterion was a very naive mistake. If he gives it up but retains his two basic assumptions, he will have to ban theories from science and regard the growth of science as an accumulation of proven basic statements. This indeed is the final stage of classical empiricism after the evaporation of the hope that facts can prove or at least disprove theories.

[2] This is no coincidence; cf. *below*, pp. 175 ff.

[3] Cf. Popper [1934], chapter VIII.

[4] For a *much* stronger case, cf. *below*, sect. 3.

To sum up: classical justificationists only admitted proven theories; neoclassical justificationists probable ones; dogmatic falsificationists realized that in either case no theories are admissible. They decided to admit theories if they are disprovable—disprovable by a finite number of observations. But even if there were such disprovable theories—those which can be contradicted by a finite number of observable facts—they are still logically too near to the empirical basis. For instance, on the terms of the dogmatic falsificationist, a theory like 'All planets move in ellipses' may be disproved by five observations; therefore the dogmatic falsificationist will regard it as scientific. A theory like 'All planets move in circles' may be disproved by four observations; therefore the dogmatic falsificationist will regard it as still more scientific. The acme of scientificness will be a theory like 'All swans are white' which is disprovable by one single observation. On the other hand, he will reject all probabilistic theories together with Newton's, Maxwell's, Einstein's theories, as unscientific, for no finite number of observations can ever disprove them.

If we accept the demarcation criterion of dogmatic falsificationism, *and* also the idea that facts can prove 'factual' propositions, we have to declare that the most important, if not all, theories ever proposed in the history of science are metaphysical, that most, if not all, of the accepted progress is pseudo-progress, that most, if not all, of the work done is irrational. If, however, still accepting the demarcation criterion of dogmatic falsificationism, we deny that facts can prove propositions, then we certainly end up in complete scepticism: then all science is undoubtedly irrational metaphysics and should be rejected. *Scientific theories are not only equally unprovable, and equally improbable, but they are also equally undisprovable.* But the recognition that not only the theoretical but *all* the propositions in science are fallible, means the total collapse of *all* forms of dogmatic justificationism as theories of scientific rationality.

(*b*) *Methodological falsificationism. The 'empirical basis'.*

The collapse of dogmatic falsificationism because of fallibilistic arguments seems to bring us back to square one. If *all* scientific statements are fallible theories, one can criticize them only for inconsistency. But then, in what sense, if any, is science empirical? If scientific theories are neither provable, nor probabilifiable, nor disprovable, then the sceptics seem to be finally right: science is no more than vain speculation and there is no such thing as progress in scientific knowledge. Can we still oppose scepticism? *Can we save scientific criticism from fallibilism*? Is it possible to have a fallibilistic theory of scientific progress? In particular, if scientific criticism is fallible, on what ground can we ever eliminate a theory?

A most intriguing answer is provided by *methodological falsificationism.* Methodological falsificationism is a brand of conventionalism; therefore in order to understand it, we must first discuss conventionalism in general.

There is an important demarcation between *'passivist' and 'activist' theories of knowledge.* 'Passivists' hold that true knowledge is Nature's imprint on a perfectly inert mind: mental *activity* can only result in bias and distortion. The most influential passivist school is classical empiricism. 'Activists' hold that we cannot read the book of Nature without mental activity, without interpreting them in the light of our expectations or theories.[1] Now *conservative 'activists'* hold that we are born with our basic expectations; with them we turn the world into 'our world' but must then live for ever in the prison of our world. The idea that we live and die in the prison of our 'conceptual frameworks' was developed primarily by Kant; pessimistic Kantians thought that the real world is for ever unknowable because of this prison, while optimistic Kantians thought that God created our conceptual framework to fit the world.[2] But *revolutionary activists* believe that conceptual frameworks can be developed and also replaced by new, *better* ones; it is *we* who create our 'prisons' and we can also, critically, demolish them.[3]

New steps from conservative to revolutionary activism were made by Whewell and then by Poincaré, Milhaud and Le Roy. Whewell held that theories are developed by trial and error—in the 'preludes to the inductive epochs'. The best ones among them are then 'proved'—during the 'inductive epochs'—by a long primarily *a priori* consideration which he called 'progressive intuition'. The 'inductive epochs' are followed by 'sequels to the inductive epochs': cumulative developments of auxiliary theories.[4] Poincaré, Milhaud and Le Roy were averse to the idea of *proof* by progressive intuition and preferred to explain the continuing historical success of Newtonian mechanics by a *methodological decision* taken by scientists: after a considerable period of initial empirical success scientists

[1] This demarcation—and terminology—is due to Popper; cf. especially his [1934], section 19 and his [1945], chapter 23 and footnote 3 to chapter 25.

[2] No version of conservative activism explained why Newton's *gravitational* theory should be invulnerable; Kantians restricted themselves to the explanation of the tenacity of Euclidean geometry and Newtonian *mechanics*. About Newtonian *gravitation* and *optics* (or other branches of science) they had an ambiguous, and occasionally inductivist position.

[3] I do not include Hegel among 'revolutionary *activists*'. For Hegel and his followers change in conceptual frameworks is a predetermined, inevitable process, where individual creativity or rational criticism plays no essential role. Those who run ahead are equally at fault as those who stay behind in this 'dialectic'. The clever man is not he who creates a better 'prison' or who demolishes critically the old one, but the one who is always in step with history. Thus dialectic accounts for change without criticism.

[4] Cf. Whewell's [1837], [1840] and [1858].

may *decide* not to allow the theory to be refuted. Once they have taken this decision, they solve (or dissolve) the apparent anomalies by auxiliary hypotheses or other 'conventionalist stratagems.'[1] This *conservative conventionalism* has, however, the disadvantage of making us unable to get out of our self-imposed prisons, once the first period of trial-and-error is over and the great decision taken. It cannot solve the problem of the elimination of those theories which have been triumphant for a long period. According to conservative conventionalism, experiments may have sufficient power to refute young theories, but not to refute old, established theories: *as science grows, the power of empirical evidence diminishes.*[2]

Poincaré's critics refused to accept his idea, that, although the scientists build their conceptual frameworks, there comes a time when these frameworks turn into prisons which cannot be demolished. This criticism gave rise to two rival schools of *revolutionary conventionalism*: Duhem's simplicism and Popper's methodological falsificationism.[3]

Duhem accepts the conventionalists' position that no physical theory ever crumbles merely under the weight of 'refutations', but claims that it still may crumble under the weight of 'continual repairs, and many tangled-up stays' when 'the worm-eaten columns' cannot support 'the tottering building' any longer[4]; then the theory loses its original simplicity and has to be replaced. But falsification is then left to subjective taste or, at best, to scientific fashion, and leaves too much leeway for dogmatic adherence to a favourite theory.[5]

Popper set out to find a criterion which is both more objective and more

[1] Cf. especially Poincaré [1891] and [1902]; Milhaud [1896]; Le Roy [1899] and [1901]. It was one of the chief philosophical merits of conventionalists to direct the limelight to the fact that any theory can be saved by 'conventionalist stratagems' from refutations. (The term 'conventionalist stratagem' is Popper's; cf. the critical discussion of Poincaré's conventionalism in his [1934], especially sections 19 and 20.)

[2] Poincaré first elaborated his conventionalism only with regard to geometry (cf. his [1891]). Then Milhaud and Le Roy generalized Poincaré's idea to cover all branches of accepted physical theory. Poincaré's [1902] starts with a strong criticism of the Bergsonian Le Roy against whom he defends the empirical (falsifiable or 'inductive') character of all physics *except for* geometry and mechanics. Duhem, in turn, criticized Poincaré: in his view there was a possibility of overthrowing even Newtonian mechanics.

[3] The *loci classici* are Duhem's [1905] and Popper's [1934]. Duhem was not a *consistent* revolutionary conventionalist. Very much like Whewell, he thought that conceptual changes are only *preliminaries* to the final—if perhaps distant—'natural classification': 'The more a theory is perfected, the more we apprehend that the logical order in which it arranges experimental laws is the reflection of an ontological order.' In particular, he refused to see Newton's mechanics *actually* 'crumbling' and characterized Einstein's relativity theory as the manifestation of a 'frantic and hectic race in pursuit of a novel idea' which 'has turned physics into a real chaos where logic loses its way and commonsense runs away frightened' (Preface—of 1914—to the second edition of his [1905]).

[4] Duhem [1905], chapter VI, section 10.

[5] For a further discussion of conventionalism, cf. *below*, pp. 184-189.

hard-hitting. He could not accept the emasculation of empiricism, inherent even in Duhem's approach, and proposed a methodology which allows experiments to be powerful even in 'mature' science. Popper's methodological falsificationism is both conventionalist and falsificationist, but he 'differs from the [conservative] conventionalists in holding that the statements decided by agreement are *not* [spatio-temporally] universal but [spatio-temporally] singular'[1]; and he differs from the dogmatic falsificationist in holding that the truth-value of such statements cannot be proved by facts but, in some cases, may be decided by agreement.[2]

The *conservative conventionalist* (or methodological justificationist, if you wish) makes unfalsifiable by *fiat* some (spatio-temporally) universal theories, which are distinguished by their explanatory power, simplicity or beauty. Our *revolutionary conventionalist* (or 'methodological falsificationist') makes unfalsifiable by *fiat* some (spatio-temporally) singular statements which are distinguishable by the fact that there exists at the time a 'relevant technique' such that 'anyone who has learned it' will be able to *decide* that the statement is 'acceptable'.[3] Such a statement may be called an 'observational' or 'basic' statement, but only in inverted commas.[4] Indeed, the very selection of all such statements is a matter of a decision, which is not based on exclusively psychological considerations. This decision is then followed by a second kind of decision concerning the separation of the set of *accepted* basic statements from the rest.

These *two decisions* correspond to the *two assumptions* of dogmatic falsificationism. But there are important differences. First, the methodological falsificationist is not a justificationist, he has no illusions about 'experimental proofs' and is fully aware of the fallibility of his decisions and the risks he is taking.

The methodological falsificationist realizes that in the 'experimental techniques' of the scientist fallible theories are involved,[5] 'in the light of which' he interprets the facts. In spite of this he 'applies' these theories, he regards them in the given context not as theories under test but as *unproblematic background knowledge* 'which we accept (tentatively) as unproblematic while we are testing the theory'.[6] He may call these theories —and the statements whose truth-value he decides in their light—'obser-

[1] Popper [1934], section 30.

[2] *In this section I discuss the 'naive' variant of Popper's methodological falsificationism. Thus, throughout the section 'methodological falsificationism' stands for 'naive methodological falsificationism'; for this 'naivety'*, cf. *below*, pp. 115-116.

[3] Popper [1934], section 27.

[4] *Op. cit.* section 28. For the non-basicness of these methodologically 'basic' statements, cf. e.g. Popper [1934] *passim* and Popper [1959*a*], p. 35, footnote *2.

[5] Cf. Popper [1934], end of section 26 and also his [1968*c*], pp. 291–2.

[6] Cf. Popper [1963], p. 390.

vational': but this is only a manner of speech which he inherited from naturalistic falsificationism.[1] The methodological falsificationist *uses our most successful theories as extensions of our senses* and widens the range of theories which can be applied in testing far beyond the dogmatic falsificationist's range of strictly observational theories. For instance, let us imagine that a big radio-star is discovered with a system of radio-star satellites orbiting it. We should like to test some gravitational theory on this planetary system—a matter of considerable interest. Now let us imagine that Jodrell Bank succeeds in providing a set of space-time co-ordinates of the planets which is inconsistent with the theory. We shall take these statements as potential falsifiers. Of course, these basic statements are not 'observational' in the usual sense but only " 'observational' ". They describe planets that neither the human eye nor optical instruments can reach. Their truth-value is arrived at by an 'experimental technique'. This 'experimental technique' is based on the 'application' of a well-corroborated theory of radio-optics. Calling these statements 'observational' is no more than a manner of saying that, in the context of his problem, that is, in testing our gravitational theory, the methodological falsificationist uses radio-optics uncritically, as 'background knowledge'. *The need for decisions to demarcate the theory under test from unproblematic background knowledge is a characteristic feature of this brand of methodological falsificationism.*[2] (This situation does not really differ from Galileo's 'observation' of Jupiter's satellites: moreover, as some of Galileo's contemporaries rightly pointed out, he relied on a virtually non-existent optical theory—which then was less corroborated, and even less articulated, than present-day radio-optics. On the other hand, calling the reports of our human eye 'observational' only indicates that we 'rely' on some vague physiological theory of human vision.[3])

This consideration shows the conventional element in granting—in a given context—the (methodologically) 'observational' status to a theory.[4] Similarly, there is a considerable conventional element in the decision concerning the actual truth-value of a basic statement which we take after we have decided which 'observational theory' to apply. One single observation may be the stray result of some trivial error: in order to reduce such risks, methodological falsificationists prescribe some safety control. The simplest such control is to repeat the experiment (it is a matter of

[1] Indeed, Popper carefully puts 'observational' in quotes; cf. his [1934], section 28.

[2] This demarcation plays a role both in the *first* and in the *fourth* type of decisions of the methodological falsificationist. (For the *fourth* decision, cf. *below*, p. 110.)

[3] For a fascinating discussion, cf. Feyerabend [1969].

[4] One wonders whether it would not be better to make a break with the terminology of naturalistic falsificationism and rechristen observational theories '*touchstone theories*'.

convention how many times); another is to 'fortify' the potential falsifier by a 'well-corroborated falsifying hypothesis'.[1]

The methodological falsificationist also points out that, as a matter of fact, these conventions are institutionalized and endorsed by the scientific community; the list of 'accepted' falsifiers is provided by the verdict of the experimental scientists.[2]

This is how the methodological falsificationist establishes his 'empirical basis'. (He uses inverted commas in order 'to give ironical emphasis' to the term.[3]) This 'basis' can be hardly called a 'basis' by justificationist standards: there is nothing proven about it—it denotes 'piles driven into a swamp'.[4] Indeed, if this 'empirical basis' clashes with a theory, the theory may be *called* 'falsified', but it is not falsified in the sense that it is disproved. Methodological 'falsification' is very different from dogmatic falsification. If a theory is falsified, it is proven false; if it is 'falsified', it may still be true. If we follow up this sort of 'falsification' by the actual 'elimination' of a theory, we may well end up by eliminating a true, and accepting a false, theory (a possibility which is thoroughly abhorrent to the old-fashioned justificationist).

Yet the methodological falsificationist advises that exactly this is to be done. The methodological falsificationist realizes that if we want to reconcile fallibilism with (non-justificationist) rationality, we *must* find a way to eliminate *some* theories. If we do not succeed, the growth of science will be nothing but growing chaos.

Therefore the methodological falsificationist maintains that '[if we want] to make the method of selection by elimination work, and to ensure that only the fittest theories survive, their struggle for life must be made severe'.[5] Once a theory has been falsified, in spite of the risk involved, it must be eliminated: '[with theories we work only] as long as they stand up to tests'.[6] The elimination must be methodologically conclusive: 'In general we regard an inter-subjectively testable falsification as final . . . A corroborative appraisal made at a later date . . . can replace a positive degree of corroboration by a negative one, but not *vice versa*'.[7] This is the

[1] Cf. Popper [1934], section 22. Many philosophers overlooked Popper's important qualification that a basic-statement has no power to refute anything without the support of a well-corroborated falsifying hypothesis.

[2] Cf. Popper [1934], section 30.

[3] Popper [1963], p. 387.

[4] Popper [1934], section 30; also cf. section 29: 'The Relativity of Basic Statements.'

[5] Popper [1957], p. 134. Popper, in other places, emphasizes that his method cannot 'ensure' the survival of the fittest. Natural selection may go wrong: the fittest may perish and monsters survive.

[6] Popper [1935].

[7] Popper [1934], section 82.

methodological falsificationist's explanation of how we get out of a rut: 'It is always the experiment which saves us from following a track that leads nowhere.'[1]

The methodological falsificationist separates rejection and disproof, which the dogmatic falsificationist had conflated.[2] He is a fallibilist but his fallibilism does not weaken his critical stance: he turns fallible propositions into a 'basis' for a hard-line policy. On these grounds he proposes a *new demarcation criterion*: only those theories—that is, non-'observational' propositions—which forbid certain 'observable' states of affairs, and therefore may be 'falsified' and rejected, are 'scientific': or, briefly, *a theory is 'scientific' (or 'acceptable') if it has an 'empirical basis'*. This criterion brings out sharply the difference between dogmatic and methodological falsificationism.[3]

This methodological demarcation criterion is much more liberal than the dogmatic one. Methodological falsificationism opens up new avenues of criticism: many more theories may qualify as 'scientific'. We have already seen that there are more 'observational' theories than observational theories,[4] and therefore there are more 'basic' statements than basic statements.[5] Furthermore, probabilistic theories may qualify now as 'scientific': although they are not falsifiable they can be easily made 'falsifiable' by an *additional* (*third type*) *decision* which the scientist can make by specifying certain rejection rules which may make statistically interpreted evidence 'inconsistent' with the probabilistic theory.[6]

[1] Popper [1934], section 82.

[2] This kind of methodological 'falsification' is, unlike dogmatic falsification (disproof), a pragmatic, methodological idea. But then what exactly should we mean by it? Popper's answer—which I am going to discard—is that methodological 'falsification' indicates an 'urgent need of replacing a falsified hypothesis by a better one' (Popper [1959*a*], p. 87, footnote *1). This shift is an excellent illustration of the process I described in my [1963–4] whereby critical discussion shifts the original *problem* without necessarily changing the old *terms*. The byproducts of such processes are *meaning-shifts*. For a further discussion, cf. *below*, p. 122, footnote 4, and p. 157, footnote 1.

[3] The demarcation criterion of the dogmatic falsificationist was: a theory is 'scientific' if it has an empirical basis (see *above*, p. 98).

[4] See *above*, pp. 98–9.

[5] Incidentally, Popper, in his [1934], does not seem to have seen this point clearly. He writes: 'Admittedly, it is possible to interpret the concept of an *observable event* in a psychologistic sense. But I am using it in such a sense that it might just as well be replaced by "an event involving position and movement of macroscopic physical bodies" '. ([1934], section 28.) In the light of our discussion, for instance, we may regard a positron passing through a Wilson chamber at time t_0 as an 'observable' event, in spite of the non-macroscopic character of the positron.

[6] Popper [1934], section 68. Indeed, this methodological falsificationism is the philosophical basis of some of the most interesting developments in modern statistics. The Neyman-Pearson approach rests completely on methodological falsificationism. Also cf. Braithwaite [1953], chapter VI. (Unfortunately, Braithwaite reinterprets Popper's demarcation criterion as separating meaningful from meaningless rather than scientific from non-scientific propositions.)

But even these three decisions are not sufficient to enable us to 'falsify' a theory which cannot explain anything 'observable' without a *ceteris paribus* clause.[1] No finite number of 'observations' is enough to 'falsify' such a theory. However, if this is the case how can one reasonably defend a methodology which claims to 'interpret natural laws or theories as . . . statements which are partially decidable, i.e. which are, for logical reasons, not verifiable but, in an asymmetrical way, falsifiable . . .'?[2] How can we interpret theories like Newton's theory of dynamics and gravitation as 'one-sidedly decidable'?[3] How can we make in such cases genuine 'attempts to weed out false theories—to find the weak points of a theory in order to reject it if it is falsified by the test'?[4] How can we draw them into the realm of rational discussion? The methodological falsificationist solves the problem by making a further (*fourth type*) *decision*: when he tests a theory together with a *ceteris paribus* clause and finds that this conjunction has been refuted, he must decide whether to take the refutation also as a refutation of the specific theory. For instance, he may accept Mercury's 'anomalous' perihelion as a refutation of the treble conjunction N_3 of Newton's theory, the known initial conditions and the *ceteris paribus* clause. Then he tests the initial conditions 'severely'[5] and may decide to relegate them into the 'unproblematic background knowledge'. This decision implies the refutation of the double conjunction N_2 of Newton's theory and the *ceteris paribus* clause. Now he has to take the crucial decision: whether to relegate also the *ceteris paribus* clause into the pool of 'unproblematic background knowledge'. He will do so if he finds the *ceteris paribus* clause well corroborated.

How can one test a *ceteris paribus* clause severely? By assuming that there *are* other influencing factors, by specifying such factors, and by testing these specific assumptions. If many of them are refuted, the *ceteris paribus* clause will be regarded as well-corroborated.

Yet the decision to 'accept' a *ceteris paribus* clause is a very risky one because of the grave consequences it implies. If it is decided to accept it as part of such background knowledge, the statements describing Mercury's perihelion from the empirical basis of N_2 are turned into the empirical basis of Newton's specific theory N_1 and what was previously a mere 'anomaly' in relation to N_1, becomes now crucial evidence against it, its falsification. (We may call an event described by a statement A an '*anomaly* in relation to a theory T' if A is a potential falsifier of the conjunction of T and a *ceteris paribus* clause but it becomes a potential falsifier of T itself after

[1] Cf. *above*, pp. 101–3.

[2] Popper [1933]. [3] Popper [1933]. [4] Popper [1957], p. 133.

[5] For a discussion of this important concept of Popperian methodology, cf. my [1968*a*], pp. 397 ff.

having decided to relegate the *ceteris paribus* clause into 'unproblematic background knowledge.'[1]) Since, for our savage falsificationist, falsifications are methodologically conclusive,[2] the fateful decision amounts to the methodological elimination of Newton's theory, making further work on it irrational. If the scientist shrinks back from such bold decisions he will 'never benefit from experience', 'believing, perhaps, that it is his business to defend a successful system against criticism as long as it is not *conclusively disproved*'.[3] He will degenerate into an apologist who may always claim that 'the discrepancies which are asserted to exist between the experimental results and the theory are only apparent and that they will disappear with the advance of our understanding'.[4] But for the falsificationist this is 'the very reverse of the critical attitude which is the proper one for the scientist',[5] and is impermissible. To use one of the methodological falsificationist's favourite expressions: the theory 'must be made to stick its neck out'.

The methodological falsificationist is in a serious plight when it comes to deciding where to draw the demarcation, even if only in a well-defined context, between the problematic and unproblematic. The plight is most dramatic when he has to make a decision about *ceteris paribus* clauses, when he has to promote one of the hundreds of 'anomalous phenomena' into a 'crucial experiment', and decide that in such a case the experiment was 'controlled'.[6]

Thus, with the help of this fourth type of decision,[7] our methodological falsificationist has finally succeeded in interpreting even theories like Newton's theory as 'scientific'.[8]

[1] For an improved 'explication', cf. *below*, p. 159, footnote 1.

[2] Cf. *above*, p. 108, text to footnotes 6 and 7.

[3] Popper [1934], section 9.

[4] *Ibid.*

[5] *Ibid.*

[6] The problem of '*controlled experiment*' may be said to be nothing else but the problem of arranging experimental conditions in such a way as to minimize the risk involved in such decisions.

[7] This type of decision belongs, in an important sense, to the same category as the first decision: it demarcates, by decision, problematic from unproblematic knowledge. Cf. *above*, p. 107, text to footnote 2.

[8] Our exposition shows clearly the complexity of the decisions needed to define the 'empirical content' of a theory—that is, the set of its potential falsifiers. 'Empirical content' depends on our *decision* as to which are our 'observational theories' and which anomalies are to be promoted to counterexamples. If one attempts to compare the empirical content of different scientific theories in order to see which is 'more scientific', then one will get involved in an enormously complex and therefore hopelessly arbitrary system of decisions about their respective classes of 'relatively atomic statements' and their 'fields of application'. (For the meaning of these (very) technical terms, cf. Popper [1934], section 38.) But such comparison is possible only when one theory supersedes another (cf. Popper, [1959*a*], p. 401, footnote 7). And even then, there may be difficulties (which would not, however, add up to irremediable 'incommensurability').

Indeed, there is no reason why he should not go yet another step. Why not decide that a theory—which even these four decisions cannot turn into an empirically falsifiable one—is falsified if it clashes with another theory which is scientific on some of the previously specified grounds and is also well-corroborated?[1] After all, if we reject one theory because one of its potential falsifiers is seen to be true in the light of an observational theory, why not reject another theory because it clashes *directly* with one that may be relegated into unproblematic background knowledge? This would allow us, by a *fifth type decision*, to eliminate even 'syntactically metaphysical' theories, that is, theories, which, like 'all-some' statements or purely existential statements,[2] because of their *logical form* cannot have spatio-temporally singular potential falsifiers.

To sum up: the methodological falsificationist offers an interesting solution to the problem of combining hard-hitting criticism with fallibilism. Not only does he offer a philosophical basis for falsification after fallibilism had pulled the carpet from under the feet of the dogmatic falsificationist, but he also widens the range of such criticism very considerably. By putting falsification in a new setting, he saves the attractive code of honour of the dogmatic falsificationist: that scientific honesty consists in specifying, in advance, an experiment such, that if the result contradicts the theory, the theory has to be given up.[3]

Methodological falsificationism represents a considerable advance beyond both dogmatic falsificationism and conservative conventionalism. It recommends risky decisions. But the risks are daring to the point of recklessness and one wonders whether there is no way of lessening them.

Let us first have a closer look at the risks involved.

Decisions play a crucial role in this methodology—as in any brand of conventionalism. Decisions however may lead us disastrously astray. The methodological falsificationist is the first to admit this. But this, he argues, is the price which we have to pay for the possibility of progress.

One has to appreciate the dare-devil attitude of our methodological falsificationist. He feels himself to be a hero who, faced with two catastrophic alternatives, dared to reflect coolly on their relative merits and choose the lesser evil. One of the alternatives was sceptical fallibilism, with its 'anything goes' attitude, the despairing abandonment of all intellectual standards, and hence of the idea of scientific progress. Nothing can be established, nothing can be rejected, nothing even communicated: the

[1] This was suggested by J. O. Wisdom: cf. his [1963].

[2] For instance: 'All metals have a solvent'; or 'There exists a substance which can turn all metals into gold'. For discussions of such theories, cf. especially Watkins [1957] and Watkins [1960]. But cf. *below*, pp. 126–7 and pp. 183–4.

[3] See *above*, p. 96.

growth of science is a growth of chaos, a veritable Babel. For two thousand years, scientists and scientifically-minded philosophers chose justificationist illusions of some kind to escape this nightmare. Some of them argued that *one has to choose between inductivist justificationism and irrationalism*: 'I do not see any way out of a dogmatic assertion that we know the inductive principle or some equivalent; the only alternative is to throw over almost everything that is regarded as knowledge by science and common sense.'[1] Our methodological falsificationist proudly rejects such escapism: he dares to measure up to the full impact of fallibilism and yet escape scepticism by a daring and risky conventionalist policy, with no dogmas. He is fully aware of the risks but insists that *one has to choose between some sort of methodological falsificationism and irrationalism*. He offers a game in which one has little hope of winning, but claims that it is still better to play than to give up.[2]

Indeed, those critics of naive falsificationism who offer no alternative method of criticism are inevitably driven to irrationalism. For instance, Neurath's muddled argument, that the falsification and ensuing elimination of a hypothesis may turn out to have been 'an obstacle in the progress of science',[3] carries no weight as long as the only alternative he seems to offer is chaos. Hempel is, no doubt, right in stressing that 'science offers various examples [when] a conflict between a highly-confirmed theory and an occasional recalcitrant experiential sentence may well be resolved by revoking the latter rather than by sacrificing the former'[4]; nevertheless he admits that he can offer no other 'fundamental standard' than that of naive falsificationism.[5] Neurath—and, seemingly, Hempel—reject falsificationism as 'pseudo-rationalism'[6]; but where is 'real rationalism'? Popper warned already in 1934 that Neurath's permissive methodology (or rather lack of methodology) would make science unempirical and therefore

[1] Russell [1943], p. 683.

[2] I am sure that some will welcome methodological falsificationism as an 'existentialist' philosophy of science.

[3] Neurath [1935], p. 356.

[4] Hempel [1952], p. 621. Agassi, in his [1966], follows Neurath and Hempel, especially pp. 16 ff. It is rather amusing that Agassi, in making this point, thinks that he is taking up arms against 'the whole literature concerning the methods of science'.

Indeed, many scientists were fully aware of the difficulties inherent in the 'confrontation of theory and facts'. (Cf. Einstein [1949], p. 27.) Several philosophers sympathetic to falsificationism emphasized that 'the process of refuting a scientific hypothesis is more complicated than it appears to be at first sight' (Braithwaite [1953], p. 20). But only Popper offered a constructive, rational solution.

[5] Hempel [1952], p. 622. Hempel's crisp 'theses on empirical certainty' do nothing but refurbish Neurath's—and some of Popper's—old arguments (against Carnap, I take it); but deplorably, he does not mention either his predecessors or his adversaries.

[6] Neurath [1935].

irrational: 'We need a set of rules to limit the arbitrariness of "deleting" (or else "accepting") a protocol sentence. Neurath fails to give any such rules and thus unwittingly throws empiricism overboard ... Every system becomes defensible if one is allowed (as everybody is, in Neurath's view) simply to "delete" a protocol sentence if it is inconvenient'.[1] Popper agrees with Neurath that all propositions are fallible; but he forcefully makes the crucial point that we cannot make progress unless we have a firm rational strategy or method to guide us when they clash.[2]

But is not the firm strategy of the brand of methodological falsificationism hitherto discussed *too firm*? Are not the decisions it advocates bound to be *too arbitrary*? Some may even claim that all that distinguishes methodological from dogmatic falsificationism is that *it pays lip-service to fallibilism*!

To criticize a theory of criticism is usually very difficult. Naturalistic falsificationism was relatively easy to refute, since it rested on an empirical psychology of perception: one could show that it was simply *false*. But how can methodological falsificationism be falsified? No disaster can ever disprove a non-justificationist theory of rationality. Moreover, how can we ever recognize an epistemological disaster? We have no means to judge whether the verisimilitude of our successive theories increases or decreases.[3] At this stage we have not yet developed a general theory of criticism even for scientific theories, let alone for theories of rationality[4]: therefore if we want to falsify our methodological falsificationism, we have to do it before having a theory of how to do it.

If we look at history of science, if we try to see how some of the most celebrated falsifications happened, we have to come to the conclusion that either some of them are plainly irrational, or that they rest on rationality principles radically different from the ones we just discussed. First of all, our falsificationist must deplore the fact that stubborn theoreticians frequently challenge experimental verdicts and have them reversed. In the falsificationist conception of scientific 'law and order' we have described there is no place for such successful appeals. Further difficulties arise from the falsification of theories to which a *ceteris paribus* clause is appended.[5]

[1] Popper [1934], section 26.

[2] Neurath's [1935] shows that he never grasped Popper's simple argument.

[3] I am using here 'verisimilitude' in Popper's sense: the difference between the truth content and falsity content of a theory. For the risks involved in estimating it, cf. my [1968*a*], especially pp. 395 ff.

[4] I tried to develop such a general theory of criticism in my [1970].

[5] The falsification of theories depends on the high degree of corroboration of the *ceteris paribus* clause. This however is not always the case. This is why the methodological falsificationist may advise us to rely on our 'scientific instinct' (Popper [1934], section 18, footnote 2) or 'hunch' (Braithwaite [1953], p. 20).

Their falsification as it occurs in actual history is *prima facie* irrational by the standards of our falsificationist. By his standards, scientists frequently seem to be irrationally slow: for instance, eighty-five years elapsed between the acceptance of the perihelion of Mercury as an anomaly and its acceptance as a falsification of Newton's theory, in spite of the fact that the *ceteris paribus* clause was reasonably well corroborated. On the other hand, scientists frequently seem to be irrationally rash: for instance, Galileo and his disciples accepted Copernican heliocentric celestial mechanics in spite of the abundant evidence against the rotation of the Earth; or Bohr and his disciples accepted a theory of light emission in spite of the fact that it ran counter to Maxwell's well-corroborated theory.

Indeed, it is not difficult to see at least two crucial characteristics common to both dogmatic and our methodological falsificationism which are clearly dissonant with the actual history of science: that (1) *a test is—or must be made—a two-cornered fight between theory and experiment so that in the final confrontation only these two face each other; and* (2) *the only interesting outcome of such confrontation is* (*conclusive*) *falsification*: '[*the only genuine*] *discoveries are refutations of scientific hypotheses*.'[1] However, history of science suggests that (1′) tests are—at least—three-cornered fights between rival theories and experiment and (2′) some of the most interesting experiments result, *prima facie*, in confirmation rather than falsification.

But if—as seems to be the case—the history of science does not bear out our theory of scientific rationality, we have two alternatives. One alternative is to abandon efforts to give a rational explanation of the success of science. Scientific method (or 'logic of discovery'), conceived as the discipline of rational appraisal of scientific theories—and of criteria of *progress*—vanishes. We, may, of course, still try to explain *changes* in 'paradigms' in terms of social psychology.[2] This is Polanyi's and Kuhn's way.[3] The other alternative is to try at least to *reduce* the conventional element in falsificationism (we cannot possibly eliminate it) and replace the

[1] Agassi [1959]; he calls Popper's idea of science '*scientia negativa*' (Agassi [1968]).

[2] It should be mentioned here that the Kuhnian sceptic is still left with what I would call the '*scientific sceptic's dilemma*': any scientific sceptic will still try to explain changes in beliefs and will regard his own psychology as a theory which is more than simple belief, which, in some sense, is 'scientific'. Hume, while trying to show up science as a mere system of beliefs with the help of his stimulus-response theory of learning, never raised the problem of whether his theory of learning applies also to his own theory of learning. In contemporary terms, we might well ask, does the popularity of Kuhn's philosophy indicate that people recognize its *truth*? In this case it would be refuted. Or does this popularity indicate that people regarded it as an attractive new fashion? In this case, it would be 'verified'. But would Kuhn like *this* 'verification'?

[3] Feyerabend who contributed probably more than anybody else to the spread of Popper's ideas, seems now to have joined the enemy camp. Cf. his intriguing [1970].

naive versions of methodological falsificationism—characterized by the theses (1) and (2) above—by a *sophisticated* version which would give a new *rationale* of falsification and thereby rescue methodology and the idea of scientific *progress*. This is Popper's way, and the one I intend to follow.

(*c*) *Sophisticated versus naive methodological falsificationism. Progressive and degenerating problemshifts.*

Sophisticated falsificationism differs from naive falsificationism both in its rules of *acceptance* (or 'demarcation criterion') and its rules of *falsification* or elimination. For the naive falsificationist any theory which can be interpreted as experimentally falsifiable, is 'acceptable' or 'scientific'.[1] For the sophisticated falsificationist a theory is 'acceptable' or 'scientific' only if it has corroborated excess empirical content over its predecessor (or rival), that is, only if it leads to the discovery of novel facts. This condition can be analysed into two clauses: that the new theory has excess empirical content ('*acceptability*'$_1$) and that some of this excess content is verified ('*acceptability*'$_2$). The first clause can be checked instantly[2] by *a priori* logical analysis; the second can be checked only empirically and this may take an indefinite time.

Again, for the naive falsificationist a theory is *falsified* by a '(fortified'[3]) 'observational' statement which conflicts with it (or rather, which he decides to interpret as conflicting with it). The sophisticated falsificationist regards a scientific theory T as falsified if and only if another theory T' has been proposed with the following characteristics: (1) T' has excess empirical content over T: that is, it predicts *novel* facts, that is, facts improbable in the light of, or even forbidden, by T;[4] (2) T' explains the previous success of T, that is, all the unrefuted content of T is contained (within the limits of observational error) in the content of T'; and (3) some of the excess content of T' is corroborated.[5]

In order to be able to appraise these definitions we need to understand their problem background and their consequences. First, we have to remember the conventionalists' methodological discovery that no experimental result can ever kill a theory: any theory can be saved from counterinstances either by some auxiliary hypothesis or by a suitable reinterpretation of its terms. Naive falsificationists solved this problem by relegating—in crucial contexts—the auxiliary hypotheses to the realm of unproblematic background knowledge, eliminating them from the deductive model of the test-

[1] Cf. *above*, p. 109. [2] But *cf. below*, pp. 155–7. [3] Cf. *above*, p. 108, text to footnote 1.

[4] I use 'prediction' in a wide sense that includes 'postdiction'.

[5] *For a detailed discussion of these acceptance and rejection rules and for references to Popper's work*, cf. my [1968*a*], pp. 375–90. For some qualifications (concerning continuity and consistency as regulative principles), cf. *below*, pp. 131–2 and 141–6.

situation and thereby *forcing* the chosen theory into logical isolation, in which it becomes a sitting target for the attack of test-experiments. But since this procedure did not offer a suitable guide for a rational reconstruction of the history of science, we may just as well completely rethink our approach. Why aim at falsification at any price? Why not rather impose certain standards on the theoretical adjustments by which one is allowed to save a theory? Indeed, some such standards have been well-known for centuries, and we find them expressed in age-old wisecracks against *ad hoc* explanations, empty prevarications, face-saving, linguistic tricks.[1] We have already seen that Duhem adumbrated such standards in terms of 'simplicity' and 'good sense'.[2] But *when* does lack of 'simplicity' in the protective belt of theoretical adjustments reach the point at which the theory *must* be abandoned?[3] In what sense was Copernican theory, for instance, 'simpler' than Ptolemaic?[4] The vague notion of Duhemian 'simplicity' leaves, as the naive falsificationist correctly argued, the decision very much to taste and fashion.[5]

Can one improve on Duhem's approach? Popper did. His solution—a sophisticated version of methodological falsificationism—is more objective and more rigorous. Popper agrees with the conventionalists that theories and factual propositions can always be harmonized with the help of auxiliary hypotheses: he agrees that the problem is how to demarcate between scientific and pseudoscientific *adjustments*, between rational and irrational changes of theory. According to Popper, saving a theory with the help of auxiliary hypotheses which satisfy certain well-defined conditions represents scientific progress; but saving a theory with the help of auxiliary hypotheses which do not, represents degeneration. Popper calls such inadmissible auxiliary hypotheses *ad hoc* hypotheses, mere linguistic devices, 'conventionalist stratagems'.[6] But then any scientific theory has to

[1] Molière, for instance, ridiculed the doctors of his *Malade Imaginaire,* who offered the *virtus dormitiva* of opium as the answer to the question as to why opium produced sleep. One might even argue that Newton's famous dictum *hypotheses non fingo* was really directed against *ad hoc* explanations—like his own explanation of gravitational forces by an aether-model in order to meet Cartesian objections. [2] Cf. *above,* p. 105.

[3] Incidentally, Duhem agreed with Bernard that experiments alone—without simplicity considerations—can decide the fate of theories in physiology. But in physics, he argued, they cannot ([1905], chapter VI, section 1).

[4] Koestler correctly points out that only Galileo created the myth that the Copernican theory was simple (Koestler [1959], p. 476); in fact, 'the motion of the earth [had not] done much to simplify the old theories, for though the objectionable equants had disappeared, the system was still bristling with auxiliary circles' (Dreyer [1906], chapter XIII). [5] Cf. *above,* p. 105.

[6] Popper [1934], sections 19 and 20. I have discussed in some detail—under the heads 'monster-barring', 'exception-barring', 'monster-adjustment'—such stratagems as they appear in informal, quasi-empirical mathematics; cf. my [1963–4].

be appraised together with its auxiliary hypotheses, initial conditions, etc., and, especially, together with its predecessors so that we may see by what sort of *change* it was brought about. Then, of course, what we appraise is a *series of theories* rather than isolated *theories*.

Now we can easily understand why we formulated the criteria of acceptance and rejection of sophisticated methodological falsificationism as we did.[1] But it may be worth while to reformulate them slightly, couching them explicitly in terms of *series of theories*.

Let us take a series of theories, T_1, T_2, T_3, . . . where each subsequent theory results from adding auxiliary clauses to (or from semantical reinterpretations of) the previous theory in order to accommodate some anomaly, each theory having at least as much content as the unrefuted content of its predecessor. Let us say that such a series of theories is *theoretically progressive* (*or* '*constitutes a theoretically progressive problemshift*') if each new theory has some excess empirical content over its predecessor, that is, if it predicts some novel, hitherto unexpected fact. Let us say that a theoretically progressive series of theories is also *empirically progressive* (*or* '*constitutes an empirically progressive problemshift*') if some of this excess empirical content is also corroborated, that is, if each new theory leads us to the actual discovery of some *new fact*.[2] Finally, let us call a problemshift *progressive* if it is both theoretically and empirically progressive, and *degenerating* if it is not.[3] We '*accept*' problemshifts as 'scientific' only if they are at least theoretically progressive; if they are not, we '*reject*' them as 'pseudoscientific'. Progress is measured by the degree to which a problemshift is progressive, by the degree to which the series of theories leads us to the discovery of novel facts. We regard a theory in the series 'falsified' when it is superseded by a theory with higher corroborated content.[4]

This demarcation between progressive and degenerating problemshifts sheds new light on the appraisal of *scientific*—*or*, *rather*, *progressive*—

[1] Cf. *above*, p. 116.

[2] If I already know P_1: 'Swan A is white', P_ω: 'All swans are white' represents no progress, because it may only lead to the discovery of such further similar facts as P_2: 'Swan B is white'. So-called 'empirical generalizations' constitute no progress. A *new* fact must be improbable or even impossible in the light of previous knowledge. Cf. *above*, p. 116, and *below*, pp. 155 ff.

[3] The appropriateness of the term 'problemshift' for a series of theories rather than of problems may be questioned. I chose it partly because I have not found a more appropriate alternative—'theoryshift' sounds dreadful—partly because theories are always problematical, they never solve all the problems they have set out to solve. Anyway, in the second half of the paper, the more natural term 'research programme' will replace 'problemshifts' in the most relevant contexts.

[4] For the 'falsification' of certain series of theories (of 'research programmes') as opposed to the 'falsification' of one theory within the series, cf. *below*, pp. 155 ff.

explanations. If we put forward a theory to resolve a contradiction between a previous theory and a counterexample in such a way that the new theory, instead of offering a content-increasing (scientific) *explanation*, only offers a content-decreasing (linguistic) *reinterpretation*, the contradiction is resolved in a merely semantical, unscientific way. *A given fact is explained scientifically only if a new fact is also explained with it.*[1]

Sophisticated falsificationism thus shifts the problem of how to appraise *theories* to the problem of how to appraise *series of theories*. Not an isolated *theory*, but only a series of theories can be said to be scientific or unscientific: to apply the term 'scientific' to one *single* theory is a category mistake.[2]

The time-honoured empirical criterion for a satisfactory theory was agreement with the observed facts. Our empirical criterion for a series of theories is that it should produce new facts. *The idea of growth and the concept of empirical character are soldered into one.*

This revised form of methodological falsificationism has many new features. First, it denies that 'in the case of a scientific theory, our decision depends upon the results of experiments. If these confirm the theory, we may accept it until we find a better one. If they contradict the theory, we reject it.'[3] It denies that 'what ultimately decides the fate of a theory is the result of a test, *i.e.* an agreement about basic statements'.[4] Contrary to naive falsificationism, *no experiment, experimental report, observation statement or well-corroborated low-level falsifying hypothesis alone can lead to falsification.*[5] *There is no falsification before the emergence of a better theory.*[6]

[1] Indeed, in the original manuscript of my [1968*a*] I wrote: 'A theory without excess corroboration has no excess explanatory power; *therefore, according to Popper, it does not represent growth and therefore it is not "scientific"; therefore, we should say, it has no explanatory power*' (p. 386). I cut out the italicized half of the sentence under pressure from my colleagues who thought it sounded too eccentric. I regret it now.

[2] Popper's conflation of 'theories' and 'series of theories' prevented him from getting the basic ideas of sophisticated falsificationism across more successfully. His ambiguous usage led to such confusing formulations as 'Marxism [as the core of a series of theories or of a "research programme"] is irrefutable' and, at the same time, 'Marxism [as a particular conjunction of this core and some specified auxiliary hypotheses, initial conditions and a *ceteris paribus* clause] has been refuted.' (Cf. Popper [1963].)

Of course, there is nothing wrong in saying that an isolated, single theory is 'scientific' if it represents an advance on its predecessor, as long as one clearly realizes that in this formulation we appraise the theory as the outcome of—and in the context of—a certain historical development.

[3] Popper [1945], vol. II, p. 233. Popper's more sophisticated attitude surfaces in the remark that 'concrete and practical consequences can be *more* directly tested by experiment' (*ibid.* my italics).

[4] Popper [1934], section 30.

[5] For the *pragmatic* character of methodological 'falsification', cf. *above*, p. 109, footnote 2.

[6] 'In most cases we have, before falsifying a hypothesis, another one up our sleeves' (Popper [1959*a*], p. 87, footnote *1). But, as our argument shows, we *must* have one. Or, as Feyerabend put it: 'The best criticism is provided by those theories which can replace the rivals they have removed' ([1965], p. 227). He notes that in *some* cases 'alternatives

But then the distinctively negative character of naive falsificationism vanishes; criticism becomes more difficult, and also positive, constructive. But, of course, if falsification depends on the emergence of better theories, on the invention of theories which anticipate new facts, then falsification is *not* simply a relation between a theory and the empirical basis, but a multiple relation between competing theories, the original 'empirical basis', and the empirical growth resulting from the competition. Falsification can thus be said to have a '*historical character*'.[1] Moreover, some of the theories which bring about falsification are frequently proposed *after* the 'counterevidence'. This may sound paradoxical for people indoctrinated with naive falsificationism. Indeed, this epistemological theory of the relation between theory and experiment differs sharply from the epistemological theory of naive falsificationism. The very term 'counterevidence' has to be abandoned in the sense that no experimental result must be interpreted directly as 'counterevidence'. If we still want to retain this time-honoured term, we have to redefine it like this: 'counterevidence to T_1' is a corroborating instance to T_2 which is either inconsistent with or independent of T_1 (with the *proviso* that T_2 is a theory which satisfactorily explains the empirical success of T_1). This shows that '*crucial counterevidence*'—or '*crucial experiments*'—can be recognized as such among the scores of anomalies only *with hindsight*, in the light of some superseding theory.[2]

Thus the crucial element in falsification is whether the *new theory* offers any novel, excess information compared with its predecessor and whether some of this excess information is corroborated. Justificationists valued 'confirming' instances of a theory; naive falsificationists stressed 'refuting' instances; for the methodological falsificationists it is the—rather rare—corroborating instances of the *excess* information which are the crucial ones; these receive all the attention. We are no longer interested in the

---

will be quite indispensable for the purpose of refutation' (*ibid.* p. 254). But according to our argument *refutation without an alternative shows nothing but the poverty of our imagination in providing a rescue hypothesis*. Also cf. *below*, p. 121, footnote 4.

[1] Cf. my [1968*a*], pp. 387 ff.

[2] In the distorting mirror of naive falsificationism, new theories which replace old refuted ones, are themselves born unrefuted. Therefore they do not believe that there is a relevant difference between anomalies and crucial counterevidence. For them, anomaly is a dishonest euphemism for counterevidence. But in actual history new theories are born refuted: they inherit many anomalies of the old theory. Moreover, frequently it is *only* the new theory which dramatically predicts that fact which will function as crucial counterevidence against its predecessor, while the 'old' anomalies may well stay on as 'new' anomalies.

All this will be still clearer when we introduce the idea of 'research programme': cf. *below*, pp. 135 and 176 ff.

thousands of trivial verifying instances nor in the hundreds of readily available anomalies: the few crucial *excess-verifying instances* are decisive.[1] This consideration rehabilitates—and reinterprets—the old proverb: *Exemplum docet, exempla obscurant.*

'Falsification' in the sense of naive falsificationism (corroborated counter-evidence) is not a *sufficient* condition for eliminating a specific theory: in spite of hundreds of known anomalies we do not regard it as falsified (that is, eliminated) until we have a better one.[2] Nor is 'falsification' in the naive sense *necessary* for falsification in the sophisticated sense: a progressive problemshift does not have to be interspersed with 'refutations'. Science can grow without any 'refutations' leading the way. Naive falsificationists suggest a linear growth of science, in the sense that theories are followed by powerful refutations which eliminate them; these refutations in turn are followed by new theories.[3] It is perfectly *possible* that theories be put forward 'progressively' in such a rapid succession that the 'refutation' of the n-th appears only as the corroboration of the $n+1$-th. The problem fever of science is raised by proliferation of rival theories rather than counterexamples or anomalies.

This shows that the slogan of *proliferation of theories* is much more important for sophisticated than for naive falsificationism. For the naive falsificationist science grows through repeated experimental overthrow of theories; new rival theories proposed before such 'overthrows' may speed up growth but are not absolutely necessary[4]; constant proliferation of theories is optional but not mandatory. For the sophisticated falsificationist

[1] *Sophisticated falsificationism adumbrates a new theory of learning;* cf. *below*, p. 123.

[2] It is clear that the theory T' may have excess corroborated empirical content over another theory T even if both T and T' are refuted. Empirical content has nothing to do with truth or falsity. Corroborated contents can also be compared irrespective of the refuted content. Thus we may see the rationality of the elimination of Newton's theory in favour of Einstein's, even though Einstein's theory may be said to have been born—like Newton's—'refuted'. We have only to remember that 'qualitative confirmation' is a euphemism for 'quantitative disconfirmation'. (Cf. my [1968*a*], pp. 384–6.)

[3] Cf. Popper [1934], section 85, p. 279 of the 1959 English translation.

[4] It is true that a certain type of *proliferation of rival theories* is allowed to play an accidental heuristic role in falsification. In many cases falsification heuristically 'depends on [the condition] that sufficiently many and sufficiently different theories are offered' (Popper [1940]). For instance, we may have a theory T which is apparently unrefuted. But it may happen that a new theory T', inconsistent with T, is proposed which equally fits the available facts: the differences are smaller than the range of observational error. In such cases the inconsistency prods us into improving our 'experimental techniques', and thus refining the 'empirical basis' so that either T or T' (or, incidentally, both) can be falsified: 'We need [a] new theory in order to find out where the old theory was deficient' (Popper [1963], p. 246). But the role of this proliferation is *accidental* in the sense that, once the empirical basis is refined, the fight is between this refined empirical basis and the theory T under test; the rival theory T' acted only as a *catalyst*. (Also cf. *above*, p. 119, footnote 6.)

proliferation of theories cannot wait until the accepted theories are 'refuted' (or until their protagonists get into a Kuhnian crisis of confidence).[1] While naive falsificationism stresses 'the urgency of replacing a *falsified* hypothesis by a better one',[2] sophisticated falsificationism stresses the urgency of replacing *any* hypothesis by a better one. Falsification cannot 'compel the theorist to search for a better theory',[3] simply because falsification cannot precede the better theory.

The problem-shift from naive to sophisticated falsificationism involves a semantic difficulty. For the naive falsificationist a 'refutation' is an experimental result which, by force of his decisions, is made to conflict with the theory under test. But according to sophisticated falsificationism one must not take such decisions before the alleged 'refuting instance' has become the confirming instance of a new, better theory. Therefore whenever we see terms like 'refutation', 'falsification', 'counterexample', we have to check in each case whether these terms are being applied in virtue of decisions by the naive or by the sophisticated falsificationist.[4]

Sophisticated methodological falsificationism offers new standards for intellectual honesty. Justificationist honesty demanded the acceptance of only what was proven and the rejection of everything unproven. Neojustificationist honesty demanded the specification of the probability of any hypothesis in the light of the available empirical evidence. The honesty of naive falsificationism demanded the testing of the falsifiable and the rejection of the unfalsifiable and the falsified. Finally, the honesty of sophisticated falsificationism demanded that one should try to look at things from different points of view, to put forward new theories which anticipate novel facts, and to reject theories which have been superseded by more powerful ones.

Sophisticated methodological falsificationism blends several different traditions. From the empiricists it has inherited the determination to learn primarily from experience. From the Kantians it has taken the activist approach to the theory of knowledge. From the conventionalists it has learned the importance of decisions in methodology.

[1] Also cf. Feyerabend [1965], pp. 254–5.

[2] Popper [1959*a*], p. 87, footnote *1.

[3] Popper [1934], section 30.

[4] Cf. also *above*, p. 109, footnote 2. [*Added in press:*] Possibly it would be better in future to abandon these terms altogether, just as we have abandoned terms like 'inductive (or experimental) proof'. Then we may call (naive) 'refutations' anomalies, and (sophisticatedly) 'falsified' theories 'superseded' ones. Our 'ordinary' language is impregnated not only by 'inductivist' but also by falsificationist dogmatism. A reform is overdue.

I should like to emphasize here a further distinctive feature of sophisticated methodological empiricism: the crucial role of excess corroboration. For the inductivist, learning about a new theory is learning how much confirming evidence supports it; about refuted theories one *learns* nothing (learning, after all, is to build up proven or probable *knowledge*). For the dogmatic falsificationist, learning about a theory is learning whether it is refuted or not; about confirmed theories one learns nothing (one cannot prove or probabilify anything), about refuted theories one learns that they are disproved.[1] For the sophisticated falsificationist, learning about a theory is primarily learning which new facts it anticipated: indeed, for the sort of Popperian empiricism I advocate, the only relevant evidence is the evidence anticipated by a theory, and *empiricalness (or scientific character) and theoretical progress are inseparably connected.*[2]

This idea is not entirely new. Leibnitz, for instance, in his famous letter to Conring in 1678, wrote: 'It is the greatest commendation of an hypothesis (next to [proven] truth) if by its help predictions can be made even about phenomena or experiments not tried.'[3] Leibnitz's view was widely accepted by scientists. But since all appraisal of a scientific theory was before Popper appraisal of its degree of justification, this position was regarded by some logicians as untenable. Mill, for instance, complains in 1843 in horror that 'it seems to be thought that an hypothesis . . is entitled to a more favourable reception, if besides accounting for all the facts previously known, it has led to the anticipation and prediction of others which experience afterwards verified'.[4] Mill had a point: this appraisal was in conflict both with justificationism and with probabilism: why should an event *prove* more, if it was anticipated by the theory than if it was known already before? As long as *proof* was the only criterion of the scientific character of a theory, Leibnitz's criterion could only be regarded as irrelevant.[5] Also, the *probability* of a theory given evidence cannot possibly be influenced, as Keynes pointed out, by *when* the evidence was produced: the probability of a theory given evidence can depend only on the theory

[1] For a defence of this theory of 'learning from experience', cf. Agassi [1969].

[2] *These remarks show that 'learning from experience' is a normative idea; therefore all purely 'empirical' learning theories miss the heart of the problem.*

[3] Cf. Leibnitz [1678]. The expression in brackets shows that Leibnitz regarded this criterion as second best and thought that the best theories are those which are proved. Thus Leibnitz's position—like Whewell's—is a far cry from fully fledged sophisticated falsificationism.

[4] Mill [1843], vol. II, p. 23.

[5] This was J. S. Mill's argument (*ibid.*). He directed it against Whewell, who thought that 'consilience of inductions' or successful prediction of improbable events *verifies* (that is, *proves*) a theory. (Whewell [1858], pp. 95–6.) No doubt, *the basic contradiction both in Whewell's and in Duhem's philosophy of science is their conflation of heuristic power and proven truth. Popper separated the two.*

and the evidence,[1] and not upon whether the evidence was produced before or after the theory.

In spite of this convincing justificationist criticism, the criterion survived among some of the best scientists, since it formulated their strong dislike of merely *ad hoc* explanations, which 'though [they] truly express the facts [they set out to explain, are] not born out by any other phenomena'.[2]

But it was only Popper who recognized that the *prima facie* inconsistency between the few odd, casual remarks against *ad hoc* hypotheses on the one hand and the huge edifice of justificationist philosophy of knowledge must be solved by demolishing justificationism and by introducing new, non-justificationist criteria for appraising scientific theories based on anti-adhocness.

Let us look at a few examples. Einstein's theory is not better than Newton's *because* Newton's theory was 'refuted' but Einstein's was not: there are many known 'anomalies' to Einsteinian theory. Einstein's theory is better than—that is, represents progress compared with—Newton's theory *anno 1916* (that is, Newton's laws of dynamics, law of gravitation, the known set of initial conditions; 'minus' the list of known anomalies such as Mercury's perihelion) *because* it explained everything that Newton's theory had successfully explained, and it explained also *to some extent* some known anomalies and, in addition, forbade events like transmission of light along straight lines near large masses about which Newton's theory had said nothing but which had been permitted by other well-corroborated scientific theories of the day; moreover, *at least some* of the unexpected excess Einsteinian content was in fact *corroborated* (for instance, by the eclipse experiments).

On the other hand, according to these sophisticated standards, Galileo's theory that the natural motion of terrestrial objects was circular, introduced no improvement since it did not forbid anything that had not been forbidden by the relevant theories he intended to improve upon (that is, by Aristotelian physics and by Copernican celestial kinematics). This theory was therefore *ad hoc* and therefore—from the heuristic point of view—valueless.[3]

A beautiful example of a theory which satisfied only the first part of Popper's criterion of progress (excess content) but not the second part (corroborated excess content) was given by Popper himself: the Bohr–

[1] Keynes [1921], p. 305. But cf. my [1968*a*], p. 394.

[2] This is Whewell's critical comment on an *ad hoc* auxiliary hypothesis in Newton's theory of light (Whewell [1857], vol. II, p. 317.)

[3] In the terminology of my [1968*a*], this theory was '*ad hoc*$_1$' (cf. my [1968*a*], p. 389, footnote 1); the example was originally suggested to me by Paul Feyerabend as a paradigm of a *valuable ad hoc theory*. But cf. *below*, p. 142, especially footnote 3.

Kramers–Slater theory of 1924. This theory was refuted in *all* its new predictions.[1]

Let us finally consider how much conventionalism remains in sophisticated falsificationism. Certainly *less* than in naive falsificationism. We need *fewer* methodological decisions. The '*fourth-type decision*' which was essential for the naive version[2] has become completely redundant. To show this we only have to realize that if a scientific theory, consisting of some 'laws of nature', initial conditions, auxiliary theories (but without a *ceteris paribus* clause) conflicts with some factual propositions we do not have to decide which—explicit or 'hidden'—part to replace. We may try to replace *any* part and only when we have hit on an explanation of the anomaly with the help of some content-increasing change (or auxiliary hypothesis), and nature corroborates it, do we move on to eliminate the 'refuted' complex. Thus sophisticated falsification is a slower but possibly safer process than naive falsification.

Let us take an example. Let us assume that the course of a planet differs from the one predicted. Some conclude that this refutes the dynamics and gravitational theory applied: the initial conditions and the *ceteris paribus* clause have been ingeniously corroborated. Others conclude that this refutes the initial conditions used in the calculations: dynamics and gravitational theory have been superbly corroborated in the last two hundred years and all suggestions concerning further factors in play failed. Yet others conclude that this refutes the underlying assumption that there were no other factors in play except for those which were taken into account: these people may possibly be motivated by the metaphysical principle that any explanation is only approximative because of the infinite complexity of the factors involved in determining any single event. Should we praise the first type as '*critical*', scold the second type as '*hack*', and condemn the third as '*apologetic*'? No. We do not need to draw any conclusions about such 'refutation'. We never reject a specific theory simply by *fiat*. If we have an inconsistency like the one mentioned, we do not have to decide which ingredients of the theory we regard as problematic and which ones as unproblematic: we regard all ingredients as problematic in the light of the conflicting accepted basic statement and try to replace all of them. If we succeed in replacing some ingredient in a 'progressive' way (that is, the replacement has more corroborated empirical content than the original), we call it 'falsified'.

We do not need the *fifth type decision* of the naive falsificationist either.

[1] In the terminology of my [1968*a*], this theory was not '*ad hoc*$_1$', but it was '*ad hoc*$_2$' (cf. my [1968*a*], p. 389, footnote 1). For a simple but artificial illustration, see *ibid.* p. 387, footnote 2. (For *ad hoc*$_3$, cf. *below*, p. 175, footnote 3.)

[2] Cf. *above*, p. 110.

In order to show this let us have a new look at the problem of the appraisal of (syntactically) metaphysical theories—and the problem of their retention and elimination. The 'sophisticated' solution is obvious. We retain a syntactically metaphysical theory as long as the problematic instances can be explained by content-increasing changes in the auxiliary hypotheses appended to it.[1] Let us take, for instance, Cartesian metaphysics C: 'in *all* natural processes *there is* a clockwork mechanism regulated by (*a priori*) animating principles.' This is syntactically irrefutable: it can clash with no—spatiotemporally singular—'basic statement'. It may, of course, clash with a refutable theory like N: 'gravitation is a force equal to $fm_1 m_2/r^2$ *which acts at a distance*'. But N will only clash with C if 'action at a distance' is interpreted literally and possibly, in addition, as representing an *ultimate* truth, irreducible to any still deeper cause. (Popper would call this an 'essentialist' interpretation.) Alternatively we can regard 'action at a distance' as a mediate cause. Then we interpret 'action at a distance' figurativ-ely, and regard it as a shorthand for some hidden mechanism of action by contact. (We may call this a 'nominalist' interpretation.) In this case we can attempt to explain N by C—Newton himself and several French physicists of the eighteenth century tried to do so. If an auxiliary theory which performs this explanation (or, if you wish, 'reduction') produces novel facts (that is, it is 'independently testable'), Cartesian metaphysics should be regarded as good, scientific, empirical metaphysics, generating a progressive problemshift. A progressive (syntactically) metaphysical theory produces a sustained progressive shift in its protective belt of auxiliary theories. If the reduction of the theory to the 'metaphysical' frame-work does not produce new empirical content, let alone novel facts, then the reduction represents a degenerating problemshift, it is a mere linguistic exercise. The Cartesian efforts to bolster up their 'metaphysics' in order to explain Newtonian gravitation is an outstanding example of such a merely linguistic reduction.[2]

Thus we do not eliminate a (syntactically) metaphysical theory if it clashes with a well-corroborated scientific theory, as naive falsificationism suggests. We eliminate it if it produces a degenerating shift in the long run and there is a better, rival, metaphysics to replace it. The methodology

[1] *We can formulate this condition with striking clarity only in terms of the methodology of research programmes to be explained in* §3*: we retain a syntactically metaphysical theory as the 'hard core' of a research programme as long as its associated positive heuristic produces a progressive problemshift in the 'protective belt' of auxiliary hypotheses*. Cf. *below*, pp. 136–7.

[2] This phenomenon was described in a beautiful paper by Whewell [1851]; but he could not explain it methodologically. Instead of recognizing the victory of the *progressive* Newtonian programme over the *degenerating* Cartesian programme, he thought this was the victory of proven truth over falsity. For details cf. my [1970]: for a general discussion of the demarcation between progressive and degenerating reduction cf. Popper [1969].

of a research programme with a 'metaphysical' core does not differ from the methodology of one with a 'refutable' core except for the logical level of the inconsistencies which are the driving force of the programme.[1]

(It has to be stressed, however, that the very choice of the logical form in which to articulate a theory depends to a large extent on our methodological decision. For instance, instead of formulating Cartesian metaphysics as an 'all-some' statement, we can formulate it as an 'all-statement': 'all natural processes are clockworks'. A 'basic statement' contradicting this would be: '*a* is a natural process and it is not clockwork'. The question is whether according to the 'experimental techniques', or rather, to the interpretative theories of the day, '*x* is not a clockwork' can be 'established' or not. Thus the rational choice of the logical form of a theory depends on the state of our knowledge; for instance, a metaphysical 'all-some' statement of today may become, with the change in the level of observational theories, a scientific 'all-statement' tomorrow. I have already argued that only series of theories and not theories should be classified as scientific or non-scientific; now I have indicated that even the logical form of a theory can only be rationally chosen on the basis of a critical appraisal of the state of the research programme in which it is embedded.)

The first, second, and third type decisions of naive falsificationism[2] however cannot be avoided, but as we shall show, the conventional element in the second decision—and also in the third—can be slightly reduced. We cannot avoid the decision which sort of propositions should be the 'observational' ones and which the 'theoretical' ones. We cannot avoid either the decision about the truth-value of some 'observational propositions'. These decisions are vital for the decision whether a problemshift is empirically progressive or degenerating.[3] But the sophisticated falsificationist may at least mitigate the arbitrariness of this second decision by allowing for an *appeal procedure*.

Naive falsificationists do not lay down any such appeal procedure. They accept a basic statement if it is backed up by a well-corroborated falsifying hypothesis,[4] and let it overrule the theory under test—even though they are well aware of the risk.[5] But there is no reason why we should not regard a falsifying hypothesis—and the basic statement it supports—as being just as problematic as a falsified hypothesis. Now how exactly can we expose the problematicality of a basic statement? On what grounds can the protagonists of the 'falsified' theory appeal and win?

Some people may say that we might go on testing the basic statement (or

[1] Cf. *above*, p. 126, footnote 1.

[2] Cf. *above*, pp. 106 and 109.

[3] Cf. *above*, p. 118.

[4] Popper [1934], section 22.

[5] Cf. e.g. Popper [1959*a*], p. 107, footnote *2. Also cf. *above*, pp. 112–14.

the falsifying hypothesis) 'by their deductive consequences' until agreement is finally reached. In this testing we deduce—in the same deductive model—further consequences from the basic statement either with the help of the theory under test or some other theory which we regard as unproblematic. Although this procedure 'has no natural end', we always come to a point when there is no further disagreement.[1]

But when the theoretician appeals against the verdict of the experimentalist, the appeal court does not normally cross-question the basic statement directly but rather questions the *interpretative theory* in the light of which its truth-value had been established.

One typical example of a series of successful appeals is the Proutians' fight against unfavourable experimental evidence from 1815 to 1911. For decades Prout's theory T ('that all atoms are compounds of hydrogen atoms and thus "atomic weights" of all chemical elements must be expressible as whole numbers') and falsifying 'observational' hypotheses, like Stas's 'refutation' R ('the atomic weight of chlorine is 35·5') confronted each other. As we know, in the end T prevailed over R.[2]

The first stage of any serious criticism of a scientific theory is to reconstruct, improve, its logical deductive articulation. Let us do this in the case of Prout's theory *vis à vis* Stas's refutation. First of all, we have to realize that in the formulation we just quoted, T and R were *not* inconsistent. (Physicists rarely articulate their theories sufficiently to be pinned down and caught by the critic.) In order to show them up as inconsistent we have to put them in the following form. T: 'the atomic weight of all pure (homogeneous) chemical elements are multiples of the atomic weight of hydrogen', and R: 'chlorine is a pure (homogeneous) chemical element and its atomic weight is 35·5'. The last statement is in the form of a falsifying hypothesis which, if well corroborated, would allow us to use basic statements of the form B: 'Chlorine X is a pure (homogeneous) chemical element and its atomic weight is 35·5'—where X is the proper name of a 'piece' of chlorine determined, say, by its space-time co-ordinates.

But how well-corroborated is R? The first component of it says that R_1: 'Chlorine X is a pure chemical element.' This was the verdict of the experimental chemist after a rigorous application of the 'experimental techniques' of the day.

Let us have a closer look at the fine-structure of R_1. In fact R_1 stands for a conjunction of two longer statements T_1 and T_2. The first statement,

[1] This is argued in Popper [1934], section 29.

[2] Agassi claims that this example shows that we may 'stick to the hypothesis in the face of known facts in the hope that the facts will adjust themselves to theory rather than the other way round' ([1966], p. 18). But *how* can facts 'adjust themselves'? Under which *particular* conditions should the theory win? Agassi gives no answer.

T_1, could be this: 'If seventeen chemical purifying procedures $p_1, p_2 \ldots p_{17}$ are applied to a gas, what remains will be pure chlorine.' T_2 is then: '*X* was subjected to the seventeen procedures $p_1, p_2 \ldots p_{17}$.' The careful 'experimenter' carefully applied all seventeen procedures: T_2 is to be accepted. But the conclusion that therefore what remained *must* be pure chlorine is a 'hard fact' only in virtue of T_1. The experimentalist, while *testing T, applied* T_1. He *interpreted* what he saw in the light of T_1: the result was R_1. *Yet in the monotheoretical model of the explanatory theory under test this interpretative theory does not appear at all.*

But what if T_1, the interpretative theory, is false? Why not 'apply' *T* rather than T_1 and claim that atomic weights *must be* whole numbers? Then *this* will be a 'hard fact' in the light of *T*, and T_1 will be overthrown. Perhaps additional new purifying procedures must be invented and applied.

The problem is then *not* when we should stick to a '*theory*' in the face of '*known facts*' and when the other way round. The problem is *not* what to do when 'theories' clash with 'facts'. Such a 'clash' is only suggested by the '*monotheoretical deductive model*'. Whether a proposition is a '*fact*' or a '*theory*' in the context of a test-situation depends on our methodological decision. 'Empirical basis of a theory' is a mono-theoretical notion, it is *relative* to some mono-theoretical deductive structure. We may use it as first approximation; but in case of 'appeal' by the theoretician, we must use a *pluralistic model.* In the pluralistic model the clash is not 'between theories and facts' but between two high-level theories: between an *interpretative theory* to provide the facts and an *explanatory theory* to explain them; and the interpretative theory may be on quite as high a level as the explanatory theory. The clash is then not any more between a logically higher-level theory and a lower-level falsifying hypothesis. The problem should not be put in terms of whether a '*refutation*' is real or not. The problem is how to repair an *inconsistency* between the 'explanatory theory' under test and the —explicit or hidden—'interpretative' theories; or, if you wish, *the problem is which theory to consider as the interpretative one which provides the 'hard' facts and which the explanatory one which 'tentatively' explains them.* In a mono-theoretical model we regard the higher-level theory as an *explanatory theory to be judged by the 'facts'* delivered from outside (by the authoritative experimentalist): in the case of a clash we reject the explanation.[1] In a pluralistic model we may decide, alternatively, to regard the higher-

[1] The decision to use some monotheoretical model is clearly vital for the naive falsificationist to enable him to reject a theory on the *sole* ground of experimental evidence. *It is in line with the necessity for him to divide sharply, at least in a test-situation, the body of science into two: the problematic and the unproblematic.* (Cf. *above* p. 107.) *It is only the theory he decides to regard as problematic which he articulates in his deductive model of criticism.*

level theory as an *interpretative theory to judge the 'facts'* delivered from outside: in case of a clash we may reject the 'facts' as 'monsters'. In a pluralistic model of testing, several theories—more or less deductively organized—are soldered together.

This argument alone would be enough to show the correctness of the conclusion, which we drew from a different earlier argument, that experiments do not simply overthrow theories, that no theory forbids a state of affairs specifiable in advance.[1] It is not that we propose a theory and Nature may shout NO; rather, we propose a maze of theories, and Nature may shout INCONSISTENT.[2]

The problem is then *shifted* from the old problem of replacing a theory refuted by 'facts' to the new problem of how to resolve inconsistencies between closely associated theories. Which of the mutually inconsistent theories should be eliminated? The sophisticated falsificationist can answer that question easily: one had to try to replace first one, then the other, then possibly both, and opt for that new set-up which provides the biggest increase in corroborated content, which provides the most progressive problemshift.[3]

Thus we have established an appeal procedure in case the theoretician wishes to question the negative verdict of the experimentalist. The theoretician may demand that the experimentalist specify his 'interpretative theory',[4] and he may then replace it—to the experimentalist's annoyance—by a better one in the light of which his originally 'refuted' theory may receive positive appraisal.[5]

[1] Cf. *above*, p. 100.

[2] Let me here answer a possible objection: 'Surely we do not need Nature to tell us that a set of theories is *inconsistent*. Inconsistency—unlike falsehood—can be ascertained without Nature's help'. But Nature's actual 'NO' in a monotheoretical methodology takes the form of a fortified 'potential falsifier', that is a sentence which, in this way of speech, we claim Nature had uttered and which is the *negation of our theory*. Nature's actual 'INCONSISTENCY' in a pluralistic methodology takes the form of a 'factual' statement couched in the light of one of the theories involved, which we claim Nature had uttered and which, if added to our proposed theories, yields an *inconsistent system*.

[3] For instance, in our earlier example (cf. *above*, p. 107 ff.) some may try to replace the gravitational theory with a new one and others may try to replace the radio-optics by a new one: we choose the way which offers the more spectacular growth, the more progressive problemshift.

[4] Criticism does not *assume* a fully articulated deductive structure: it creates it. (Incidentally, this is the main message of my [1963–4].)

[5] A classical example of this pattern is Newton's relation to Flamsteed, the first Astronomer Royal. For instance, Newton visited Flamsteed on 1 September 1694, when working full time on his lunar theory; told him to reinterpret some of his data since they contradicted his own theory; and he explained to him exactly how to do it. Flamsteed obeyed Newton and wrote to him on 7 October: 'Since you went home, I examined the observations I employed for determining the greatest equations of the earth's orbit, and considering the moon's places at the times . . . ' I find that (*if, as you intimate, the earth inclines*

But even this appeal procedure cannot do more than *postpone* the conventional decision. For the verdict of the appeal court is not infallible either. When we decide whether it is the replacement of the 'interpretative' or of the 'explanatory' theory that produces novel facts, we again must take a decision about the acceptance or rejection of basic statements. But then we have only *postponed*—and possibly *improved*—the decision, not avoided it.[1] The difficulties concerning the empirical basis which confronted 'naive' falsificationism cannot be avoided by 'sophisticated' falsificationism either. Even if we regard a theory as 'factual', that is, if our slow-moving and limited imagination cannot offer an alternative to it (as Feyerabend used to put it), we have to make, at least occasionally and temporarily, decisions about its truth-value. *Even then, experience still remains, in an important sense, the 'impartial arbiter'*[2] *of scientific controversy.* We cannot get rid of the problem of the 'empirical basis', if we want to learn from experience[3]: but we can make our learning less dogmatic—but also less fast and less dramatic. By regarding some observational theories as problematic we may make our methodology more flexible: but we cannot articulate and include *all* 'background knowledge' (or 'background ignorance'?) into our critical deductive model. This process is bound to be piecemeal and some conventional line must be drawn at any given time.

There is one objection even to the sophisticated version of methodological falsificationism which cannot be answered without some concession to Duhemian 'simplicism'. The objection is the so-called 'tacking paradox'. According to our definitions, adding to a theory completely disconnected low-level hypotheses may constitute a 'progressive shift'. It is difficult to eliminate such makeshift shifts without demanding that 'the additional assertions must be connected with the contradicting assertion *more intimately* than by mere conjunction'.[4] This, of course, is a sort of simplicity

on that side the moon then is) you may abate abt 20″ from it . . .' Thus Newton constantly criticized and corrected Flamsteed's observational theories. Newton taught Flamsteed, for instance, a better theory of the refractive power of the atmosphere; Flamsteed accepted this and corrected his original 'data'. One can understand the constant humiliation and slowly increasing fury of this great observer, having his data criticized and improved by a man who, on his own confession, made no observations himself: it was this feeling—I suspect—which led finally to a vicious personal controversy.

[1] The same applies to the third type of decision. If we reject a stochastic hypothesis only for one which, in our sense, supersedes it, the exact form of the 'rejection rules' becomes *less* important.

[2] Popper [1945], vol. II, chapter 23, p. 218.

[3] Agassi is then wrong in his thesis that 'observation reports may be accepted as false and hence the problem of the empirical basis is thereby disposed of' (Agassi [1966], p. 20).

[4] Feyerabend [1965], p. 226.

requirement which would assure the continuity in the series of theories which can be said to constitute *one* problemshift.

This leads us to further problems. For one of the crucial features of sophisticated falsificationism is that it replaces the concept of *theory* as the basic concept of the logic of discovery by the concept of *series of theories. It is a succession of theories and not one given theory which is appraised as scientific or pseudo-scientific.* But the members of such series of theories are usually connected by a remarkable *continuity* which welds them into *research programmes.* This *continuity*—reminiscent of Kuhnian 'normal science'—plays a vital role in the history of science; the main problems of the logic of discovery cannot be satisfactorily discussed except in the framework of a *methodology of research programmes.*

3. A METHODOLOGY OF SCIENTIFIC RESEARCH PROGRAMMES

I have discussed the problem of objective appraisal of scientific growth in terms of progressive and degenerating problemshifts in series of scientific theories. The most important such series in the growth of science are characterized by a certain *continuity* which connects their members. This continuity evolves from a genuine research programme adumbrated at the start.[1] The programme consists of methodological rules: some tell us what paths of research to avoid (*negative heuristic*), and others what paths to pursue (*positive heuristic*).

Even science as a whole can be regarded as a huge research programme with Popper's supreme heuristic rule: 'devise conjectures which have more empirical content than their predecessors.' Such methodological rules may be formulated, as Popper pointed out, as metaphysical principles.[2] For instance, the *universal* anti-conventionalist rule against exception-barring may be stated as the metaphysical principle: 'Nature does not allow exceptions'. This is why Watkins called such rules 'influential metaphysics'.[3]

But what I have primarily in mind is not science as a whole, but rather *particular* research programmes, such as the one known as 'Cartesian metaphysics'. Cartesian metaphysics, that is, the mechanistic theory of the

[1] One may point out that the negative and positive heuristic gives a rough (implicit) definition of the 'conceptual framework' (and consequently of the language). The recognition that the history of science is the history of research programmes rather than of theories may therefore be seen as a partial vindication of the view that the history of science is the history of conceptual frameworks or of scientific languages.

[2] Popper [1934], sections 11 and 70. I use 'metaphysical' as a technical term of naive falsificationism: a contingent proposition is 'metaphysical' if it has no 'potential falsifiers'.

[3] Watkins [1958]. Watkins cautions that 'the logical gap between statements and prescriptions in the metaphysical-methodological field is illustrated by the fact that a person may reject a [metaphysical] doctrine in its fact-stating form while subscribing to the prescriptive version of it' (*Ibid.* pp. 356–7).

universe—according to which the universe is a huge clockwork (and system of vortices) with push as the only cause of motion—functioned as a powerful heuristic principle. It discouraged work on scientific theories—like [the 'essentialist' version of] Newton's theory of action at a distance—which were inconsistent with it (*negative heuristic*). On the other hand, it encouraged work on auxiliary hypotheses which might have saved it from apparent counterevidence—like Keplerian ellipses (*positive heuristic*).[1]

(*a*) *Negative heuristic: the 'hard core' of the programme.*

All scientific research programmes may be characterized by their '*hard core*'. The negative heuristic of the programme forbids us to direct the *modus tollens* at this 'hard core'. Instead, we must use our ingenuity to articulate or even invent 'auxiliary hypotheses', which form a *protective belt* around this core, and we must redirect the *modus tollens* to *these*. It is this protective belt of auxiliary hypotheses which has to bear the brunt of tests and get adjusted and re-adjusted, or even completely replaced, to defend the thus-hardened core. A research programme is successful if all this leads to a progressive problemshift; unsuccessful if it leads to a degenerating problemshift.

The classical example of a successful research programme is Newton's gravitational theory: possibly the most successful research programme ever. When it was first produced, it was submerged in an ocean of 'anomalies' (or, if you wish, 'counterexamples'[2]), and opposed by the observational theories supporting these anomalies. But Newtonians turned, with brilliant tenacity and ingenuity, one counter-instance after another into corroborating instances, primarily by overthrowing the original observational theories in the light of which this 'contrary evidence' was established. In the process they themselves produced new counter-examples which they again resolved. They 'turned each new difficulty into a new victory of their programme'.[3]

In Newton's programme the negative heuristic bids us to divert the *modus tollens* from Newton's three laws of dynamics and his law of gravitation. This 'core' is 'irrefutable' by the methodological decision of its protagonists: anomalies must lead to changes only in the 'protective' belt of auxiliary, 'observational' hypothesis and initial conditions.[4]

I have given a contrived micro-example of a progressive Newtonian

[1] For this Cartesian research programme, cf. Popper [1958] and Watkins [1958], pp. 350–1.

[2] For the clarification of the concepts of 'counterexample' and 'anomaly' cf. *above*, p. 110, and especially *below*, p. 159, footnote 1.

[3] Laplace [1796], livre IV, chapter ii.

[4] The actual hard core of a programme does not actually emerge fully armed like Athene from the head of Zeus. It develops slowly, by a long, preliminary process of trial and error. In this paper this process is not discussed.

problemshift.[1] If we analyse it, it turns out that each successive link in this exercise predicts some new fact; each step represents an increase in empirical content: the example constitutes a *consistently progressive theoretical shift*. Also, each prediction is in the end verified; although on three subsequent occasions they may have seemed momentarily to be 'refuted'.[2] While 'theoretical progress' (in the sense here described) may be verified immediately,[3] 'empirical progress' cannot, and in a research programme we may be frustrated by a long series of 'refutations' before ingenious and lucky content-increasing auxiliary hypotheses turn a chain of defeats—*with hindsight*—into a resounding success story, either by revising some false 'facts' or by adding novel auxiliary hypotheses. We may then say that we must require that each step of a research programme be consistently content-increasing: that each step constitute a *consistently progressive theoretical problemshift*. All we need in addition to this is that at least every now and then the increase in content should be seen to be retrospectively corroborated: the programme as a whole should also display an *intermittently progressive empirical shift*. We do not demand that each step produce *immediately* an *observed* new fact. Our term '*intermittently*' gives sufficient *rational* scope for dogmatic adherence to a programme in face of *prima facie* 'refutations'.

The idea of 'negative heuristic' of a scientific research programme rationalizes classical conventionalism to a considerable extent. We may rationally decide not to allow 'refutations' to transmit falsity to the hard core as long as the corroborated empirical content of the protecting belt of auxiliary hypotheses increases. But our approach differs from Poincaré's justificationist conventionalism in the sense that, unlike Poincaré's, we maintain that if and when the programme ceases to anticipate novel facts, its hard core might have to be abandoned: that is, *our* hard core, unlike Poincaré's, may crumble under certain conditions. In this sense we side with Duhem who thought that such a possibility must be allowed for[4]; but for Duhem the reason for such crumbling is purely *aesthetic*,[5] while for us it is mainly *logical and empirical*.

(*b*) *Positive heuristic: the construction of the 'protective belt' and the relative autonomy of theoretical science.*

Research programmes, besides their negative heuristic, are also characterized by their positive heuristic.

[1] Cf. *above*, pp. 100–1. For *real* examples, cf. my [1970].

[2] The 'refutation' was each time successfully diverted to 'hidden lemmas'; that is, to lemmas emerging, as it were, from the *ceteris paribus* clause.

[3] But cf. *below*, pp. 155–7. [4] Cf. *above*, p. 105. [5] *Ibid.*

Even the most rapidly and consistently progressive research programmes can digest their 'counter-evidence' only piecemeal: anomalies are never completely exhausted. But it should not be thought that yet unexplained anomalies—'puzzles' as Kuhn might call them—are taken in random order, and the protective belt built up in an eclectic fashion, without any preconceived order. The order is usually decided in the theoretician's cabinet, independently of the *known* anomalies. Few theoretical scientists engaged in a research programme pay undue attention to 'refutations'. They have a long-term research policy which anticipates these refutations. This research policy, or order of research, is set out—in more or less detail—in the *positive heuristic* of the research programme. The negative heuristic specifies the 'hard core' of the programme which is 'irrefutable' by the methodological decision of its protagonists; the positive heuristic consists of a partially articulated set of suggestions or hints on how to change, develop the 'refutable variants' of the research-programme, how to modify, sophisticate, the 'refutable' protective belt.

The positive heuristic of the programme saves the scientist from becoming confused by the ocean of anomalies. The positive heuristic sets out a programme which lists a chain of ever more complicated *models* simulating reality: the scientist's attention is riveted on building his models following instructions which are laid down in the positive part of his programme. He ignores the *actual* counterexamples, the available '*data*'.[1] Newton first worked out his programme for a planetary system with a fixed point-like sun and one single point-like planet. It was in this model that he derived his inverse square law for Kepler's ellipse. But this model was forbidden by Newton's own third law of dynamics, therefore the model had to be replaced by one in which both sun and planet revolved round their common centre of gravity. This change was not motivated by any observation (the data did not suggest an 'anomaly' here) but by a theoretical difficulty in developing the programme. Then he worked out the programme for more planets as if there were only heliocentric but no interplanetary forces. Then he worked out the case where the sun and planets were not mass-points but mass-*balls*. Again, for this change he did not *need* the observation of an anomaly; infinite density was forbidden by an (inarticulated) touchstone theory, therefore planets *had* to be extended. This change involved considerable mathematical difficulties, held up Newton's work—and delayed the publication of the *Principia* by more than a decade. Having solved this 'puzzle',

[1] If a scientist (or mathematician) has a positive heuristic, he refuses to be drawn into observation. He will 'lie down on his couch, shut his eyes and forget about the data'. (Cf. my [1963–4], especially pp. 300 ff., where there is a detailed case study of such a programme.) Occasionally, of course, he will ask Nature a shrewd question: he will then be encouraged by Nature's *YES*, but not discouraged by its *NO*.

he started work on *spinning balls* and their wobbles. Then he admitted interplanetary forces and started work on *perturbations*. At this point he started to look more anxiously at the facts. Many of them were beautifully explained (qualitatively) by this model, many were not. It was then that he started to work on *bulging* planets, rather than round planets, etc.

Newton despised people who, like Hooke, stumbled on a first naive model but did not have the tenacity and ability to develop it into a research programme, and who thought that a first version, a mere aside, constituted a 'discovery'. He held up publication until his programme had achieved a remarkable progressive shift.[1]

Most, if not all, Newtonian 'puzzles', leading to a series of new variants superseding each other, were forseeable at the time of Newton's first naive model and no doubt Newton and his colleagues *did* forsee them: Newton must have been fully aware of the blatant falsity of his first variants.[2] Nothing shows the existence of a positive heuristic of a research programme clearer than this fact: this is why one speaks of 'models' in research programmes. A '*model*' is a set of initial conditions (possibly together with some of the observational theories) which one knows is *bound* to be replaced during the further development of the programme, and one even knows, more or less, how. This shows once more how irrelevant 'refutations' of any specific variant are in a research programme: their existence is fully expected, the positive heuristic is there as the strategy both for predicting (producing) and digesting them. Indeed, if the positive heuristic is clearly spelt out, the difficulties of the programme are mathematical rather than empirical.[3]

One may formulate the 'positive heuristic' of a research programme as a 'metaphysical' principle. For instance one may formulate Newton's programme like this: 'the planets are essentially gravitating spinning-tops of roughly spherical shape'. This idea was never *rigidly* maintained: the planets are not *just* gravitational, they have also, for example, electromagnetic characteristics which may influence their motion. Positive heur-

[1] Reichenbach, following Cajori, gives a different explanation of what delayed Newton in the publication of his *Principia*: 'To his disappointment he found that the observational results disagreed with his calculations. Rather than set any theory, however beautiful, before the facts, Newton put the manuscript of his theory into his drawer. Some twenty years later, after new measurements of the circumference of the earth had been made by a French expedition, Newton saw that the figures on which he had based his test were false and that the improved figures agreed with his theoretical calculation. It was only after this test that he published his law . . . The story of Newton is one of the most striking illustrations of the method of modern science' (Reichenbach [1951], pp. 101–2). Feyerabend criticizes Reichenbach's account (Feyerabend [1965], p. 229), but does not give an alternative *rationale*.

[2] For a further discussion of Newton's research programme, cf. my [1970].

[3] For this point cf. Truesdell [1960].

istic is thus in general more flexible than negative heuristic. Moreover, it occasionally happens that when a research programme gets into a degenerating phase, a little revolution or a *creative shift* in its positive heuristic may push it forward again.[1] It is better therefore to separate the 'hard core' from the more flexible metaphysical principles expressing the positive heuristic.

Our considerations show that the positive heuristic forges ahead with almost complete disregard of 'refutations': it may seem that it is the '*verifications*'[2] rather than the refutations which provide the contact points with reality. Although one must point out that any 'verification' of the $n+1$-th version of the programme is a refutation of the n-th version, we cannot deny that *some* defeats of the subsequent versions are always foreseen: it is the 'verifications' which keep the programme going, recalcitrant instances notwithstanding.

We may appraise research programmes, even after their 'elimination', for their *heuristic power*: how many new facts did they produce, how great was 'their capacity to explain their refutations in the course of their growth'?[3]

(We may also appraise them for the stimulus they gave to mathematics. The real difficulties for the theoretical scientist arise rather from the *mathematical difficulties* of the programme than from anomalies. The greatness of the Newtonian programme comes partly from the development—by Newtonians—of classical infinitesimal analysis which was a crucial precondition of its success.)

Thus the methodology of scientific research programmes accounts for the *relative autonomy of theoretical science:* a historical fact whose rationality cannot be explained by the earlier falsificationists. Which problems scientists working in powerful research programmes rationally choose, is determined by the positive heuristic of the programme rather than by psychologically worrying (or technologically urgent) anomalies. The anomalies are listed but shoved aside in the hope that they will turn, in due course, into corroborations of the programme. Only those scientists have to rivet their attention on anomalies who are either engaged in trial-and-error exercises[4] or who work in a degenerating phase of a research programme when the positive heuristic ran out of steam. (All this, of course, must sound repugnant to naive falsificationists who hold that once

[1] Soddy's contribution to Prout's programme or Pauli's to Bohr's (old quantum theory) programme are typical examples of such creative shifts.

[2] A 'verification' is a corroboration of excess content in the expanding programme. But, of course, a 'verification' does not *verify* a programme: it shows only its heuristic power.

[3] Cf. my [1963–4], pp. 324–30. Unfortunately in 1963–4 I had not yet made a clear terminological distinction between theories and research programmes, and this impaired my exposition of a research programme in informal, quasi-empirical mathematics. There are fewer such shortcomings in my [1971].

[4] Cf. *below*, p. 175.

a theory is 'refuted' by experiment (by *their* rule book), it is irrational (and dishonest) to develop it further: one has to replace the old 'refuted' theory by a new, unrefuted one.)

(*c*) *Two illustrations: Prout and Bohr.*

The dialectic of positive and negative heuristic in a research programme can best be illuminated by examples. Therefore I am now going to sketch a few aspects of two spectacularly successful research programmes: Prout's programme[1] based on the idea that all atoms are compounded of hydrogen atoms and Bohr's programme based on the idea that light-emission is due to electrons jumping from one orbit to another within the atoms.

(*In writing a historical case study, one should, I think, adopt the following procedure:* (1) *one gives a rational reconstruction;* (2) *one tries to compare this rational reconstruction with actual history and to criticize both one's rational reconstruction for lack of historicity and the actual history for lack of rationality. Thus any historical study must be preceded by a heuristic study: history of science without philosophy of science is blind. In this paper it is not my purpose to go on seriously to the second stage.*)

(*c 1*) *Prout: a research programme progressing in an ocean of anomalies.*

Prout, in an anonymous paper of 1815, claimed that the atomic weights of all pure chemical elements were whole numbers. He knew very well that anomalies abounded, but said that these arose because chemical substances as they ordinarily occurred were *impure*: that is, the relevant 'experimental techniques' of the time were unreliable, or, to put it in our terms, the contemporary 'observational' theories in the light of which the truth values of the basic statements of his theory were established, were false.[2] The champions of Prout's theory therefore embarked on a major venture: to overthrow those theories which supplied the counter-evidence to their thesis. For this they had to revolutionize the established analytical chemistry of the time and correspondingly revise the experimental techniques with which pure elements were to be separated.[3]

[1] Already mentioned *above*, pp. 128–9.

[2] Alas, all this is rational reconstruction rather than actual history. Prout denied the existence of any anomalies. For instance, he claimed that the atomic weight of chlorine was exactly 36.

[3] Prout was aware of some of the basic methodological features of his programme. Let us quote the first lines of his [1815]: 'The author of the following essay submits it to the public with the greatest diffidence... He trusts, however, that its importance will be seen, and that some one will undertake to examine it, and thus verify or refute its conclusions. If these should be proved erroneous, still new facts may be brought to light, or old ones better established, by the investigation; but if they should be verified, a new and interesting light will be thrown upon the whole science of chemistry.'

Prout's theory, as a matter of fact, defeated the theories previously applied in purification of chemical substances one after the other. Even so, some chemists became tired of the research programme and gave it up, since the successes were still far from adding up to a final victory. For instance, Stas, frustrated by some stubborn, recalcitrant instances, concluded in 1860 that Prout's theory was 'without foundations'.[1] But others were more encouraged by the progress than discouraged by the lack of complete success. For instance, Marignac immediately retorted that 'although [he is satisfied that] the experiments of Monsieur Stas are perfectly exact, [there is no proof] that the differences observed between his results and those required by Prout's law cannot be explained by the imperfect character of experimental methods'.[2] As Crookes put it in 1886: 'Not a few chemists of admitted eminence consider that we have here [in Prout's theory] an expression of the truth, masked by some residual or collateral phenomena which we have not yet succeeded in eliminating.'[3] That is, there had to be some *further* false hidden assumption in the 'observational' theories on which 'experimental techniques' for chemical purification were based and with the help of which atomic weights were calculated: in Crookes's view even in 1886 'some present atomic weights merely represented a mean value'.[4] Indeed, Crookes went on to put this idea in a scientific (content-increasing) form: he proposed concrete new theories of 'fractionation', a new 'sorting Demon'.[5] But, alas, his new observational theories turned out to be as false as they were bold and, being unable to anticipate any new fact, they were eliminated from the (rationally reconstructed) history of science. As it turned out a generation later, there was a very basic hidden assumption which failed the researchers: that two pure elements must be separable by *chemical* methods. The idea that two different pure elements may behave identically in all *chemical* reactions but can be separated by *physical* methods, required a change, a '*stretching*', of the concept of 'pure element' which constituted a change—a *concept-stretching expansion*—of the research programme itself.[6] This revolutionary highly *creative shift* was taken only by Rutherford's school[7]; and then 'after

[1] Clerk Maxwell was on Stas's side: he thought it was impossible that there should be two kinds of hydrogen, 'for if some [molecules] were of slightly greater mass than others, we have the means of producing a separation between molecules of different masses, one of which would be somewhat denser than the other. As this cannot be done, we must admit [that all are alike]' (Maxwell [1871]).

[2] Marignac [1860].

[3] Crookes [1886].

[4] *Ibid.*

[5] Crookes [1886], p. 491.

[6] For 'concept-stretching', cf. my [1963–4], part IV.

[7] The shift is anticipated in Crookes's fascinating [1888] where he indicates that the solution should be sought in a new demarcation between 'physical' and 'chemical'. But the anticipation remained philosophical; it was left to Rutherford and Soddy to develop it, after 1910, into a scientific theory.

many vicissitudes and the most convincing apparent disproofs, the hypothesis thrown out so lightly by Prout, an Edinburgh physician, in 1815, has, a century later, become the corner-stone of modern theories of the structure of atoms'.[1] However, this creative step was in fact only a side-result of progress in a different, indeed, distant research programme; Proutians, lacking this *external* stimulus, never dreamt of trying, for instance, to build powerful centrifugal machines to separate elements.

(When an 'observational' or 'interpretative' theory finally gets eliminated, the 'precise' measurements carried out within the discarded framework may look—with hindsight—rather foolish. Soddy made fun of 'experimental precision' for its own sake: 'There is something surely akin to if not transcending tragedy in the fate that has overtaken the life work of that distinguished galaxy of nineteenth-century chemists, rightly revered by their contemporaries as representing the crown and perfection of accurate scientific measurement. Their hard won results, for the moment at least, appears as of as little interest and significance as the determination of the average weight of a collection of bottles, some of them full and some of them more or less empty.'[2])

Let us stress that in the light of the methodology of research programmes here proposed there never was any rational reason to *eliminate* Prout's programme. Indeed, the programme produced a beautiful, progressive shift, even if, in between, there were considerable hitches.[3] Our sketch shows how a research programme can challenge a considerable bulk of accepted scientific knowledge: it is planted, as it were, in an inimical environment which, step by step, it can override and transform.

Also, the actual history of Prout's programme illustrates only too well how much the progress of science was hindered and slowed down by justificationism and by naive falsificationism. (The opposition to atomic theory in the nineteenth century was fostered by both.) An elaboration of this particular influence of bad methodology on science may be a rewarding research programme for the historian of science.

(*c* 2) *Bohr: a research programme progressing on inconsistent foundations.*

A brief sketch of Bohr's research programme of light emission (in *early* quantum physics) will illustrate further—and even expand—our thesis.[4]

[1] Soddy [1932], p. 50.

[2] *Ibid.*

[3] These hitches inevitably induce many individual scientists to shelve or altogether jettison the programme and join other research programmes where the positive heuristic happens to offer at the time cheaper successes: the history of science cannot be *fully* understood without mob-psychology. (Cf. *below*, pp. 177–80.)

[4] This section may again strike the historian as more a caricature than a sketch; but I hope it serves its purpose. (Cf. *above*, p. 138.) Some statements are to be taken not with a grain, but with tons, of salt.

The story of Bohr's research programme can be characterized by: (1) its initial problem; (2) its negative and positive heuristic; (3) the problems which it attempted to solve in the course of its development; and (4) its degeneration point (or, if you wish, 'saturation point') and, finally, (5) the programme by which it was superseded.

The background problem was the riddle of how Rutherford atoms (that is, minute planetary systems with electrons orbiting round a positive nucleus) can remain stable; for, according to the well-corroborated Maxwell–Lorentz theory of electromagnetism they should collapse. But Rutherford's theory was well corroborated too. Bohr's suggestion was to ignore for the time being the inconsistency and consciously develop a research programme whose 'refutable' versions were inconsistent with the Maxwell–Lorentz theory.[1] He proposed five postulates as the *hard core* of his programme: '(1) that energy radiation [within the atom] is not emitted (or absorbed) in the continuous way assumed in the ordinary electrodynamics, but only during the passing of the systems between different "stationary" states. (2) That the dynamical equilibrium of the systems in the stationary states is governed by the ordinary laws of mechanics, while these laws do not hold for the passing of the systems between the different states. (3) That the radiation emitted during the transition of a system between two stationary states is homogeneous, and that the relation between the frequency ν and the total amount of energy emitted E is given by $E = h\nu$, where h is Planck's constant. (4) That the different stationary states of a simple system consisting of an electron rotating round a positive nucleus are determined by the condition that the ratio between the total energy, emitted during the formation of the configuration, and the frequency of revolution of the electron is an entire multiple of $\frac{1}{2}h$. Assuming that the orbit of the electron is circular, this assumption is equivalent with the assumption that the angular momentum of the electron round the nucleus is equal to an entire multiple of $h/2\pi$. (5) That the "permanent" state of any atomic system, i.e. the state in which the energy emitted is maximum, is determined by the condition that the angular momentum of every electron round the centre of its orbit is equal to $h/2\pi$.'[2]

We have to appreciate the crucial methodological difference between the inconsistency introduced by Prout's programme and that introduced by Bohr's. Prout's research programme declared war on the analytical chemistry of his time: its positive heuristic was designed to overthrow it and replace it. But Bohr's research programme contained no analogous design:

[1] This, of course, is a further argument against J. O. Wisdom's thesis that metaphysical theories can be refuted by a conflicting well corroborated scientific theory (Wisdom [1963]) Also, cf. *above*, p. 112, text to footnote 1, and pp. 126–7.

[2] Bohr [1913*a*], p. 874.

its positive heuristic, even if it had been completely successful, would have left the inconsistency with the Maxwell–Lorentz theory unresolved.[1] To suggest such an idea required even greater courage than Prout's; the idea crossed Einstein's mind but he found it unacceptable, and rejected it.[2] Indeed, *some of the most important research programmes in the history of science were grafted on to older programmes with which they were blatantly inconsistent.* For instance, Copernican astronomy was 'grafted' on to Aristotelian physics, Bohr's programme on to Maxwell's. Such 'grafts' are irrational for the justificationist and for the naive falsificationist, neither of whom can countenance growth on inconsistent foundations. Therefore they are usually concealed by *ad hoc* stratagems—like Galileo's theory of circular inertia or Bohr's correspondence, and, later, complementarity principle—the only purpose of which is to hide the 'deficiency'.[3] As the young grafted programme strengthens, the peaceful co-existence comes to an end, the symbiosis becomes competitive and the champions of the new programme try to replace the old programme altogether.

It may well have been the success of his 'grafted programme' which later misled Bohr into believing that such fundamental inconsistencies in research programmes can and should be put up with *in principle*, that they do not present any serious problem and one merely has to get used to them. Bohr tried in 1922 to lower the standards of scientific criticism; he argued that '*the most* that one can demand of a theory [i.e. programme] is that the classification [it establishes] can be pushed so far that it can contribute to the development of the field of observation by the prediction of *new* phenomena.'[4]

(This statement by Bohr is similar to d'Alembert's when faced with the inconsistency in the foundations of infinitesimal theory: '*Allez en avant et la foi vous viendra.*' According to Margenau, 'it is understandable that, in the excitement over its success, men overlooked a malformation in the theory's architecture; for Bohr's atom sat like a baroque tower upon the Gothic base of classical electrodynamics.'[5] But as a matter of fact, the 'malformation' was not 'overlooked': everybody was aware of it, only they ignored it—more or less—during the progressive phase of the programme.[6]

[1] Bohr held at this time that the Maxwell–Lorentz theory would *eventually* have to be replaced (Einstein's photon theory had already indicated this need).

[2] Hevesy [1913]; cf. also *above*, p. 136, text to footnote 1.

[3] In our methodology there is no need for such protective *ad hoc* stratagems. But, on the other hand, they are harmless as long as they are clearly seen as problems, not as solutions.

[4] Bohr [1922]; my italics.

[5] Margenau [1950], p. 311.

[6] Sommerfeld ignored it more than Bohr: cf. *below*, p. 150, footnote 4.

Our methodology of research programmes shows the rationality of this attitude but it also shows the irrationality of the defence of such 'malformations' once the progressive phase is over.

It should be said here that in the thirties and forties Bohr abandoned his demand for 'new phenomena' and was prepared to 'proceed with the immediate task of co-ordinating the multifarious evidence regarding atomic phenomena, which accumulated from day to day in the exploration of this new field of knowledge'.[1] This indicates that Bohr, by this time, had fallen back on 'saving the phenomena', while Einstein sarcastically insisted that 'every theory is true provided that one suitably associates its symbols with observed quantities'.[2])

But *consistency*—in a strong sense of the term[3]—*must remain an important regulative principle* (over and above the requirement of progressive problemshift); and inconsistencies *must* be seen as problems. The reason is simple. If science aims at truth, it must aim at consistency; if it resigns consistency, it resigns truth. To claim that 'we must be modest in our demands',[4] that we must resign ourselves to—weak or strong—inconsistencies, remains a methodological vice. On the other hand, this does not mean that the discovery of an inconsistency—or of an anomaly—must *immediately* stop the development of a programme: it may be rational to put the inconsistency into some temporary, *ad hoc* quarantine, and carry on with the positive heuristic of the programme. This has been done even in mathematics, as the examples of the early infinitesimal calculus and of naive set theory show.[5]

[1] Bohr [1949], p. 206.

[2] Quoted in Schrödinger [1958], p. 170.

[3] Two propositions are inconsistent if their conjunction has no model, that is, there is no interpretation of their descriptive terms in which the conjunction is true. But in informal discourse we use more formative terms than in formal discourse: some descriptive terms are given a fixed interpretation. In this informal sense two propositions may be (weakly) inconsistent given the standard interpretations of some characteristic terms even if formally, in some unintended interpretation, they may be consistent. For instance, the first theories of electron spin were inconsistent with the special theory of relativity if 'spin' was given its ('strong') standard interpretation and thereby treated as a formative term; but the inconsistency disappears if 'spin' is treated as an uninterpreted descriptive term. The reason why we should not give up standard interpretations too easily is that such emasculation of meanings may emasculate the positive heuristic of the programme. (On the other hand, such meaning shifts may be in some cases progressive: cf. *above*, p. 126.)

For the shifting demarcation between formative and descriptive terms in informal discourse, cf. my [1963–4], 9(*b*), especially p. 335, footnote 1.

[4] Bohr [1922], last paragraph.

[5] Naive falsificationists tend to regard this liberalism as a *crime against reason*. Their main argument runs like this: 'If one were to accept contradictions, then one would have to give up any kind of scientific activity: it would mean a complete breakdown of science. This can be shown by proving that *if two contradictory statements are admitted, any*

(From this point of view, Bohr's 'correspondence principle' played an interesting double role in his programme. On the one hand it functioned as an important heuristic principle which suggested many new scientific hypotheses which, in turn, led to novel facts, especially in the field of the intensity of spectrum lines.[1] On the other hand it functioned also as a defence-mechanism, which 'endeavoured to utilize to the utmost extent the concepts of the classical theories of mechanics and electrodynamics, in spite of the contrast between these theories and the quantum of action',[2] instead of emphasizing the urgency of a unified programme. In this second role it reduced the degree of problematicality of the programme.[3])

Of course, the research programme of quantum theory as a whole was a 'grafted programme' and therefore repugnant to physicists with deeply conservative views like Planck. There are two extreme and equally irrational positions with regard to a grafted programme.

The *conservative position* is to halt the new programme until the basic inconsistency with the old programme is somehow repaired: it is irrational to work on inconsistent foundations. The 'conservatives' will concentrate on eliminating the inconsistency by explaining (approximately) the postulates of the new programme in terms of the old programme: they find it irrational to go on with the new programme without a successful *reduction* of the kind mentioned. Planck himself chose this way. He did not succeed, in spite of the decade of hard work he invested in it.[4] Therefore Laue's remark that his lecture on 14 December 1900, was the 'birthday of the quantum theory' is not quite true: that day was the birthday of Planck's reduction programme. The decision to go *ahead* with temporarily inconsistent foundations was taken by Einstein in 1905, but even he wavered in 1913, when Bohr forged forward again.

statement whatever must be admitted; for from a couple of contradictory statements any statement whatever can be validly inferred . . . A theory which involves a contradiction is therefore entirely useless *as a theory* ' (Popper [1940]). In fairness to Popper, one has to stress that he is here arguing against Hegelian dialectic, in which inconsistency becomes a *virtue*; and he is absolutely right when he points out its dangers. But Popper never analysed patterns of empirical (or non-empirical) progress on inconsistent foundations; indeed, in section 24 of his [1934] he makes consistency and falsifiability mandatory requirements for any scientific theory. I discuss this problem in more detail in my [1970].

[1] Cf. e.g. Kramers [1923].

[2] Bohr [1923].

[3] Born, in his [1954], gives a vivid account of the correspondence principle which strongly supports this double appraisal: 'The art of guessing correct formulae, which deviate from the classical ones, yet contain them as a limiting case . . . was brought to a high degree of perfection.'

[4] For the fascinating story of this long series of frustrating failures, cf. Whittaker, [1953], pp. 103–4. Planck himself gives a dramatic description of these years: 'My futile attempts to fit the elementary quantum of action into the classical theory continued for a number of years, and they cost me a great deal of effort. Many of my colleagues saw in this something bordering on a tragedy . . .' (Planck [1947]).

The *anarchist position* concerning grafted programmes is to extol anarchy in the foundations as a virtue and regard [weak] inconsistency either as some basic property of nature or as an ultimate limitation of human knowledge, as some of Bohr's followers did.

The *rational position* is best characterized by Newton's, who faced a situation which was to a certain extent similar to the one discussed. Cartesian push-mechanics, on which Newton's programme was originally grafted, was (weakly) inconsistent with Newton's theory of gravitation. Newton worked both on his positive heuristic (successfully) *and* on a reductionist programme (unsuccessfully), and disapproved both of Cartesians who, like Huyghens, thought that it was not worth wasting time on an 'unintelligible' programme and of some of his rash disciples who, like Cotes, thought that the inconsistency presented no problem.[1]

The rational position with regard to 'grafted' programmes is then to exploit their heuristic power without resigning oneself to the fundamental chaos on which it is growing. On the whole, this attitude dominated old, pre-1925 quantum theory. In the new, post-1925 quantum theory the 'anarchist' position became dominant and modern quantum physics, in its 'Copenhagen interpretation', became one of the main standard bearers of philosophical obscurantism. In the *new* theory Bohr's notorious 'complementarity principle' enthroned [weak] inconsistency as a basic factual final feature of nature, and merged subjectivist positivism and antilogical dialectic and even ordinary language philosophy into one unholy alliance. After 1925 Bohr and his associates introduced a new and unprecedented lowering of critical standards for scientific theories. This led to a defeat of reason within modern physics and to an anarchist cult of incomprehensible chaos. Einstein protested: 'The Heisenberg–Bohr tranquillizing philosophy—or religion?—is so delicately contrived that, for the time begin, it provides a gentle pillow for the true believer'.[2] On the other hand,

[1] Cf. my [1970]. Of course, a reductionist programme is scientific only if it explains more than it has set out to explain; otherwise the reduction is *not* scientific (cf. Popper [1969]). If the reduction does not produce new empirical content, let alone novel facts, then the reduction represents a degenerating problemshift—it is a mere linguistic exercise. The Cartesian efforts to bolster up their metaphysics in order to be able to interpret Newtonian gravitation in its terms, is an outstanding example for such merely linguistic reduction. Cf. *above*, p. 126, footnote 2.

[2] Einstein [1928]. Among the critics of the Copenhagen 'anarchism' we should mention —besides Einstein—Popper, Landé, Schrödinger, Margenau, Blokhinzev, Bohm, Fényes and Jánossy. For a defence of the Copenhagen interpretation, cf. Heisenberg [1955]; for a hard-hitting recent criticism, cf. Popper [1967]. Feyerabend in his [1968–9], makes use of some inconsistencies and waverings in Bohr's position for a crude apologetic falsification of Bohr's philosophy. Feyerabend misrepresents Popper's, Landé's and Margenau's critical attitude to Bohr, gives insufficient emphasis to Einstein's opposition, and seems to have forgotten completely that in some of his earlier papers he was more Popperian than Popper on this issue.

Einstein's *too* high standards may well have been the reason that prevented him for discovering (or perhaps only from publishing) the Bohr model and wave mechanics.

Einstein and his allies have not won the battle. Physics textbooks are nowadays full of statements like this: 'The two viewpoints, quanta and electromagnetic field strengths, are complementary in the sense of Bohr. This complementarity is one of the great achievements of natural philosophy in which the Copenhagen interpretation of the epistemology of quantum theory has resolved the age-old conflict between the corpuscular and the wave theories of light. From the reflection and rectilinear propagation properties of Hero of Alexandria in the first century A.D., right through to the interference and wave properties of Young and Maxwell in the nineteenth century, this controversy raged. The quantum theory of radiation during the past half century, in a striking Hegelian manner, has *completely* resolved the dichotomy'.[1]

Let us now return to the logic of discovery of *old* quantum theory and, in particular, concentrate on its *positive heuristic*. Bohr's plan was to work out first the theory of the hydrogen atom. His first model was to be based on a fixed proton-nucleus with an electron in a circular orbit; in his second model he wanted to calculate an elliptical orbit in a fixed plane; then he intended to remove the clearly artificial restrictions of the fixed nucleus and fixed plane; after this he thought of taking the possible spin of the electron into account,[2] and then he hoped to extend his programme to the structure of complicated atoms and molecules and to the effect of electromagnetic fields on them, etc., etc. All this was planned right at the start: the idea that atoms are analogous to planetary systems adumbrated a long, difficult but optimistic programme and clearly indicated the policy of research.[3] 'It looked at this time—in the year 1913—as if the authentic

[1] Power [1964], p. 31 (my italics). '*Completely*' is meant here literally. As we read in *Nature* (**222**, 1969, pp. 1034–5): 'It is absurd to think that any fundamental element of [quantum] theory can be false . . . The arguments that *scientific* results are always temporary, cannot hold. It is the *philosophers*' conceptions of modern physics that are temporary, because they have not yet realized how profoundly the discoveries of quantum physics affect the whole of epistemology. . . . The assertion that ordinary language is the ultimate source of the unambiguousness of physical description is verified most convincingly by the observational conditions in quantum physics.'

[2] This is rational reconstruction. As a matter of fact, Bohr accepted this idea only in his [1926].

[3] Besides this analogy, there was another basic idea in Bohr's positive heuristic: the 'correspondence principle'. This was indicated by him as early as 1913 (cf. the second of his five postulates quoted above on p. 141), but he developed it only later when he used it as a guiding principle in solving some problems of the later, sophisticated models (like the intensities and states of polarization). The peculiarity of this second part of his positive heuristic was that Bohr did not believe its metaphysical version: he thought it was a temporary rule until the replacement of classical electromagnetics (and possibly mechanics).

key to the spectra had at last been found, as if only time and patience would be needed to resolve their riddles completely.'[1]

Bohr's celebrated first paper of 1913 contained the initial step in the research programme. It contained his first model (I shall call it M_1) which already predicted facts hitherto unpredicted by any previous theory: the wavelengths of hydrogen's line emission spectrum. Though some of these wavelengths were known before 1913—the Balmer series (1885) and the Paschen series (1908)—Bohr's theory predicted much more than these two known series. And tests soon corroborated its novel content: one additional Bohr series was discovered by Lyman in 1914, another by Brackett in 1922 and yet another by Pfund in 1924.

Since the Balmer and the Paschen series were known before 1913, some historians present the story as an example of a Baconian 'inductive ascent': (1) the chaos of spectrum lines, (2) an 'empirical law' (Balmer), (3) the theoretical explanation (Bohr). This certainly looks like the three 'floors' of Whewell. But the progress of science would hardly have been delayed had we lacked the laudible trials and errors of the ingenious Swiss school-teacher: the speculative mainline of science, carried forward by the bold speculations of Planck, Rutherford, Einstein and Bohr would have produced Balmer's results deductively, as test-statements of their theories, without Balmer's so-called 'pioneering'. In the rational reconstruction of science there is little reward for the pains of the discoverers of 'naive conjectures'.[2]

As a matter of fact, Bohr's problem was not to explain Balmer's and Paschen's series, but to explain the paradoxical stability of the Rutherford atom. Moreover, Bohr had not even heard of these formulae before he wrote the first version of his paper.[3]

Not all the novel content of Bohr's first model M_1 was corroborated. For instance, Bohr's M_1 claimed to predict all the lines in the hydrogen emission spectrum. But there was experimental evidence for a hydrogen

[1] Davisson [1937]. A similar euphoria was experienced by MacLaurin in 1748 over Newton's programme: Newton's 'philosophy being founded on experiment and demonstration, cannot fail till reason or the nature of things are changed . . . [Newton] left to posterity little more to do, but observe the heavens, and compute after his models' (MacLaurin [1748], p. 8).

[2] I use here 'naive conjecture' as a technical term in the sense of my [1963–4]. For a case study and detailed criticism of the myth of the 'inductive basis' of science (natural or mathematical) cf. *ibid.* section 7, especially pp. 298–307. There I show that Descartes's and Euler's 'naive conjecture' that for all polyhedra $V-E+F=2$ was irrelevant and superfluous for the later development; as further examples one may mention that Boyle's and his successors' labours to establish $pv = RT$ was irrelevant for the later theoretical development (except for developing some experimental techniques), as Kepler's three laws may have been superfluous for the Newtonian theory of gravitation.

For further discussion of this point cf. *below*, p. 175.

[3] Cf. Jammer [1966], pp. 77 ff.

series where according to Bohr's M_1 there should have been none. The anomalous series was the Pickering–Fowler ultraviolet series.

Pickering discovered this series in 1896 in the spectrum of the star ζ Puppis. Fowler, after having discovered its first line also in the sun in 1898, produced the whole series in a discharge tube containing hydrogen and helium. True, it could be argued that the monster-line had nothing to do with the hydrogen—after all, the sun and ζ Puppis contain many gases and the discharge tube also contained helium. Indeed, the line could *not* be produced in a pure hydrogen tube. But Pickering's and Fowler's 'experimental technique', that led to a falsifying hypothesis of Balmer's law, had a plausible, although never severely tested, theoretical background: (*a*) their series had the same convergence number as the Balmer series and therefore was taken to be a hydrogen series and (*b*) Fowler gave a plausible explanation why helium could not possibly be responsible for producing the series.[1]

Bohr was not, however, very impressed by the 'authoritative' experimental physicists. He did not question their 'experimental precision' or the 'reliability of their observations', but questioned their observational theory. Indeed, he proposed an alternative. He first elaborated a new model (M_2) of his research programme: the model of ionized helium, with a double proton orbited by an electron. Now this model predicts an ultraviolet series in the spectrum of ionized helium which coincides with the Pickering–Fowler series. This constituted a rival theory. Then he suggested a 'crucial experiment': he predicted that Fowler's series can be produced, possibly with even stronger lines, in a tube which is filled with a mixture of helium and chlorine. Moreover, Bohr explained to the experimentalists, without even looking at their apparatus, the catalytic role of the hydrogen in Fowler's experiment and of chlorine in the experiment he suggested.[2] Indeed, he was right.[3] Thus the first apparent defeat of the research programme was turned into a resounding victory.

[1] Fowler [1912]. Incidentally his 'observational' theory was provided by 'Rydberg's theoretical investigations' which 'in the absence of strict experimental proof [he] regarded as justifying [his experimental] conclusion' (p. 65). But his theoretician colleague, Professor Nicholson, referred three months later to Fowler's findings as 'laboratory confirmations of Rydberg's theoretical deduction' (Nicholson [1913]). This little story, I think, bears out my pet thesis that most scientists tend to understand little more *about* science than fish about hydrodynamics.

In the Report of the Council to the Ninety-third Annual General Meeting of the Royal Astronomical Society, Fowler's 'observation in laboratory experiments' of new 'hydrogen lines which have so long eluded the efforts of the physicists' is described as 'an advance of great interest' and as 'a triumph of well-directed experimental work'.

[2] Bohr [1913*b*].

[3] Evans [1913]. For a similar example of a theoretical physicist teaching a refutation-keen experimentalist what he—the experimentalist—had really observed, cf. *above*, p. 130, footnote 5.

The victory, however, was immediately questioned. Fowler acknowledged that his series was not a hydrogen, but a helium series. But he pointed out that Bohr's monster-adjustment[1] still failed: the wavelengths in the Fowler series differ significantly from the values predicted by Bohr's M_2. Thus the series, although it does not refute M_1, still refutes M_2, and because of the close connection between M_1 and M_2, it undermines M_1![2]

Bohr brushed off Fowler's argument: *of course* he never meant M_2 to be taken too seriously. His values were based on a crude calculation based on the electron orbiting round a fixed nucleus; but *of course* it orbits round the common centre of gravity; *of course*, as is done when treating two-body problems, one has to substitute reduced mass for mass: $m'_e = m_e/[1+(m_e/m_n)]$.[3] This modified model was Bohr's M_3. And Fowler himself had to admit that Bohr was again right.[4]

The apparent refutation of M_2 turned into a victory for M_3; and it was clear that M_2 and M_3 would have been developed within the research programme—perhaps even M_{17} or M_{20}—without *any* stimulus from observation or experiment. It was at this stage that Einstein said of Bohr's theory: 'It is one of the greatest discoveries.'[5]

Bohr's research programme then went on as planned. The next step was to calculate elliptical orbits. This was done by Sommerfeld in 1915, but with the (unexpected) result that the increased number of possible steady orbits did *not* increase the number of possible energy levels, so there seemed to be no possibility of a crucial experiment between the elliptical and circular theory. However, electrons orbit the nucleus with very high velocity so that when they accelerate their mass should change noticeably if Einsteinian mechanics is true. Indeed, calculating such relativistic corrections, Sommerfeld got a new array of energy levels and thus the 'finestructure' of the spectrum.

The switch to this new relativistic model required much more mathematical skill and talent than the development of the first few models. Sommerfeld's achievement was primarily mathematical.[6]

[1] Monster-adjustment: turning a counterexample, in the light of some new theory, into an example. Cf. my [1963–4], pp. 127 ff. But Bohr's 'monster-adjustment' was empirically 'progressive': it predicted a new fact (the appearance of the 4686 line in tubes containing no hydrogen).

[2] Fowler [1913a].

[3] Bohr [1913c]. This monster-adjustment was also 'progressive': Bohr predicted that Fowler's observations must be slightly imprecise and the Rydberg 'constant' must have a fine structure.

[4] Fowler [1913b]. But he sceptically noted that Bohr's programme had not yet explained the spectrum lines of *un-ionized*, ordinary helium. However, he soon abandoned his scepticism and joined Bohr's research programme (Fowler [1914]).

[5] Cf. Hevesy [1913]: 'When I told him of the Fowler spectrum, the big eyes of Einstein looked still bigger and he told me: "Then it is one of the greatest discoveries." '

[6] For the vital mathematical aspects of research programmes, cf. *above*, p. 137.

Curiously, the doublets of the hydrogen spectrum had already been discovered in 1891 by Michelson.[1] Moseley pointed out immediately after Bohr's first publication that 'it fails to account for the second weaker line found in each spectrum'.[2] Bohr was not upset: he was convinced that the positive heuristic of his research programme would, *in due course*, explain and even correct Michelson's observations.[3] And so it did. Sommerfeld's theory was, of course, inconsistent with Bohr's first versions; the fine-structure experiments—with the old observations corrected!—provided the crucial evidence in its favour. Many defeats of Bohr's first models were turned by Sommerfeld and his Munich school into victories for Bohr's research programme.

It is interesting that just as Einstein got worried and slowed down in the middle of the spectacular progress of quantum physics by 1913, Bohr got worried and slowed down by 1916; and just as Bohr had, by 1913 taken the initiative from Einstein, Sommerfeld had taken the initiative from Bohr by 1916. The difference between the atmosphere of Bohr's Copenhagen school and Sommerfeld's Munich school was conspicuous: 'In Munich one used more concrete formulations and was therefore more easily understood; one had been successful in the systematization of spectra and in the use of the vector model. In Copenhagen, however, one believed that an adequate language for the new [phenomena] had not yet been found, one was reticent in the face of too definite formulations, one expressed oneself more cautiously and more in general terms, and was therefore much more difficult to understand.'[4]

Our sketch shows how a progressive shift may lend credibility—and *rationale*—to an inconsistent programme. Born, in his obituary of Planck, describes this process forcefully: 'Of course the mere introduction of the quantum of action does not yet mean that a *true* Quantum Theory has been established . . . The difficulties which the introduction of the quantum of action into the well-established classical theory has encountered from the outset have already been indicated. They have gradually increased rather than diminished; and although research in its forward march has in the meantime passed over some of them, the remaining gaps

[1] Michelson [1891–2], especially pp. 287–9. Michelson does not even mention Balmer.

[2] Moseley [1914].

[3] Sommerfeld [1916], p. 68.

[4] Hund [1961]. This is discussed at some length in Feyerabend [1968–9], pp. 83–7 But Feyerabend's paper is heavily biased. The main aim of his paper is to play down Bohr's methodological anarchism and show that Bohr *opposed* the Copenhagen interpretation of the *new* (post-1925) quantum programme. In order to do so, Feyerabend, on the one hand, overemphasizes Bohr's unhappiness about the inconsistency of the *old* (pre-1925) quantum programme and, on the other hand, makes too much of the fact that Sommerfeld cared less for the problematicality of the inconsistent foundations of the *old* programme than Bohr.

in the theory are the more distressing to the conscientious theoretical physicist. In fact, what in Bohr's theory served as the basis of the laws of action consists of certain hypotheses which a generation ago would doubtless have been flatly rejected by every physicist. That within the atom certain quantized orbits (i.e. picked out on the quantum principle) should play a special role could well be granted; somewhat less easy to accept is the further assumption that the electrons moving on these curvilinear orbits, and therefore accelerated, radiate no energy. But that the sharply defined frequency of an emitted light quantum should be different from the frequency of the emitting electron would be regarded by a theoretician who had grown up in the classical school as monstrous and almost inconceivable. But numbers [or, rather, *progressive problemshifts*] decide, and in consequence the tables have been turned. While originally it was a question of fitting in with as little strain as possible a new and strange element into an existing system which was generally regarded as settled, *the intruder, after having won an assured position, now has assumed the offensive*; and it now appears certain that it is about to blow up the old system at some point. The only question now is, at what point and to what extent this will happen.'[1]

One of the most important points one learns from studying research programmes is that relatively few experiments are really important. The heuristic guidance the theoretical physicist receives from tests and 'refutations' is usually so trivial that large-scale testing—or even bothering too much with the data already available—may well be a waste of time. In most cases we need no refutations to tell us that the theory is in urgent need of replacement: the positive heuristic of the programme drives us forward anyway. Also, to give a stern 'refutable interpretation' to a fledgling version of a programme is dangerous methodological cruelty. The first versions may even 'apply' only to non-existing 'ideal' cases; it may take decades of theoretical work to arrive at the first novel facts and still more time to arrive at *interestingly testable* versions of the research programmes, at the stage when refutations are no longer forseeable in the light of the programme itself.

The dialectic of research programmes is then not necessarily an alternating series of speculative conjectures and empirical refutations. The interaction between the development of the programme and the empirical checks may be very varied—which pattern is actually realized depends only on historical accident. Let us mention three typical variants.

(1) Let us imagine that each of the first three consecutive versions, H_1, H_2, H_3 predict some new facts successfully but others unsuccessfully, that is

[1] Born [1948], p. 180; my italics.

each version is both corroborated *and* refuted in turn. Finally H_4 is proposed which predicts some novel facts but stands up to the severest tests. The problemshift is progressive, and also we have a beautiful Popperian alternation of conjectures and refutations.[1] People will admire this as a classical example of theoretical and experimental work going hand in hand.

(2) Another pattern could have been a lone Bohr (possibly without Balmer preceding him), working out H_1, H_2, H_3, H_4 but self-critically withholding publication until H_4. Then H_4 is tested: all the evidence will turn up as corroborations of H_4, the first (and only) published hypothesis. The theoretician—at his desk—is here seen to work far ahead of the experimenter: we have a period of relative autonomy of theoretical progress.

(3) Let us now imagine that *all* the empirical evidence mentioned in these three patterns is already there at the time of the invention of H_1, H_2, H_3, H_4. In this case H_1, H_2, H_3, H_4 will not represent an empirically progressive problemshift and therefore, although all the evidence supports his theories, the scientist has to work on further in order to prove the scientific value of his programme.[2] Such a state of affairs may be brought about either by the fact that an older research programme (which has been challenged by the one leading to H_1, H_2, H_3, H_4) had already produced all these facts—or by the fact that too much government money lay around for collecting data about spectrum lines and hacks stumbled upon all the data. However the latter case is extremely unlikely, for, as Cullen used to say, 'the number of false facts, afloat in the world, infinitely exceeds that of the false theories'[3]; in most such cases the research programme will clash with the available 'facts', the theoretician will look into the 'experimental techniques' of the experimentalist, and having overthrown and replaced his observational theories will correct his facts thereby producing *novel* ones.[4]

[1] In the first three patterns we do not involve complications like successful appeals against the verdict of the experimental scientists.

[2] This shows that if exactly the same theories and the same evidence is rationally reconstructed in different time orders, they may constitute either a progressive or a degenerative shift. Also cf. my [1968*a*], p. 387.

[3] Cf. McCulloch [1825], p. 21. For a strong argument on how extremely unlikely such a pattern is, see *below*, pp. 156–7.

[4] Perhaps it should be mentioned that manic data collection—and 'too much' precision —prevents even the formation of naive 'empirical' hypotheses like Balmer's. Had Balmer known of Michelson's fine-spectra, would he have ever found his formula? Or, had Tycho de Brahe's data been more precise, would Kepler's elliptical law ever have been put forward? The same applies to the naive first version of the general gas law, etc. The Descartes–Euler conjecture on polyhedra might never have been made but for the scarcity of data; cf. my [1963–4], pp. 298 ff.

After this methodological excursion, let us return to Bohr's programme. Not all developments in the programme were foreseen and planned when the positive heuristic was first sketched. When some curious gaps appeared in Sommerfeld's sophisticated models (some predicted lines never did appear), Pauli proposed a deep auxiliary hypothesis (his 'exclusion principle') which accounted not only for the known gaps but reshaped the shell theory of the periodic system of elements and anticipated facts then unknown.

I do not wish to give here an elaborate account of the development of Bohr's programme. But its detailed study from the methodological viewpoint is a veritable goldmine: its marvellously fast progress—on inconsistent foundations!—was breathtaking, the beauty, originality and empirical success of its auxiliary hypotheses, put forward by scientists of brilliance and even genius, was unprecedented in the history of physics.[1] Occasionally the next version of the programme required only a trivial improvement, like the replacement of mass by reduced mass. Occasionally, however, to arrive at the next version required new sophisticated mathematics, like the mathematics of the many-body problem, or new sophisticated physical auxiliary theories. The additional mathematics or physics was either dragged in from some part of extant knowledge (like relativity theory) or invented (like Pauli's exclusion principle). In the latter case we have a 'creative shift' in the positive heuristic.

But even this great programme came to a point where its heuristic power petered out. *Ad hoc* hypotheses multiplied and could not be replaced by content-increasing explanations. For instance, Bohr's theory of molecular (band) spectra predicted the following formula for diatomic molecules:

$$\nu = \frac{h}{8\pi^2 I}[(m+1)^2 - m^2]$$

But the formula was refuted. Bohrians replaced the term m^2 by $m(m+1)$: this fitted the facts but was sadly *ad hoc*.

Then came the problem of some unexplained doublets in alkali spectra. Landé explained them in 1924 by an *ad hoc* 'relativistic splitting rule', Goudsmit and Uhlenbeck in 1925 by electron spin. If Landé's explanation was *ad hoc*, Goudsmit's and Uhlenbeck's was also inconsistent with special relativity theory: surface points on the largish electron had to travel

[1] 'Between the appearance of Bohr's great trilogy in 1913 and the advent of wave mechanics in 1925, a large number of papers appeared developing Bohr's ideas into an impressive theory of atomic phenomena. It was a collective effort and the names of the physicists contributing to it make up an imposing roll-call: Bohr, Born, Klein, Rosseland, Kramers, Pauli, Sommerfeld, Planck, Einstein, Ehrenfest, Epstein, Debye, Schwarzschild, Wilson ...' (Ter Haar [1967], p. 43).

faster than light, and the electron had even to be bigger than the whole atom.[1] Considerable courage was needed to propose it. (Kronig got the idea earlier but refrained from publishing it because he thought it was inadmissible.[2])

But temerity in proposing wild inconsistencies did not reap any more rewards. The programme lagged behind the discovery of 'facts'. Undigested anomalies swamped the field. With ever more sterile inconsistencies and ever more *ad hoc* hypotheses, the degenerating phase of the research programme had set in: it started—to use one of Popper's favourite phrases —'to lose its empirical character'.[3] Also many problems, like the theory of perturbations, could not even be expected to be solved within it. A rival research programme soon appeared: wave mechanics. Not only did the new programme, even in its first version (de Broglie, 1924), explain Planck's and Bohr's quantum conditions; it also led to an exciting new fact, to the Davisson–Germer experiment. In its later, ever more sophisticated versions it offered solutions to problems which had been completely out of the reach of Bohr's research programme, and explained the *ad hoc* later theories of Bohr's programme by theories satisfying high methodological standards. Wave mechanics soon caught up with, vanquished and replaced Bohr's programme.

De Broglie's paper came at the time when Bohr's programme was degenerating. But this was mere coincidence. One wonders what would have happened if de Broglie had written and published his paper in 1914 instead of 1924.

(*d*) *A new look at crucial experiments: the end of instant rationality.*

It would be wrong to assume that one must stay with a research programme until it has exhausted all its heuristic power, that one must not introduce a rival programme before everybody agrees that the point of degeneration has probably been reached. (Although one can understand the irritation of a physicist when, in the middle of the progressive phase of a research programme, he is confronted by a proliferation of vague metaphysical theories

[1] A footnote in their paper reads: 'It should be observed that [according to our theory] the peripheral velocity of the electron would considerably exceed the velocity of light' (Uhlenbeck and Goudsmit [1925]).

[2] Jammer [1966], pp. 146–8 and 151.

[3] For a vivid description of this degenerating phase of Bohr's programme, cf. Margenau [1950], pp. 311–3.

In the progressive phase of a programme the main heuristic stimulus comes from the positive heuristic: anomalies are largely ignored. In the degenerating phase the heuristic power of the programme peters out. In the absence of a rival programme this situation may be reflected in the psychology of the scientists by an unusual hypersensitivity to anomalies and by a feeling of a Kuhnian 'crisis'.

stimulating no empirical progress.[1]) One must never allow a research programme to become a *Weltanschauung*, or a sort of *scientific rigour*, setting itself up as an arbiter between explanation and non-explanation, as mathematical rigour sets itself up as an arbiter between proof and non-proof. Unfortunately this is the position which Kuhn tends to advocate: indeed, what he calls 'normal science' is nothing but a research programme that has achieved monopoly. But, as a matter of fact, research programmes have achieved complete monopoly only rarely and then only for relatively short periods, in spite of the efforts of some Cartesians, Newtonians and Bohrians. *The history of science has been and should be a history of competing research programmes (or, if you wish, 'paradigms'), but it has not been and must not become a succession of periods of normal science: the sooner competition starts, the better for progress.* 'Theoretical pluralism' is better than 'theoretical monism': on this point Popper and Feyerabend are right and Kuhn is wrong.[2]

The idea of competing scientific research programmes leads us to the problem: *how are research programmes eliminated*? It has transpired from our previous considerations that a degenerating problemshift is no more a sufficient reason to eliminate a research programme than some old-fashioned 'refutation' or a Kuhnian 'crisis'. *Can there be any objective* (as opposed to socio-psychological) *reason to reject a programme, that is, to eliminate its hard core and its programme for constructing protective belts*? Our answer, in outline, is that such an objective reason is provided by a rival research programme which explains the previous success of its rival and supersedes it by a further display of *heuristic power*.[3]

However, the criterion of 'heuristic power' strongly depends on how we construe '*factual novelty*'. Until now we have assumed that it is immediately ascertainable whether a new theory predicts a novel fact or not.[4] But *the novelty of a factual proposition can frequently be seen only after a long period has elapsed.* In order to show this, I shall start with an example.

[1] This is what must have irritated Newton most in the 'sceptical proliferation of theories' by Cartesians. Cf. my [1970].

[2] Nevertheless there is something to be said for at least *some* people sticking to a research programme until it reaches its 'saturation point'; a new programme is then challenged to account for the full success of the old. It is no argument against this that the rival may, when it was first proposed, already have explained all the success of the first programme; the growth of a research programme cannot be predicted—it may stimulate important unforeseeable auxiliary theories of its own. Also, if a version A_n of a research programme P_1 is mathematically equivalent to a version A_m of a rival P_2, one should develop both: their heuristic strength can still be very different.

[3] I use '*heuristic power*' here as a technical term to characterize the power of a research programme to anticipate theoretically novel facts in its growth. I could of course use '*explanatory power*': cf. *above*, p. 119, footnote 1.

[4] Cf. *above*, p. 116, text to footnote 2, and p. 134, text to footnote 3.

Bohr's theory logically implied Balmer's formula for hydrogen lines as a consequence.[1] Was this a novel fact? One might have been tempted to deny this, since after all, Balmer's formula was well-known. But this is a half-truth. Balmer merely 'observed' B_1: that *hydrogen lines obey the Balmer formula.* Bohr predicted B_2: that *the differences in the energy levels in different orbits of the hydrogen electron obey the Balmer formula.* Now one may say that B_1 already contains all the purely 'observational' content of B_2. But to say this presupposes that there can be a pure 'observational level', untainted by theory, and impervious to theoretical change. In fact, B_1 was accepted only because the optical, chemical and other theories *applied* by Balmer were well corroborated and accepted as interpretative theories; and these theories could always be questioned. It might be argued that we can 'purge' even B_1 of its theoretical presuppositions, and arrive at what Balmer really 'observed', which might be expressed in the more modest assertion, B_0: that *the lines emitted in certain tubes in certain well-specified circumstances* (*or in the course of a 'controlled experiment'*[2]) *obey the Balmer formula.* Now some of Popper's arguments show that we can *never* arrive at any hard 'observational' rock-bottom in this way; 'observational' theories can easily be shown to be involved in B_0.[3] On the other hand, given that Bohr's programme after a long progressive development, had shown its heuristic power, its hard core would itself have become well corroborated[4] and therefore qualified as an 'observational' or interpretative theory. But then B_2 will be seen not as a mere theoretical reinterpretation of B_1, but as a *new fact* in its own right.

These considerations lend new emphasis to the hindsight element in our appraisals and lead to a further liberalization of our standards. A new research programme which has just entered the competition may start by explaining 'old facts' in a novel way but may take a very long time before it is seen to produce 'genuinely novel' facts. For instance, the kinetic theory of heat *seemed* to lag behind the results of the phenomenological

[1] Cf. *above*, p. 147.

[2] Cf. *above*, p. 111, footnote 6.

[3] One of Popper's arguments is particularly important: 'There is a widespread belief that the statement "I see that this table here is white", possesses some profound advantage over the statement "This table here is white", from the point of view of epistemology. But from the point of view of evaluating its possible objective tests, the first statement, in speaking about me, does not appear more secure than the second statement, which speaks about the table here' ([1934], section 27). Neurath makes a characteristically blockheaded comment on this passage: 'For us such protocol statements have the advantage of *having more stability*. One may retain the statement: "People in the 16th century saw fiery swords in the sky" while crossing out "There were fiery swords in the sky" ' (Neurath [1935], p. 362).

[4] *This remark, incidentally, defines a 'degree of corroboration' for the 'irrefutable' hard cores of research programmes. Newton's theory (in isolation) had no empirical content, yet it was, in this sense, highly corroborated.*

theory for decades before it finally overtook it with the Einstein–Smoluchowski theory of Brownian motion in 1905. After this, what had previously seemed a speculative reinterpretation of old facts (about heat, etc.) turned out to be a discovery of novel facts (about atoms).

All this suggests that we must not discard a budding research programme simply because it has so far failed to overtake a powerful rival. We should not abandon it if, supposing its rival were not there, it would constitute a progressive problemshift.[1] *And we should certainly regard a newly interpreted fact as a new fact, ignoring the insolent priority claims of amateur fact collectors. As long as a budding research programme can be rationally reconstructed as a progressive problemshift, it should be sheltered for a while from a powerful established rival.*[2]

These considerations, on the whole, stress the importance of methodological tolerance, and leave the question of how research programmes are eliminated still unanswered. The reader may even suspect that laying this much stress on fallibility liberalizes or, rather, softens up, our standards to the extent that we will be landed with radical scepticism. Even the celebrated '*crucial experiments*' will then have no force to overthrow a research programme; anything goes.[3]

But this suspicion is unfounded. *Within* a research programme '*minor crucial experiments*' between subsequent versions are quite common. Experiments easily 'decide' between the n-th and $n+1$-th scientific version, since the $n+1$-th is not only inconsistent with the n-th, but also supersedes it. If the $n+1$-th version has more corroborated content in the light of the *same* programme and in the light of the *same* well corroborated observational theories elimination is a relatively routine affair (only relatively, for even here this decision may be subject to appeal). Appeal procedures too are occasionally easy: in many cases the challenged observational theory, far from being well corroborated, is in fact an inarticulate, naive, 'hidden' assumption; it is only the challenge which reveals the existence of this hidden assumption, and brings about its articulation, testing and downfall. Time and again, however, the observational theories are themselves embedded in some research programme and then the appeal procedure

[1] Incidentally, in the methodology of research programmes, the pragmatic meaning of 'rejection' [of a programme] becomes crystal clear: it means *the decision to cease working on it.*

[2] Some might regard—cautiously—this sheltered period of development as '*prescientific*' (or 'theoretical'); and be prepared only when it starts producing 'genuinely novel' facts to recognize its truly *scientific* (or 'empirical') character—but then their recognition will have to be retroactive.

[3] Incidentally, *this conflict between fallibility and criticism can be rightly said to be the main problem—and driving force—of the Popperian research programme in the theory of knowledge.*

leads to a clash between two research programmes: in such cases we may need a '*major crucial experiment*'.

When two research programmes compete, their first 'ideal' models usually deal with different aspects of the domain (for example, the first model of Newton's semi-corpuscular optics described light-refraction, the first model of Huyghens's wave optics light-interference). As the rival research programmes expand, they gradually encroach on each other's territory and the n-th version of the first will be blatantly, dramatically inconsistent with the m-th version of the second.[1] An experiment is repeatedly performed, and as a result, the first is defeated in *this battle*, while the second wins. But *the war* is not over: any research programme is allowed a few such defeats. All its needs for a comeback is to produce an $n+1$-th (or $n+k$-th) content-increasing version and a verification of some of its novel content.

If such a comeback, after sustained effort, is not forthcoming, the war is lost and the original experiment is seen, *with hindsight*, to have been 'crucial'. But especially if the defeated programme is a young, fast-developing programme, and if we decide to give sufficient credit to its 'pre-scientific' successes, allegedly crucial experiments dissolve one after the other in the wake of its forward surge. Even if the defeated programme is an old, established and 'tired' programme, near its 'natural saturation point',[2] it may continue to resist for a long time and hold out with ingenious content-increasing innovations even if these are unrewarded with empirical success. It is very difficult to defeat a research programme supported by talented, imaginative scientists. Alternatively, stubborn defenders of the defeated programme may offer *ad hoc* explanations of the experiments or a shrewd *ad hoc* 'reduction' of the victorious programme to the defeated one. But such efforts we should reject as unscientific.[3]

Our considerations explain why crucial experiments are seen to be crucial only decades later. Kepler's ellipses were generally admitted as crucial evidence for Newton and against Descartes only about one hundred years after Newton's claim. The anomalous behaviour of Mercury's perihelion

[1] An especially interesting case of such competition is *competitive symbiosis*, when a new programme is grafted on to an old one which is inconsistent with it; cf. *above*, p. 142

[2] There is not such thing as a *natural* 'saturation point'; in my [1963–4], especially on pp. 327–8, I was more of a Hegelian, and I thought there was; now I use the expression with an ironical emphasis. There is no predictable or ascertainable limitation on human imagination in inventing new, content-increasing theories or on the 'cunning of reason' (*List der Vernunft*) in rewarding them with some empirical success even if they are false or even if the new theory has less verisimilitude—in Popper's sense—than its predecessor. (Probably all scientific theories ever uttered by men will be false: they still may be rewarded by empirical successes and even have increasing verisimilitude.)

[3] For an example, cf. *above*, p. 126, footnote 2.

was known for decades as one of the many yet unsolved difficulties in Newton's programme; but only the fact that Einstein's theory explained it better transformed a dull anomaly into a brilliant 'refutation' of Newton's research programme.[1] Young claimed that his double-slit experiment of 1802 was a crucial experiment between the corpuscular and the wave programmes of optics; but his claim was only acknowledged much later, after Fresnel developed the wave programme much further 'progressively' and it became clear that the Newtonians could not match its heuristic power. The anomaly, which had been known for decades, received the honorific title of refutation, the experiment the honorific title of 'crucial experiment' only after a long period of uneven development of the two rival programmes. Brownian motion was for nearly a century in the middle of the battlefield before it was *seen* to defeat the phenomenological research programme and turn the war in favour of the atomists. Michelson's 'refutation' of the Balmer series was ignored for a generation until Bohr's triumphant research programme backed it up.

It may be worthwhile to discuss in detail some examples of experiments whose 'crucial' character became evident only retrospectively. First I shall take the celebrated Michelson–Morley experiment of 1887 which allegedly falsified the ether theory and 'led to the theory of relativity', then the Lummer–Pringsheim experiments which allegedly falsified the classical theory of radiation and 'led to the quantum theory'.[2] Finally I shall discuss an experiment which many physicists thought would turn out to decide against the conservation laws but which, in fact, ended up as their most triumphant corroboration.

(*d 1*) *The Michelson–Morley experiment.*

Michelson first devised an experiment in order to test Fresnel's and Stokes's contradictory theories about the influence of the motion of the earth on the ether,[3] during his visit to Helmholtz's Berlin institute in 1881. According to Fresnel's theory, the earth moves through an ether at rest, but the ether within the earth is *partially* carried along with the earth; Fresnel's theory therefore entailed that the velocity of the ether outside the

[1] *Thus an anomaly in a research programme is a phenomenon which we regard as something to be explained in terms of the programme. More generally, we may speak, following Kuhn, about 'puzzles': a 'puzzle' in a programme is a problem which we regard as a challenge to that particular programme. A 'puzzle' can be resolved in three ways: by solving it within the original programme (the anomaly turns into an example); by neutralizing it, i.e. solving it within an independent, different programme (the anomaly disappears); or, finally, by solving it within a rival programme (the anomaly turns into a counterexample).*

[2] Cf. Popper [1934], section 30.

[3] Cf. Fresnel [1818], Stokes [1845] and [1846]. For an excellent brief exposition cf. Lorentz [1895].

earth relative to the earth was positive (i.e. Fresnel's theory implied the existence of an 'ether wind'). According to Stokes's theory, the ether was dragged along by the earth and immediately on the surface of the earth, the velocity of the ether was equal to that of the earth: therefore its relative velocity was zero (i.e. there was no ether wind on the surface). Stokes originally thought that the two theories were observationally equivalent: for instance, with suitable auxiliary assumptions both theories explained the aberration of light. But Michelson claimed that his 1881 experiment was a crucial experiment between the two and that it *proved* Stokes's theory.[1] He claimed that the velocity of the earth relative to the ether is far less than Fresnel's theory would have it. Indeed, he concluded that from his experiment' 'the *necessary conclusion* follows that the hypothesis [of a stationary ether] is erroneous. This conclusion *directly contradicts* the explanation of the phenomenon of aberrration which . . . presupposes that the earth moves through the ether, the latter remaining at rest'.[2] As often happens, Michelson the experimenter was then taught a lesson by a theoretician. Lorentz, the leading theoretical physicist of the period, in what Michelson later described as 'a very searching analysis . . . of the entire experiment',[3] showed that Michelson 'misinterpreted' the facts and that what he observed did *not* in fact contradict the hypothesis of the stationary ether. Lorentz showed that Michelson's calculations were wrong; Fresnel's theory predicted only half of the effect Michelson had calculated. Lorentz concluded that Michelson's experiment did *not* refute Fresnel's theory, and that it certainly did not prove Stokes's theory either. Lorentz went on to show that Stokes's theory was inconsistent: that it assumed the ether at the earth's surface to be at rest with regard to the latter *and* required that the relative velocity have a potential; but these two conditions are incompatible. But even if Michelson *had* refuted *one* theory of the stationary ether, the programme is untouched: one can easily devise several other versions of the ether programme, which predict very small values for the ether winds and he, Lorentz, immediately produced one. This theory was testable and Lorentz proudly submitted it to the verdict of experiment.[4] Michelson, jointly with Morley, took up the challenge. The relative velocity of the earth to the ether again seemed to be zero, in conflict with Lorentz's theory. By this time, Michelson had become more cautious in interpreting his data and even thought of the possibility that the solar system as a whole might have moved in the opposite direction to the earth; therefore he decided to repeat the experiment

[1] This transpires, obliquely, from the concluding section of his [1881].

[2] Michelson [1881], p. 128. My italics. [3] Michelson and Morley [1887], p. 335.

[4] Lorentz [1886]. For the inconsistency of Stokes's theory also cf. his [1892*b*].

'at intervals of three months and thus avoid all uncertainty'.[1] Michelson, in his second paper, does not talk any more about 'necessary conclusions' and 'direct contradictions'. He only thinks that from his experiment 'it appears, from all that precedes, *reasonably certain* that if there be any relative motion between the earth and the luminiferous ether, it must be *small*; quite small enough entirely to refute *Fresnel's* explanation of aberration'.[2] Thus in this paper Michelson still claims to have refuted Fresnel's theory (and also Lorentz's new theory); but there is not a word about his old 1881 claim that he refuted 'the theory of stationary ether' in general. (Indeed, he believed that in order to do so, he would have to test the ether wind also at high altitudes, 'at the top of an isolated mountain peak, for instance'.[3])

While some ether-theorists—like Kelvin—did not trust Michelson's 'experimental skill',[4] Lorentz pointed out that, in spite of Michelson's naive claim, even his *new* experiment 'furnishes no evidence for the question for which it was undertaken'.[5] One can regard Fresnel's theory perfectly well as an *interpretative* theory, which interprets facts, rather than is refutable by them, and then, Lorentz showed, 'the significance of the Michelson–Morley experiment lies rather in the fact that it can teach us something about *the changes in the dimensions*'[6]: the dimensions of bodies is affected by their movement through the ether. Lorentz elaborated this 'creative shift' within Fresnel's programme with great ingenuity and thereby claimed to have 'removed the contradiction between Fresnel's theory and Michelson's result'.[7] But he admitted that 'since the nature of the molecular forces is entirely unknown to us, it is impossible to test the hypothesis'[8]: *at least for the time being* it could predict no novel facts.[9]

[1] Michelson and Morley [1887], p. 341. But Pearce Williams points out that he never did. (Pearce Williams [1968], p. 34.)

[2] *Ibid.* p. 341. My italics.

[3] Michelson and Morley [1887]. This remark shows that Michelson realized that his 1887 experiment was completely consistent with an ether wind higher up. Max Born, in his [1920], that is, thirty-three years later, asserted that from the 1887 experiment 'we *must* conclude that the ether wind does not exist'. (My italics).

[4] Kelvin said in the 1900 International Congress of Physics that 'the only cloud in the clear sky of the [ether] theory was the null result of the Michelson–Morley experiment' (cf. Miller [1925]) and immediately persuaded Morley and Miller, who were there, to repeat the experiment.

[5] Lorentz [1892*a*].

[6] *Ibid.* My italics.

[7] Lorentz [1895].

[8] Lorentz [1892*b*].

[9] Fitzgerald at the same time, independently of Lorentz, produced a testable version of this 'creative shift' which was quickly refuted by Trouton's, Rayleigh's and Brace's experiments: it was theoretically but not empirically progressive. Cf. Whittaker [1947], p. 53 and Whittaker [1953], pp. 28–30.

There is a widespread view that Fitzgerald's theory was *ad hoc*. What contemporary physicists meant was that the theory was *ad hoc*$_2$ (cf. *above*, p. 125, footnote 1): that there was '*no independent* [*positive*] *evidence*' for it. (Cf. e.g. Larmor [1904], p. 624.) Later,

In the meanwhile, in 1897, Michelson carried out his long planned experiment to measure the velocity of ether wind on mountain tops. He found none. Since he had thought earlier that he had proved Stokes's theory which predicted an ether wind higher up, he was dumbfounded. If Stokes's theory was still correct, the gradient of the velocity of the ether had to be very small. Michelson had to conclude that 'the earth's influence upon the ether extended to distances of the order of the earth's diameter'.[1] He thought that this was an 'improbable' result, and decided that in 1887 he had drawn the wrong conclusion from his experiment: it was Stokes's theory which had to be rejected and Fresnel's which had to be accepted; and he decided that he would accept *any* reasonable auxiliary hypothesis to have it saved, including Lorentz's 1892 theory.[2] He *now* seemed to prefer the Fitzgerald–Lorentz contraction and by 1904 his colleagues at Case were trying to find out whether this contraction varies with different materials.[3]

While most physicists tried to interpret Michelson's experiments within the framework of the ether programme, Einstein, unaware of Michelson, Fitzgerald and Lorentz, but stimulated primarily by Mach's criticism of Newtonian mechanics, arrived at a new, progressive research programme.[4] This new programme not only 'predicted' and explained the outcome of the Michelson–Morley experiment but also predicted a huge array of previously undreamt-of facts, which obtained dramatic corroborations. It was *only then*, twenty-five years later, that the Michelson–Morley experiment came to be seen as 'the greatest negative experiment in the history of science'.[5] But this could not be seen instantly. Even if the experiment was negative, it was not clear, negative exactly to *what*? Moreover, Michelson in 1881 thought that it was also *positive*: he held that he had *refuted* Fresnel's

---

under Popper's influence the term '*ad hoc*' was primarily used in the sense of *ad hoc*$_1$, that there was *no independent test* possible for it. But, as the refuting experiments show, it is a mistake to claim, as Popper does, that Fitzgerald's theory was *ad hoc*$_1$ (cf. Popper [1934], section 20). This shows again how important it is to separate *ad hoc*$_1$, and *ad hoc*$_2$.

When Grünbaum, in his [1959*a*], pointed out Popper's mistake, Popper admitted it but replied that Fitzgerald's theory was certainly *more ad hoc* than Einstein's (Popper [1959*b*]), and that this provides yet another '... excellent example of "degrees of *ad-hocness*" and of one of the main theses of [his] book—that *degrees* of *ad-hocness* are related (inversely) to degrees of testability and significance'. But the difference is *not* simply a matter of degrees of a unique *ad-hocness* which can be measured by testability. Also cf. *below*, p. 175.

[1] Michelson [1897], p. 478.

[2] Lorentz, indeed, immediately commented: 'While [Michelson] considers so far-reaching an influence of the earth improbable, I should, on the contrary, *expect* it' (Lorentz [1897]; my italics).

[3] Morley and Miller [1904].

[4] There has been a considerable controversy about the historico-heuristic background of Einstein's theory, in the light of which this statement may turn out to be false.

[5] Bernal [1965], p. 530. For Kelvin, in 1905, it was only a 'cloud in the clear sky': cf. *above*, p. 161, footnote 4.

but had *verified* Stokes's theory. Michelson himself and then Fitzgerald and Lorentz explained the result also *positively* within the ether programme.[1] As it is with all experimental results, its negativity for the old programme was established *only later*, by the slow accumulation of *ad hoc* attempts to account for it within the degenerating old programme and by the gradual establishment of a new *progressive* victorious programme in which it has become a positive instance. But the possibility of the rehabilitation of some part of the 'degenerating' old programme could never be rationally excluded.

Only an extremely difficult and—indefinitely—long process can establish a research programme as superseding its rival; and it is unwise to use the term 'crucial experiment' too rashly. Even when a research programme is seen to be swept away by its predecessor, it is not swept away by some 'crucial' experiment; and even if some such crucial experiment is later called in doubt, the new research programme cannot be stopped without a powerful progressive upsurge of the old programme.[2] The negativity—and importance—of the Michelson–Morley experiment lies primarily in the progressive shift in the *new* research programme to which it came to lend powerful support, and its 'greatness' is only a reflection of the greatness of the two *programmes* involved.

It would be interesting to give a detailed analysis of the rival shifts involved in the waning fortunes of the ether theory. But under the influence of naive falsificationism the most interesting degenerating phase in the ether theory after Michelson's 'crucial experiment' is simply ignored by most Einsteinians. They believe that the Michelson–Morley experiment single-handedly defeated the ether theory, the tenacity of which was only due to obscurantist conservatism. On the other hand, this post-Michelson period of the ether theory is not scrutinized *critically* by the anti-Einsteinians, who believe that the ether theory suffered no setback whatsoever: what is good in Einstein's theory was essentially in Lorentz's ether theory and Einstein's victory is only due to positivist fashion. But, in fact, Michelson's long series of experiments from 1881 to 1935, conducted in order to test subsequent versions of the ether programme provides a fascinating

[1] Indeed, Chwolson's excellent physics textbook said in 1902 that the probability of the ether hypothesis borders on certainty. (Cf. Einstein [1909], p. 817.)

[2] Polanyi tells us with *gusto* how, in 1925, in his presidential address to the American Physical Society, Miller announced that Michelson's and Morley's reports notwithstanding, he had 'overwhelming evidence' for an ether-drift; yet the audience remained committed to Einstein's theory. Polanyi draws the conclusion that no ' "objectivist" framework' can account for the scientist's acceptance or rejection of theories (Polanyi [1958], pp. 12–14). But my reconstruction makes the tenacity of the Einsteinian research programme in the face of alleged contrary evidence a completely *rational* phenomenon and thereby undermines Polanyi's 'post-critical'-mystical message.

example of a degenerating problemshift.[1] (But problemshifts may get out of degenerating troughs. It is well known that Lorentz's ether theory can easily be strengthened in such a way that it becomes, in an interesting sense, equivalent with Einstein's no-ether theory.[2] The ether may, in the context of a major 'creative shift', still return.[3])

The fact that we heed hindsight to evaluate experiments explains why, between 1881 and 1886, Michelson's experiment was not even mentioned in the literature. Indeed, when a French physicist, Potier, pointed out to Michelson his 1881 mistake, Michelson decided not to publish a correction note. He explains the reason for this decision in a letter to Rayleigh in March 1887: 'I have repeatedly tried to interest my scientific friends in this experiment without avail, and the reason for my never publishing the correction (I am ashamed to confess it) was that I was discouraged at the slight attention the work received, and did not think it worthwhile.'[4] This letter, incidentally, was a reply to a letter from Rayleigh which drew Michelson's attention to Lorentz's paper. This letter triggered off the 1887 experiment. But even after 1887, and even after 1905, the Michelson–Morley experiment was not yet generally regarded as disproving the existence of the ether, and with good reason. This may explain why Michelson was awarded his Nobel Prize (in 1907), not for 'refuting the ether theory', but 'for his optical precision *instruments* and the spectro-scopic and methodological investigations carried out with their aid'[5]; and why the Michelson–Morley experiment was not even mentioned in the presentation speeches. Michelson, in his *Nobel Lecture*, did not mention it; and he kept quiet

[1] *One typical sign of the degeneration of a programme which is not discussed in this paper is the proliferation of contradictory 'facts'. Using a false theory as an interpretative theory, one may get—without committing any 'experimental mistake'—contradictory factual propositions, inconsistent experimental results.* Michelson, who stuck to the ether to the bitter end, was primarily frustrated by the inconsistency of the 'facts' he arrived at by his ultra-precise measurements. His 1887 experiment 'showed' that there was no ether wind on the earth's surface. But aberration 'showed' that there was. Moreover, his own 1925 experiment (either never mentioned or, as in Jaffé's [1960], misrepresented) also 'proved' that there was one (cf. Michelson and Gale [1925] and, for a sharp criticism, Runge [1925]).

[2] Cf. e.g. Ehrenfest [1913], pp. 17–18, quoted and discussed by Dorling in his [1968]. But one should not forget that *two specific theories, while being mathematically (and observationally) equivalent, may still be embedded into different rival research programmes, and the power of the positive heuristic of these programmes may well be different.* This point has been overlooked by proposers of such equivalence proofs (a good example is the equivalence proof between Schrödinger's and Heisenberg's approach to quantum physics). Also cf. *above*, p. 155, footnote 2.

[3] Cf. e.g. Dirac [1951]: 'If one reexamines the question in the light of present-day knowledge, one finds that the aether is no longer ruled out by relativity, and good reasons can now be advanced for postulating an aether.' Also cf. the concluding paragraph of Rabi [1961] and Prokhovnik [1967].

[4] Shankland [1964], p. 29.

[5] My italics.

about the fact that although he might have originally devised his instruments to measure precisely the velocity of light, he was compelled to improve them for testing some specific ether theories and that the 'precision' of his 1887 experiment was largely motivated by Lorentz's theoretical criticism: a fact which standard contemporary literature never mentions.[1]

Finally, one tends to forget that even if the Michelson–Morley experiment had shown an 'ether wind', Einstein's programme might have been victorious nonetheless. When Miller, an ardent champion of the classical ether programme, published his sensational claim that the Michelson–Morley experiment was sloppily conducted and in fact there *was* an ether wind, the news correspondent of *Science* crowed that 'Professor Miller's results knock out the relativity theory radically'.[2] In Einstein's view, however, even if Miller had reported the true state of affairs '[only] the *present form* of relativity theory' would have to be abandoned.[3] In fact, Synge pointed out that Miller's results, even if taken at their face value, do not conflict with Einstein's theory: only Miller's explanation of them does. One can easily replace the extant auxiliary theory of rigid bodies by a new, Gardner–Synge theory, and then Miller's results are fully digested within Einstein's programme.[4]

(*d 2*) *The Lummer–Pringsheim experiments.*

Let us discuss another alleged crucial experiment. Planck claimed that Lummer's and Pringsheim's experiments, which '*refuted*' Wien's and Rayleigh's and Jeans's laws of radiation at the turn of the century, '*led to*'— or 'even brought about'—the quantum theory.[5] But again the role of these experiments is much more complicated and is very much in line with our approach. It is not simply that Lummer's and Pringsheim's experiments put an end to the classical approach but were neatly explained by quantum physics. On the one hand, some early versions of quantum theory by Einstein *entail* Wien's law and therefore were no less refuted by

[1] Einstein himself tended to believe that Michelson devised his interferometer in order to test Fresnel's theory. (Cf. Einstein [1931].) Incidentally, Michelson's early experiments on spectrum lines—like his [1881–2]—were also relevant to the ether theories of his day. Michelson over-emphasized his success in 'precise measurements' only when he was frustrated by his lack of success in evaluating their relevance for theories. Einstein, who disliked precision for its own sake, asked him why he devoted so much energy to it. Michelson's answer was 'because he found it fun.' (Cf. Einstein [1931].)

[2] *Science* [1925].

[3] Einstein [1927]. My italics.

[4] Synge [1952–4].

[5] Planck [1929]. Popper, in his [1934], section 30, Gamow in his [1966] (p. 37), take over this locution. Of course, observation statements do not 'lead' to some uniquely determined theory.

Lummer's and Pringsheim's experiments than the classical theory.[1] On the other hand, several classical explanations of the Planck formula were offered. For instance, at the 1913 meeting of the British Association for the Advancement of Science, there was a special meeting on radiation, attended among others by Jeans, Rayleigh, J. J. Thomson, Larmor, Rutherford, Bragg, Poynting, Lorentz, Pringsheim and Bohr. Pringsheim and Rayleigh were studiedly neutral about quantum theoretical speculations, but Professor Love 'represented the older views, and maintained the possibility of explaining facts about radiation without adopting the theory of quanta. He criticized the application of the equi-partition of energy theory, on which part of the quantum theory rests. The evidence for the quantum theory of most weight is the agreement with experiment of Planck's formula for the emissivity of a black body. From the mathematical point of view, there may be many more formulae which would agree equally well with the experiments. A formula due to A. Korn was dealt with, which gave results over a wide range, showing just about as good agreement with experiment as the Planck formula. In further contention that *the resources of ordinary theory are not exhausted*, he pointed out that it may be possible to extend the calculation for the emissivity of a thin plate due to Lorentz to other cases. For this calculation no simple analytical expression represents the results over the whole range of wavelengths, and it may well be that in the general case no simple formula exists which is applicable to all wavelengths. Planck's formula may, in fact, be nothing more than an empirical formula.'[2] One example of classical explanations was due to Callendar: 'The disagreement with experiment of Wien's well-known formula for the partition of energy in full radiation, is readily explained if we assume that it represents only the intrinsic energy. The corresponding value of the pressure is very easily deduced by reference to Carnot's principle, as Lord Rayleigh has indicated. The formula which I have proposed (*Phil. Mag.*, October 1913) is simply the sum of the pressure and energy-density thus obtained, and gives very satisfactory agreement with experiment, both for radiation and specific heat. I prefer it to Planck's formula (among other reasons) on the ground that the latter cannot be reconciled with the classical thermodynamics, and involves the conception of a *quantum*, or indivisible unit of action, which is unthinkable. The corresponding physical magnitude on my theory, which I have elsewhere called a molecule of caloric, is not necessarily indivisible, but bears a very simple relation to the intrinsic energy of an atom, which is all that is

[1] Cf. Ter Haar [1967], p. 18. A budding research programme usually starts by explaining already refuted 'empirical laws'—and this, in the light of my approach, may be *rationally* regarded as a success.

[2] *Nature* [1913–14], p. 306; my italics.

required to explain the facts that radiation may in special cases be emitted in atomic units which are multiples of a particular magnitude.'[1]

These quotations may have been tediously long but at least they show again convincingly the absence of instant crucial experiments. Lummer's and Pringsheim's refutations did not eliminate the classical approach to the radiation problem. The situation can be better described by pointing out that Planck's original '*ad hoc*' formula[2]—which fitted (and corrected) Lummer's and Pringsheim's data—could be explained *progressively* within the new quantum theoretical programme,[3] while neither his '*ad hoc*' formula, nor its 'semi-empirical' rivals could be explained within the classical programme except at the price of a degenerating problemshift. The 'progressive' development, incidentally, hinged on a 'creative shift': the replacement (by Einstein) of the Boltzman–Maxwell by the Bose–Einstein statistics.[4] The progressiveness of the new development was abundantly clear: in Planck's version it predicted correctly the value of the Boltzman–Planck constant and in Einstein's version it predicted a stunning series of further novel facts.[5] But before the invention of the new—but sadly *ad hoc*—auxiliary hypotheses in the old programme, before the unfolding of the new programme, and before the discovery of the new facts indicating a progressive problemshift in the latter, the objective relevance of the Lummer–Pringsheim experiments was very limited.

[1] Callendar [1914].

[2] I am referring to Planck's formula as given in his [1900*a*] in which he admitted that after having tried for a long time to prove that 'Wien's law must be *necessarily* true', the 'law' was refuted. So he switched from proving lofty eternal laws to 'constructing completely arbitrary expressions'. But of course any physical theory turns out to be 'completely arbitrary' by justificationist standards. In fact, Planck's arbitrary formula contradicted—and victoriously corrected—contemporary empirical evidence. (Planck told this part of the story in his scientific autobiography.) Of course, in an important sense, Planck's *original* radiation formula was 'arbitrary', 'formal', '*ad hoc*': it was a rather isolated formula which was not part of a research programme. (Cf. *below*, p. 175, footnote 3.) As he himself put it: 'Even if the absolutely precise validity of the radiation formula is taken for granted, so long as it had merely the standing of a law disclosed by a lucky intuition, it could not be expected to possess more than a formal significance. For this reason, on the very day when I formulated this law, I began to devote myself to the task of investing it with a true physical meaning' ([1947], p. 41). But the primary importance of 'investing the formula with a physical meaning'—not necessarily '*true* physical meaning'—is that such interpretation frequently leads to a suggestive research programme and *growth*.

[3] First by Planck himself, in his [1900*b*] which 'founded' the research programme of quantum theory.

[4] This had already been done by Planck, but only inadvertently, as it were by mistake. Cf. Ter Haar [1967], p. 18. Indeed, one role of Pringsheim's and Lummer's results was to stimulate the critical analysis of the informal deductions in the quantum theory of radiation, deductions which were loaded with vital 'hidden lemmas' articulated only in the later development. A most important step in this 'articulating process' was Ehrenfest's [1911].

[5] Cf. e.g. Joffé's 1910 list (Joffé [1911], p. 547).

(*d 3*) *Beta-decay versus conservation laws.*

Finally, I shall tell a story of an experiment which very nearly, but not quite, became 'the greatest negative experiment in the history of science'. The story again illustrates the supreme difficulties of deciding exactly *what* one learns from experience, what it 'proves' and what it 'disproves'. The piece of experience under scrutiny will be Chadwick's 'observation' of beta decay in 1914. The story shows how an experiment may first be regarded as presenting a routine puzzle within a research programme, then nearly promoted to the rank of 'crucial experiment', and then again downgraded to presenting a (*new*) routine puzzle, all this depending on the *whole* changing theoretical and empirical landscape. Most conventional accounts are confused by these changes and prefer to falsify history.[1]

When Chadwick discovered the continuous spectrum of radioactive beta-emission in 1914, nobody thought that this curious phenomenon had anything to do with conservation laws. Two ingenious rival explanations were offered in 1922, both within the framework of the atomic physics of the day, one by L. Meitner, the other by C. D. Ellis. According to Miss Meitner, the electrons were partly primary electrons from the nucleus, partly secondary electrons from the electron shell. According to Mr Ellis, they were all primary electrons. Both theories contained sophisticated auxiliary hypotheses, but both predicted novel facts. The predicted facts contradicted each other and the experimental testimony supported Ellis against Meitner.[2] Miss Meitner appealed; the experimental 'appeal court' refused to support her, but ruled that one crucial auxiliary hypothesis in Ellis's theory had to be rejected.[3] The result of the contest was a draw.

Still nobody would have thought that Chadwick's experiment defied the law of conservation of energy, had not Bohr and Kramers arrived exactly at the time of the Ellis–Meitner controversy at the idea that a consistent theory could be developed only if they renounced the principle of conservation of energy in single processes. One of the main features of the fascinating Bohr–Kramers–Slater theory in 1924 was that the classical laws of conservation of energy and momentum were replaced by statistical ones.[4] This theory (or, rather, 'programme') was immediately 'refuted'

[1] A notable partial exception is Pauli's account (Pauli [1958]). In what follows I am trying both to correct Pauli's story and to show that its rationality can be easily seen in the light of our approach.

[2] Ellis and Wooster [1927].

[3] Meitner and Orthmann [1930].

[4] Slater co-operated only reluctantly in sacrificing the conservation principle. He wrote to van der Waerden in 1964: 'As you suspected, the idea of statistical conservation of energy and momentum was put into the theory by Bohr and Kramers, quite against my better judgment.' Van der Waerden does his amusing best to exonerate Slater from the terrible crime of being responsible for a false theory (van der Waerden [1967], p. 13).

and none of its consequences corroborated; indeed, it was never sufficiently developed to explain beta-decay. But in spite of the immediate abandonment of this programme (not simply because of its 'refutations' by the Compton–Simon and Bothe-Geiger experiments but because of the emergence of a powerful rival: the Heisenberg–Schrödinger programme[1]), Bohr remained convinced that the non-statistical conservation laws would finally have to be abandoned and that the beta-decay anomaly would never be explained until these laws were replaced; at which time beta-decay would be seen as a crucial experiment against the conservation laws. Gamow tells us how Bohr tried to use the idea of non-conservation of energy in beta-decay for an ingenious explanation of the seemingly eternal production of energy in stars.[2] Only Pauli, in his Mephistophelian urge to defy the Lord, remained conservative[3] and devised, in 1930, his neutrino theory in order to explain beta decay and in order to save the principle of conservation of energy. He communicated his idea in a jocular letter to a conference in Tübingen—he himself preferred to stay in Zürich to attend a ball.[4] He first mentioned it in a public lecture in 1931 in Pasadena, but he did not allow the lecture to be published because he felt 'unsure' about it. Bohr, at that time (in 1932), still thought that—at least in nuclear physics—one may have 'to renounce the very idea of energy balance'.[5] Pauli finally decided to publish his talk on the neutrino which he delivered to the 1933 Solvay conference, in spite of the fact that 'the reception at the Congress, except for two young physicists, was sceptical'.[6] But Pauli's theory had some methodological merits. It saved not only the principle of conservation of energy but also the principle of conservation of spin and statistics: it explained not only the beta-decay spectrum but, at the same time, the 'nitrogen anomaly'.[7] By Whewellian standards this 'consilience of inductions' should have been sufficient to establish the

[1] Popper is wrong to suggest that these 'refutations' were sufficient to bring about the downfall of this theory. (Popper [1963], p. 242.)

[2] Gamow [1966], pp. 72–4. Bohr never published this theory (it was untestable as it stood) but 'it looked'—writes Gamow—'as if he would not be greatly surprised if it were true'. Gamow does not date this unpublished theory but it seems that Bohr entertained it in 1928–9 when Gamow was working in Copenhagen.

[3] Cf. the amusing play 'Faust' produced in Bohr's institute in 1932; published by Gamow as an appendix to his [1966].

[4] Cf. Pauli [1958], p. 160.

[5] Bohr [1932]. Ehrenfest too sided firmly with Bohr against the neutrino. Chadwick's discovery of the neutron in 1932 only slightly shook their opposition: they still dreaded the idea of a particle which has neither charge nor, possibly, even (rest) mass, but only 'disembodied' spin.

[6] Wu [1966].

[7] For a fascinating discussion of the open problems presented by the beta-decay and by the nitrogen anomaly, cf. Bohr's Faraday Lecture in 1930, read before, but published after, Pauli's solution (Bohr [1930], especially pp. 380–3).

respectability of Pauli's theory. But on our criteria, the successful prediction of some *novel* fact was needed. This too was provided by Pauli's theory. For Pauli's theory had an interesting observable consequence: if it was right, the β-spectra had to have a clear upper bound. This question was *at the time* undecided, but Ellis and Mott became interested[1] and soon, Ellis's student, Henderson, showed that the experiments supported Pauli's programme.[2] Bohr was not impressed. He knew that if a major programme based on *statistical* conservation of energy ever got going, the growing belt of auxiliary hypotheses would take proper care of the most negative-looking evidence.

Indeed, in these years most leading physicists thought that in nuclear physics the laws of conservation of energy and momentum break down.[3] The reason was stated clearly by Lise Meitner who admitted defeat only in 1933: 'All the attempts to uphold the validity of the law of conservation of energy also for *single* processes demanded a second process [in the beta-decay]. But no such process was found . . .'[4]: that is, the conservation programme for the nucleus showed an empirically degenerating problemshift. There were several ingenious attempts to account for the continuous beta-emission spectrum without assuming a 'thief particle'.[5] These attempts were discussed with great interest,[6] but they were abandoned because they failed to establish a progressive shift.

At this point, Fermi entered on the scene. In 1933–4 he reinterpreted the beta-emission problem in the framework of the research programme of the new quantum theory. Thus he initiated a small new research programme of the neutrino (which later grew into the programme of weak interactions). He calculated some first crude models.[7] Although his theory did not yet predict any new fact, he made it clear that this was only a matter of some further work.

Two years passed and Fermi's promise was still not fulfilled. But the new programme of quantum physics developed fast, at least as far as the non-nuclear phenomena were concerned. Bohr became convinced that some of the basic original ideas of the Bohr–Kramers–Slater programme were now firmly embedded in the new quantum programme and that the

[1] Ellis and Mott [1933].

[2] Henderson [1934].

[3] Mott [1933], p. 823. Heisenberg, in his celebrated [1932], in which he introduced the proton-neutron model of the nucleus, pointed out that 'because of the breakdown of the conservation of energy in the beta-decay one cannot give a unique definition of the binding energy of the electron within the neutron' (p. 164).

[4] Meitner [1933], p. 132.

[5] E.g. Thomson [1929] and Kudar [1929–30].

[6] For a most interesting discussion cf. Rutherford, Chadwick and Ellis [1930], pp. 335–6.

[7] Fermi [1933] and [1934].

new programme solved the intrinsic theoretical problems of the old quantum programme without touching the conservation laws. Therefore Bohr followed Fermi's work with sympathy, and in 1936, in an unusual sequence of events, gave it, by our standards prematurely, public support.

In 1936 Shankland devised a new test of rival theories of photon scattering. His results seemed to support the discarded Bohr–Kramers–Slater theory and undermine the reliability of experiments which, more than a decade earlier, refuted it.[1] Shankland's paper created a sensation. Those physicists who abhorred the new trend were quick to hail Shankland's experiment. Dirac, for instance, immediately welcomed back the 'refuted' Bohr–Kramers–Slater programme, wrote a very sharp article against the 'so-called quantum electrodynamics' and demanded 'a profound alteration in current theoretical ideas, involving a departure from the conservation laws [in order] to get a satisfactory relativistic quantum mechanics'.[2] In the article Dirac suggested again that beta-decay may well turn out to be a piece of crucial evidence against the conservation laws and made fun of the 'new unobservable particle, the neutrino, specially postulated by some investigators in an attempt formally to preserve conservation of energy by assuming the unobservable particle to carry off the balance'.[3] Immediately afterwards Peierls joined the discussion. Peierls suggested that Shankland's experiment may turn out to refute even the statistical conservation of energy. He added: 'That, too, seems satisfactory, once detailed conservation has been abandoned.'[4]

In Bohr's Copenhagen institute, Shankland's experiments were immediately repeated and discarded. Jacobsen, a colleague of Bohr reported this in a letter to *Nature*. Jacobsen's results were accompanied by a letter from Bohr himself, who firmly came out against the rebels, and in defence of Heisenberg's new quantum programme. In particular, he came out in defence of the neutrino against Dirac: 'It may be remarked that the grounds for serious doubts as regards the strict validity of the conservation laws in the problem of the emission of β-rays from atomic nuclei are now largely removed by the suggestive agreement between the rapidly increasing experimental evidence regarding β-ray phenomena and the consequences of the neutrino hypotheses of Pauli so remarkably developed in Fermi's theory.'[5]

Fermi's theory, in its first versions, had no striking empirical success. Indeed, even the available data, especially in the case of *RaE*, on which beta emission research then centred, sharply contradicted Fermi's 1933–4 theory. He wanted to deal with these in the second part of his paper which,

[1] Shankland [1936]. [2] Dirac [1936]. [3] Dirac [1936].
[4] Peierls [1936]. [5] Bohr [1936].

however, was never published. Even if one construes Fermi's 1933–4 theory as a first version of a flexible programme, by 1936 one could not possibly detect any serious sign of a progressive shift.[1] But Bohr wanted to put his *authority* behind Fermi's daring application of Heisenberg's new big programme to the nucleus; and since Shankland's experiment and Dirac's and Peierls's attack brought the beta-decay into the focus of the criticism of the new big programme, he over-praised Fermi's neutrino programme which promised to fill in a sensitive gap. No doubt, the later development spared Bohr from a dramatic humilation: the programmes based on conservation principles progressed, while no progress was made in the rival camp.[2]

The moral of this story is again that the status of an experiment as 'crucial' depends on the status of the theoretical competition in which it is embedded. As the fortunes of the competing camps wax or wane, the interpretation and appraisal of the experiment may change.

Our scientific folklore however is impregnated with theories of instant rationality. The story which I described is falsified in most accounts and reconstructed in terms of some wrong theory of rationality. Even the very best popular expositions teem with such falsifications. Let me mention two examples.

In one paper we learn this about beta-decay: 'When this situation was faced for the first time, the alternatives seemed grim. Physicists *either* had to accept a breakdown of the law of energy conservation, *or* they had to suppose the existence of a new and unseen particle. Such a particle, emitted along with the proton and the electron in the disintegration of the neutron, could save the central pillar of physics by carrying off the missing energy. This was in the early 1930s, when the introduction of a new particle was

[1] Several physicists between 1933 and 1936 offered alternatives or proposed *ad hoc* changes of Fermi's theory; cf. e.g. Becke and Sitte [1933], Bethe and Peierls [1934], Konopinski and Uhlenbeck [1934]. Wu and Moszkowski write in 1966 that 'the Fermi theory [i.e. programme] of β-decay is *now* known to predict with remarkable accuracy both the relation between the rate of β-decay and the energy of disintegration, and also the shape of β-spectra'. But they stress that 'at the very beginning the Fermi theory unfortunately met an unfair test. Until the time when artificial radioactive nuclei could be copiously produced, RaE was the only candidate that beautifully fulfilled many experimental requirements as a β source for the investigation of its spectrum shape. How could we have known then that the β spectrum of RaE would turn out to be only a very special case, one whose spectrum has, in fact, been understood only very recently. Its peculiar energy dependence defied what was expected of the simple Fermi theory of β decay and greatly slackened the pace of the theory's [i.e. programme's] initial progress' (Wu and Moszkowski [1966], p. 6).

[2] It is very doubtful whether Fermi's neutrino programme was progressive or degenerating even between 1936 and 1950; and after 1950 the verdict is still not crystal clear. But this I shall try to discuss in some other occasion. (Incidentally, Schrödinger stood up for the statistical interpretation of the conservation principles in spite of his crucial role in the development of new quantum physics; cf. his [1958].)

not the casual matter it is today. Nevertheless, *after only the briefest vacillation*, physicists chose the second alternative.'[1] Of course, even the *discussed* alternatives were many more than two and the 'vacillation' was certainly not 'the briefest'.

In a well-known textbook of philosophy of science we learn that (1) 'the law (or principle) of the conservation of energy was seriously challenged by experiments on beta-ray decay whose outcome could not be denied'; that (2) 'nevertheless, the law was not abandoned, and the existence of a new kind of entity (called a "neutrino") was assumed in order to bring the law into concordance with experimental data'; and that (3) 'the rationale for this assumption is that the rejection of the conservation law would deprive a large part of our physical knowledge of its systematic coherence'.[2] But all the three points are wrong. (1) is wrong because no law can be 'seriously challenged' by experiments only; (2) is wrong because new *scientific* hypotheses are assumed not simply in order to patch up gaps between data and theory but in order to predict novel facts; and (3) is wrong because at the time it seemed that *only* the rejection of the conservation law would secure the 'systematic coherence' of our physical knowledge.

(d 4) Conclusion. The requirement of continuous growth.

There are no such things as crucial experiments, at least not if these are meant to be experiments which can *instantly* overthrow a research programme. In fact, when one research programme suffers defeat and is superseded by another one, we may—*with long hindsight*—call an experiment crucial if it turns out to have provided a spectacular corroborating instance for the victorious programme and a failure for the defeated one (in the sense that it was never 'explained progressively'—or, briefly, 'explained'[3]—within the defeated programme). But scientists, of course, do not always judge heuristic situations correctly. A rash scientist may *claim* that his experiment defeated a programme, and parts of the scientific community may even, rashly, accept his claim. But if a scientist in the 'defeated' camp puts forward a few years later a scientific explanation of the allegedly 'crucial experiment' within (or consistent with) the allegedly defeated programme, *the honorific title may be withdrawn and the 'crucial experiment' may turn from a defeat into a new victory for the programme.*

Examples abound. There were many experiments in the eighteenth century which were, as a matter of historico-sociological fact, widely accepted as 'crucial' evidence against Galileo's law of free fall, and Newton's theory of gravitation. In the nineteenth century there were several 'crucial

[1] Treiman [1959]; my italics. [2] Nagel [1961], pp. 65–6.

[3] Cf. *above*, p. 119, footnote 1.

experiments' based on measurements of light velocity which 'disproved' the corpuscular theory and which turned out later to be erroneous in the light of relativity theory. These 'crucial experiments' were later deleted from the justificationist textbooks as manifestations of shameful short-sightedness or even of envy. (Recently they reappeared in some new textbooks, this time to illustrate the inescapable irrationality of scientific fashions.) However, in those cases in which ostensibly 'crucial experiments' were indeed *later* borne out by the defeat of the programme, historians charged those who resisted them with stupidity, jealousy, or unjustified adulation of the father of the research programme in question. (Fashionable 'sociologists of knowledge'—or 'psychologists of knowledge'—tend to explain positions in purely social or psychological terms when, as a matter of fact, they are determined by rationality principles. A typical example is the explanation of Einstein's opposition to Bohr's complementarity principle on the ground that 'in 1926 Einstein was forty-seven years old. Forty-seven may be the prime of life, but not for physicists'.[1])

In the light of this paper, the utopian idea of instant rationality becomes a hallmark of most brands of epistemology. Justificationists wanted scientific theories to be proved even before they were published; probabilists hoped a machine could flash up instantly the value (degree of confirmation) of a theory, given the evidence; naive falsificationists hoped that elimination at least was the instant result of the verdict of *experiment*.[2] I hope I have shown that *all these theories of instant rationality—and instant learning—fail*. The case studies of this section show that rationality works much slower than most people tend to think, and, even then, fallibly. Minerva's owl flies at dusk. I also hope I have shown that the *continuity* in science, the *tenacity* of some theories, the rationality of a certain

[1] Bernstein [1961], p. 129. In order to appraise progressive and degenerating elements in rival problemshifts one must understand the *ideas* involved. But the sociology of knowledge frequently serves as a successful cover for illiteracy: most sociologists of knowledge do not understand—or even care for—the ideas; they watch the socio-psychological patterns of behaviour. Popper used to tell a story about a 'social psychologist', Dr. X, studying scientists' group behaviour. He went into a physics seminar to study the psychology of science. He observed the 'emergence of a leader', the 'rallying round effect' in some and the 'defence-reaction' in others, the correlation between age, sex and aggressive behaviour, etc. (Dr. X claimed to have used some sophisticated small-sample techniques of modern statistics.) At the end of the enthusiastic account Popper asked Dr. X: 'What was the *problem* the group was discussing?' Dr. X was surprised: 'Why do you ask? I did not listen to the *words*! Anyway, what has *that* to do with the psychology of knowledge?'

[2] Of course, naive falsificationists may take some time to reach the 'verdict of experiment': the experiment has to be repeated and critically considered. But once the discussion ends up in an agreement among the experts, and thus a 'basic statement' becomes 'accepted', and it has been decided which specific theory was hit by it, the naive falsificationist will have little patience with those who still 'prevaricate'.

amount of dogmatism, can only be explained if we construe science as a battleground of research programmes rather than of isolated theories. One can understand very little of the growth of science when our paradigm of a chunk of scientific knowledge is an isolated theory like 'All swans are white', standing aloof, without being embedded in a major research programme. *My account implies a new criterion of demarcation between 'mature science', consisting of research programmes, and 'immature science', consisting of a mere patched up pattern of trial and error.*[1] For instance, we may have a conjecture, have it refuted and then rescued by an auxiliary hypothesis which is not *ad hoc* in the senses which we had earlier discussed. It may predict novel facts some of which may even be corroborated.[2] Yet one may achieve such 'progress' with a patched up, arbitrary series of disconnected theories. Good scientists will not find such makeshift progress satisfactory; they may even reject it as not genuinely scientific. They will call such auxiliary hypotheses merely 'formal', 'arbitrary', 'empirical', 'semi-empirical', or even '*ad hoc*'.[3]

Mature science consists of research programmes in which not only novel facts but, in an important sense, also novel auxiliary theories, are anticipated; mature science—unlike pedestrian trial-and-error—has 'heuristic power'. Let us remember that in the positive heuristic of a powerful programme there is, right at the start, a general outline of how to build the protective belts: this heuristic power generates *the autonomy of theoretical science.*[4]

This *requirement of continuous growth* is my rational reconstruction of the widely acknowledged requirement of 'unity' or 'beauty' of science. It highlights the weakness of *two*—apparently very different—types of theorizing. First, it shows up the weakness of programmes which, like Marxism or Freudism, are, no doubt, 'unified', which give a major sketch of the sort of auxiliary theories they are going to use in absorbing anomalies, but which

[1] The elaboration of this demarcation in the two following paragraphs was improved in the press, following invaluable discussions with Paul Meehl in Minneapolis in 1969.

[2] Earlier I distinguished, following Popper, two criteria of *adhocness*. I called *ad hoc*$_1$ theories which had no excess content over their predecessors (or competitors) that is, which did not predict any *novel* facts; I called *ad hoc*$_2$ theories which predicted novel facts but completely failed: none of their excess content got corroborated (cf. *above*, p. 124, footnote 3, and p. 125, footnote 1).

[3] Planck's radiation formula—given in his [1900*a*]—is a good example: cf. *above*, p. 167, footnote 2. We may call such hypotheses which are not *ad hoc*$_1$, not *ad hoc*$_2$, but still unsatisfactory in the sense specified in the text, *ad hoc*$_3$. These three—unfailingly pejorative—usages of *ad hoc* may provide a satisfactory entry in the *Oxford English Dictionary*.

It is intriguing to note that 'empirical' and 'formal' are both used as synonyms for our *ad hoc*$_3$.

Meehl, in his brilliant [1967], reports that in contemporary psychology—especially in social psychology—many alleged 'research programmes' in fact consist of chains of such *ad hoc*$_3$ stratagems.

[4] Cf. *above*, p. 137.

unfailingly devise their actual auxiliary theories in the wake of facts without, at the same time, anticipating others. (What *novel* fact has Marxism *predicted* since, say, 1917?) Secondly, it hits patched-up, unimaginative series of pedestrian 'empirical' adjustments which are so frequent, for instance, in modern social psychology. Such adjustments may, with the help of so-called 'statistical techniques', make some 'novel' predictions and may even conjure up some irrelevant grains of truth in them. But this theorizing has no unifying idea, no heuristic power, no continuity. They do not add up to a genuine research programme and are, on the whole, worthless.[1]

My account of scientific rationality, although based on Popper's, leads away from some of his general ideas. I endorse to some extent both Le Roy's conventionalism with regard to theories and Popper's conventionalism with regard to basic propositions. In this view scientists (and as I have shown, mathematicians too[2]) are not irrational when they tend to ignore counterexamples or as they prefer to call them, 'recalcitrant' or 'residual' instances, and follow the sequence of problems as prescribed by the positive heuristic of their programme, and elaborate—and apply—their theories regardless.[3] Contrary to Popper's falsificationist morality, scientists frequently and *rationally* claim 'that the experimental results are not reliable, or that the discrepancies which are asserted to exist between the experimental results and the theory are only apparent and

[1] After reading Meehl [1967] and Lykken [1968] one wonders whether the function of statistical techniques in the social sciences is not primarily to provide a machinery for producing phoney corroborations and thereby a semblance of 'scientific progress' where, in fact, there is nothing but an increase in pseudo-intellectual garbage. Meehl writes that 'in the physical sciences, the usual result of an improvement in experimental design, instrumentation, or numerical mass of data, is to increase the difficulty of the "observational hurdle" which the physical theory of interest must successfully surmount; whereas, in psychology and some of the allied behaviour sciences, the usual effect of such improvement in experimental precision is to provide an easier hurdle for the theory to surmount'. Or, as Lykken put it: 'Statistical significance [in psychology] is perhaps the least important attribute of a good experiment; it is never a sufficient condition for claiming that a theory has been usefully corroborated, that a meaningful empirical fact has been established, or that an experimental report ought to be published.' It seems to me that most theorizing condemned by Meehl and Lykken may be *ad hoc*$_3$. Thus the methodology of research programmes might help us in devising laws for stemming this intellectual pollution which may destroy our cultural environment even earlier than industrial and traffic pollution destroys our physical environment.

[2] Cf. my [1963–4].

[3] Thus the *methodological* asymmetry between universal and singular statements vanishes. We may adopt either by convention: in the 'hard core' we decide to 'accept' universal, in the 'empirical basis' singular, statements. The *logical* asymmetry between universal and singular statements is fatal only for the dogmatic inductivist who wants to learn only from hard experience and logic. The conventionalist can, of course, 'accept' this *logical* asymmetry: he does not have to be (although he *may* be) also an inductivist. He 'accepts' some universal statements, but not because he claims to deduce (or induce) them from singular ones.

that they will disappear with the advance of our understanding'.[1] When doing so, they may *not* be 'adopting the very reverse of that critical attitude which . . . is the proper one for the scientist'.[2] Indeed, Popper is right in stressing that 'the dogmatic attitude of sticking to a theory as long as possible is of considerable significance. Without it we could never find out what is in a theory—we should give the theory up before we had a real opportunity of finding out its strength; and in consequence no theory would ever be able to play its role of bringing order into the world, of preparing us for future events, of drawing our attention to events we should otherwise never observe'.[3] Thus the 'dogmatism' of 'normal science' does not prevent growth as long as we combine it with the Popperian recognition that there is good, progressive normal science and that there is bad, degenerating normal science, and as long as we retain the *determination* to eliminate, under certain objectively defined conditions, some research programmes.

The dogmatic attitude in science—which would explain its stable periods—was described by Kuhn as a prime feature of 'normal science'.[4] But Kuhn's conceptual framework for dealing with continuity in science is socio-psychological: mine is normative. I look at continuity in science through 'Popperian spectacles'. Where Kuhn sees 'paradigms', I *also* see rational 'research programmes'.

4. THE POPPERIAN VERSUS THE KUHNIAN RESEARCH PROGRAMME

Let us now sum up the Kuhn–Popper controversy.

We have shown that Kuhn is right in objecting to naive falsificationism, and also in stressing the *continuity* of scientific growth, the *tenacity* of some scientific theories. But Kuhn is wrong in thinking that by discarding naive falsificationism he has discarded thereby all brands of falsificationism. Kuhn objects to the entire Popperian research programme, and he excludes *any* possibility of a rational reconstruction of the growth of science. In a

[1] Popper [1934], section 9. [2] *Ibid.*

[3] Popper [1940], first footnote. We find a similar remark in his [1963], p. 49. But these remarks are in *prima facie* contradiction with some of his remarks in [1934] (quoted *above*, p. 111), and therefore may only be interpreted as signs of a growing awareness by Popper of an undigested anomaly in his own research programme.

[4] Indeed, my demarcation criterion between mature and immature science can be interpreted as a Popperian absorption of Kuhn's idea of 'normality' as a hallmark of [mature] science; and it also improved on our earlier argument against regarding such highly falsifiable statements as scientific. (Cf. *above*. p. 102.)

Incidentally, this demarcation between mature and immature science appears already in my [1961] and [1963–4], where I called the former 'deductive guessing' and the latter 'naive trial and error'. (See e.g. [1963–4], section 7(c): 'Deductive guessing versus naive guessing.')

succint comparison of Hume, Carnap and Popper, Watkins points out that the growth of science is inductive and irrational according to Hume, inductive and rational according to Carnap, non-inductive and rational according to Popper.[1] But Watkins's comparison can be extended by adding that it is non-inductive and irrational according to Kuhn. *In Kuhn's view there can be no logic, but only psychology of discovery.*[2] For instance, in Kuhn's conception, anomalies, inconsistencies *always* abound in science, but in 'normal' periods the dominant paradigm secures a pattern of growth which is eventually overthrown by a 'crisis'. There is no particular rational cause for the appearance of a Kuhnian 'crisis'. 'Crisis' is a psychological concept; it is a contagious panic. Then a new 'paradigm' emerges, incommensurable with its predecessor. There are no rational standards for their comparison. Each paradigm contains its own standards. The crisis sweeps away not only the old theories and rules but also the standards which made us respect them. The new paradigm brings a totally new rationality. There are no super-paradigmatic standards. The change is a bandwagon effect. Thus *in Kuhn's view scientific revolution is irrational, a matter for mob psychology*.

The reduction of philosophy of science to psychology of science did not start with Kuhn. An earlier wave of 'psychologism' followed the breakdown of justificationism. For many, justificationism represented the only possible form of rationality: the end of justificationism meant the end of rationality. The collapse of the thesis that scientific theories are provable, that the progress of science is cumulative, made justificationists panic. If 'to discover is to prove', but nothing is provable, then there can be no discoveries, only discovery-claims. Thus disappointed justificationists—ex-justificationists—thought that the elaboration of rational standards was a hopeless enterprise and that all one can do is to study—and imitate—the Scientific Mind, as it is exemplified in famous scientists. After the collapse of Newtonian physics, Popper elaborated new, non-justificationist critical standards. Now some of those who had already learned of the collapse of justificationist rationality now learned, mostly by hearsay, of Popper's colourful slogans which suggested naive falsificationism. Finding them untenable, they identified the collapse of naive falsificationism with the end of rationality itself. The elaboration of rational standards was again regarded as a hopeless enterprise; the best one can do is to study, they thought once again, the Scientific Mind.[3] Critical philosophy was to be replaced by what Polanyi called a 'post-critical' philosophy. But the Kuhnian research

[1] Watkins [1968], p. 281.

[2] Kuhn [1965]. But this position is already implicit in his [1962].

[3] Incidentally, just as some earlier ex-justificationists led the wave of sceptical irrationalism, so now some ex-falsificationists lead the *new* wave of sceptical irrationalism and anarchism. This is best exemplified in Feyerabend [1970].

programme contains a new feature: we have to study not the mind of the individual scientist but the mind of the Scientific Community. Individual psychology is now replaced by social psychology; imitation of the great scientists by submission to the collective wisdom of the community.

But Kuhn overlooked Popper's sophisticated falsificationism and the research programme he initiated. Popper replaced the central problem of classical rationality, *the old problem of foundations*, with *the new problem of fallible-critical growth*, and started to elaborate objective standards of this growth. In this paper I have tried to develop his programme a step further. I think this small development is sufficient to escape Kuhn's strictures.[1]

The reconstruction of scientific progress as proliferation of rival research programmes and progressive and degenerative problemshifts gives a picture of the scientific enterprise which is in many ways different from the picture provided by its reconstruction as a succession of bold theories and their dramatic overthrows. Its main aspects were developed from Popper's ideas and, in particular, from his ban on 'conventionalist', that is, content-decreasing, stratagems. The main difference from Popper's original version is, I think, that in my conception criticism does not—and must not—kill as fast as Popper imagined. *Purely negative, destructive criticism, like 'refutation' or demonstration of an inconsistency does not eliminate a programme. Criticism of a programme is a long and often frustrating process and one must treat budding programmes leniently.*[2] One may, of course, show up the degeneration of a research programme, but it is only *constructive criticism* which, with the help of rival research programmes, can achieve real successes; and dramatic spectacular results become visible only with hindsight and rational reconstruction.

Kuhn certainly showed that the psychology of science can reveal important and, indeed, sad truths. But the psychology of science is not autonomous; for *the—rationally reconstructed—growth of science takes place*

[1] Indeed, as I had already mentioned, *my concept of a 'research programme' may be construed as an objective, 'third world' reconstruction of Kuhn's socio-psychological concept of paradigm'*: thus the Kuhnian 'Gestalt-switch' can be performed without removing one's Popperian spectacles.

(I have not dealt with Kuhn's and Feyerabend's claim that theories cannot be eliminated on any *objective* grounds because of the 'incommensurability' of rival theories. Incommensurable theories are neither inconsistent with each other, nor comparable for content. But we can *make* them, by a dictionary, inconsistent and their content comparable. If we want to eliminate a programme, we need some methodological determination. This determination is the heart of methodological falsificationism; for instance, no result of statistical sampling is ever inconsistent with a statistical theory unless we *make them* inconsistent with the help of Popperian rejection rules, cf. *above*, p. 109.)

[2] The reluctance of economists and other social scientists to accept Popper's methodology may have been partly due to the destructive effect of naive falsificationism on budding research programmes.

essentially in the world of ideas, in Plato's and Popper's 'third world', in the world of articulated knowledge which is independent of knowing subjects.[1] *Popper's research programme* aims at a description of this objective scientific *growth*.[2] *Kuhn's research programme* seems to aim at a description of *change* in the ('normal') scientific mind (whether individual or communal).[3] But the mirror-image of the third world in the mind of the individual—even in the mind of the 'normal'—scientists is usually a caricature of the original; and to describe this caricature without relating it to the third-world original might well result in a caricature of a caricature. One cannot understand the history of science without taking into account the interaction of the three worlds.

APPENDIX

POPPER, FALSIFICATIONISM AND THE 'DUHEM-QUINE THESIS'

Popper began as a dogmatic falsificationist in the 1920s; but he soon realized the untenability of this position and published nothing before he invented *methodological falsificationism*. This was an entirely new idea in the philosophy of science and it clearly originates with Popper, who put it forward as a solution to the difficulties of dogmatic falsificationism. Indeed, the conflict between the theses that science is both critical and

[1] The *first* world is the material world, the *second* is the world of consciousness, the *third* is the world of propositions, truth, standards: the world of objective knowledge. The modern *loci classici* on this subject are Popper [1968*a*] and Popper [1968*b*]; also, cf. Toulmin's impressive programme set out in his [1967]. It should be mentioned here that many passages of Popper [1934] and even of [1963] sound like descriptions of a psychological contrast between the Critical Mind and the Inductivist Mind. But Popper's psychologistic terms can be, to a large extent, reinterpreted in third-world terms. Cf. Musgrave [1971].

[2] In fact, Popper's research programme extends beyond science. The concepts of 'progressive' and 'degenerating' problemshifts, the idea of proliferation of theories can be generalized to any sort of rational discussion and thus serve as tools for a general theory of criticism; cf. my [1970].

[3] *Actual* state of minds, beliefs, etc., belong to the second world; states of the *normal* mind belong to a limbo between the second and third. The study of actual scientific minds belongs to *psychology*; the study of the 'normal' (or 'healthy' etc.) mind belongs to a *psychologistic philosophy of science*. There are *two kinds of psychologistic philosophies of science*. According to one kind there can be no philosophy of science: only a psychology of individual scientists. According to the other kind there is a psychology of the 'scientific', 'ideal' or 'normal' mind: this turns philosophy of science into a psychology of this ideal mind and, in addition, offers a psychotherapy for turning one's mind into an ideal one. I discuss this second kind of psychologism in detail in my [1970]. Kuhn does not seem to have noticed this distinction.

fallible is one of the central problems in Popperian philosophy. While Popper offered a coherent formulation and criticism of dogmatic falsificationism, he never made a sharp distinction between naive and sophisticated falsificationism. In an earlier paper,[1] I distinguished three Poppers: *Popper*$_0$, *Popper*$_1$ and *Popper*$_2$. Popper$_0$ is the dogmatic falsificationist who never published a word: he was invented—and 'criticized'—first by Ayer and then by many others.[2] This paper will, I hope, finally kill this ghost. Popper$_1$ is the naive falsificationist, Popper$_2$ the sophisticated falsificationist. The *real* Popper developed from dogmatic to a naive version of methodological falsificationism in the twenties; he arrived at the '*acceptance rules*' *of sophisticated falsificationism* in the fifties. The transition was marked by his adding to the original requirement of testability the 'second' requirement of 'independent testability',[3] and then the 'third' requirement that some of these independent tests should result in corroborations.[4] But the real Popper never abandoned his earlier (naive) *falsification rules*. He has demanded, until this day, that '*criteria of refutation* have to be laid down beforehand: it must be agreed, which observable situations, if actually observed, mean that the theory is refuted'.[5] He still construes 'falsification' as the result of a duel between theory and observation, without another, better theory *necessarily* being involved. The real Popper has never explained in detail the appeal procedure by which some 'accepted basic statements' may be eliminated. Thus the real Popper consists of Popper$_1$ together with some elements of Popper$_2$.

The idea of a demarcation between progressive and degenerating problemshifts, as discussed in this paper, is based on Popper's work:

[1] Cf. my [1968*b*].

[2] Ayer seems to have been the first to attribute dogmatic falsificationism to Popper. (Ayer also invented the myth that according to Popper 'definite confutability' was a criterion not only of the empirical but also of the meaningful character of a proposition: cf. his [1936], chapter 1, p. 38 of the second edition.) Even today, many philosophers (cf. Juhos [1966] or Nagel [1967]) criticize the strawman Popper$_0$. Medawar, in his [1967], called *dogmatic* falsificationism 'one of the strongest ideas' in Popper's methodology. Nagel, reviewing Medawar's book, criticized Medawar for 'endorsing' what he too believes to be 'Popper's claims' (Nagel [1967], p. 70). Nagel's criticism convinced Medawar that 'the act of falsification is not immune to human error' (Medawar [1969], p. 54). But Medawar and Nagel misread Popper: his *Logik der Forschung* is the strongest ever criticism of dogmatic falsificationism.

One may take a charitable view of Medawar's mistake: for brilliant scientists whose speculative talent was thwarted under the tyranny of an inductivist logic of discovery, falsificationism, even in its dogmatic form, was bound to have a tremendous liberating effect. (Besides Medawar, another Nobel Prize winner, Eccles, learned from Popper to replace his original caution by bold falsifiable speculation: cf. his [1964], pp. 274–5.)

[3] Popper [1957].

[4] Popper [1963], pp. 242 ff.

[5] Popper [1963], p. 38, footnote 3.

indeed this demarcation is almost identical with his celebrated demarcation criterion between science and metaphysics.[1]

Popper originally had only the *theoretical* aspect of problemshifts in mind, which is hinted at in section 20 of his [1934] and developed in his [1957].[2] He added a discussion of the *empirical* aspect of problemshifts only later, in his [1963].[3] However, Popper's ban on 'conventionalist stratagems' is in some respects too strong, in others too weak. It is too *strong*, for, according to Popper, a new version of a progressive programme *never* adopts a content-decreasing stratagem to absorb an anomaly, it *never* says things like 'all bodies are Newtonian except for seventeen anomalous ones'. But since unexplained anomalies always abound, I allow such formulations; an explanation is a step forward (that is, 'scientific') if it explains at least *some* previous anomalies which were not explained 'scientifically' by its predecessor. As long as anomalies are regarded as genuine (though not necessarily urgent) problems, it does not matter much whether we dramatize them as 'refutations' or de-dramatize them as 'exceptions': the difference *then* is only a linguistic one. (This degree of tolerance of *ad hoc* stratagems allows us to progress even on inconsistent foundations. Problemshifts may then be progressive in spite of inconsistencies.[4]) However, Popper's ban on content-decreasing stratagems is also too *weak*: it cannot deal, for instance, with the 'tacking paradox',[5] and does not ban *ad hoc*$_3$ stratagems.[6] These can be eliminated only by the requirement that *the auxiliary hypotheses should be formed in accordance with the positive heuristic of a genuine research programme.* This new requirement brings us to the problem of *continuity in science.*

[1] If the reader is in doubt about the authenticity of my reformulation of Popper's demarcation criterion, he should re-read the relevant parts of Popper [1934] with Musgrave [1968] as a guide. Musgrave wrote his [1968] against Bartley who, in his [1968], mistakenly attributed to Popper the demarcation criterion of naive falsificationism, as formulated *above*, p. 109.

[2] In his [1934], Popper was primarily concerned with a ban on *surreptitious ad hoc* adjustments. Popper (Popper$_1$) demands that the design of a potentially negative crucial experiment must be presented together with the theory, and then the verdict of the experimental jury humbly accepted. It follows that conventionalist stratagems, which *after* the verdict give a retrospective twist to the original theory in order to escape the verdict, are *eo ipso* ruled out. But if we admit the refutation and *then* reformulate the theory with the help of an *ad hoc* stratagem, we may admit it as a '*new*' theory; and if it is testable, then Popper$_1$ accepts it for new criticism: 'Whenever we find that a system has been rescued by a conventionalist stratagem, we shall test it afresh, and reject it, as circumstances may require' (Popper [1934], section 20).

[3] For details, cf. my [1968*a*], especially pp. 388–90.

[4] Cf. *above*, pp. 142 ff. This tolerance is rarely, if ever, found in textbooks of scientifiic method.

[5] Cf. *above*, p. 131.

[6] Cf. *above*, p. 175, footnote 3.

The problem of *continuity* in science was raised by Popper and his followers long ago. When I proposed my theory of growth based on the idea of competing research programmes, I again followed, and tried to improve, Popperian tradition. Popper himself, in his [1934], had already stressed the heuristic importance of 'influential metaphysics',[1] and was regarded by some members of the Vienna Circle as a champion of dangerous metaphysics.[2] When his interest in the role of metaphysics revived in the 1950s, he wrote a most interesting 'Metaphysical Epilogue' about 'metaphysical research programmes' to his *Postscript: After Twenty Years*—in galleys since 1957.[3] But Popper associated tenacity not with *methodological irrefutability* but rather with *syntactical irrefutability*. By 'metaphysics' he meant syntactically specifiable statements like 'all-some' statements and purely existential statements. No basic statements could conflict with them because of their logical form. For instance, 'for all metals there is a solvent' would, in this sense, be 'metaphysical', while Newton's theory of gravitation, taken in isolation, would not be.[4] Popper, in the 1950s, also raised the problem of how to criticize metaphysical theories and suggested

[1] Cf. e.g. his [1934], end of section 4; also cf. his [1968*c*], p. 93. One should remember that such importance was denied to metaphysics by Comte and Duhem. The people who did most to reverse the anti-metaphysical tide in the philosophy and the historiography of science were Burtt, Popper and Koyré.

[2] Carnap and Hempel were trying, in their reviews of the book, to defend Popper against this charge (cf. Carnap [1935] and Hempel [1937]). Hempel wrote: '[Popper] stresses strongly certain features of his approach which are common with the approach of somewhat metaphysically oriented thinkers. It is to be hoped that this valuable work will not be misinterpreted as if it meant to allow for a new, perhaps even logically defensible, metaphysics.'

[3] A passage of this *Postscript* is here worth quoting: 'Atomism is an . . . excellent example of a non-testable metaphysical theory whose influence upon science exceeded that of many testable theories . . . The latest and greatest so far was the programme of Faraday, Maxwell, Einstein, de Broglie, and Schrödinger, of conceiving the world . . . in terms of continuous fields . . . Each of these metaphysical theories functioned, long before it became testable, as a programme for science. It indicated the direction in which satisfactory explanatory theories of science may be found, and it made possible something like an appraisal of the depth of a theory. In biology, the theory of evolution, the theory of the cell, and the theory of bacterial infection, have all played similar parts, at least for a time. In psychology, sensualism, atomism (that is, the theory that all experiences are composed of last elements, such as, for example, sense data) and psycho-analysis should be mentioned as metaphysical research programmes . . . Even purely existential assertions have sometimes proved suggestive and even fruitful in the history of science even if they never became part of it. Indeed, few metaphysical theories exerted a greater influence upon the development of science than the purely metaphysical one: "There exists a substance which can turn base metals into gold (that is, a philosopher's stone)", although it is non-falsifiable, was never verified, and is now believed by nobody.'

[4] Cf. especially Popper [1934], section 66. In the 1959 edition he added a clarifying footnote (footnote *2) in order to stress that in *metaphysical* "all-some' statements the existential quantifier must be interpreted as 'unbounded'; but, of course, he had made this absolutely clear already in section 15 of the original text.

solutions.[1] Agassi and Watkins published several interesting papers on the role of this sort of 'metaphysics' in science, which all connected 'metaphysics' with the continuity of scientific progress.[2] My treatment differs from theirs first because I go much further than they in blurring the demarcation between [Popper's] 'science' and [Popper's] 'metaphysics': I do not even use the term 'metaphysical' any more. I only talk about *scientific* research programmes whose hard core is irrefutable, not because of syntactical but because of methodological reasons which have nothing to do with logical form. Secondly, separating sharply the *descriptive problem* of the psychologico-historical role of metaphysics from the *normative problem* of how to distinguish progressive from degenerating research programmes, I elaborate the latter problem further than they had done.

Finally, I should like to discuss the '*Duhem-Quine thesis*', and its relation to falsificationism.[3]

According to the 'Duhem-Quine thesis', given sufficient imagination, any theory (whether consisting of one proposition or of a finite conjunction of many) can be permanently saved from 'refutation' by some suitable adjustment in the background knowledge in which it is embedded. As Quine put it: 'Any statement can be held true come what may, if we make drastic enough adjustments elsewhere in the system . . . Conversely, by the same token, no statement is immune to revision.'[4] Moreover, the 'system' is nothing less than 'the whole of science'. 'A recalcitrant experience can be accommodated by any of various alternative reëvaluations in various alternative quarters of the total system [including the possibility of reëvaluating the recalcitrant experience itself].'[5]

This thesis has two very different interpretations. In its *weak interpretation* it only asserts the impossibility of a direct experimental hit on a narrowly specified theoretical target and the logical possibility of shaping science in indefinitely many different ways. The weak interpretation hits only dogmatic, not methodological, falsificationism: it only denies the possibility of a *disproof* of any *separate* component of a theoretical system.

In its *strong interpretation* the Duhem–Quine thesis excludes any *rational* selection rule among the alternatives; this version is inconsistent with all forms of methodological falsificationism. The two interpretations have not been clearly separated, although the difference is methodologically vital. Duhem seems to have held only the weak interpretation: for

[1] Cf. especially his [1963], pp. 198–9 (first published in 1958).

[2] Cf. Watkins [1957] and [1958] and Agassi [1962] and [1964].

[3] This concluding part of the *Appendix* was added in print.

[4] Quine [1953], chapter ii.

[5] *Ibid.* The clause in the square brackets is mine.

him the selection is a matter of 'sagacity': we must always make the right choices in order to get nearer to 'natural classification'.[1] On the other hand, Quine, in the tradition of the American pragmatism of James and Lewis, seems to hold a position very near to the strong interpretation.[2]

Let us now have a closer look at the weak Duhem–Quine thesis. Let us take a 'recalcitrant experience' expressed in an 'observation statement' O' which is inconsistent with a conjunction of theoretical (and 'observational') statements $h_1, h_2 \ldots h_n, I_1, I_2 \ldots I_n$, where h_i are theories and I_i the corresponding initial conditions. In the 'deductive model', $h_1 \ldots h_n$, $I_1 \ldots I_n$ logically imply O; but O' is observed which implies *not*-O. Let us also assume that the premisses are independent and are all necessary for deducing O.

In this case we may restore consistency by altering *any* of the sentences in our deductive model. For instance, let h_1 be: 'whenever a thread is loaded with a weight exceeding that which characterizes the tensile strength of the thread, then it will break'; let h_2 be: 'the weight characteristic for this thread is 1 *lb*.'; let h_3 be: 'the weight put on this thread was 2 *lbs*'. Let, finally, h_0 be: 'an iron weight of 2 *lbs* was put on the thread located in the space-time position P and it did not break'. One may solve the problem in many ways. To give a few examples: (1) We reject h_1; we replace the expression 'is loaded with a weight' by 'is pulled by a force'; we introduce a new initial condition: there was a hidden magnet (or hitherto unknown force) located in the laboratory ceiling. (2) We reject h_2; we propose that the tensile strength *does* depend on how moist threads are; the tensile strength of the actual thread, since it got moist, was 2 *lbs*. (3) We reject h_3; the weight was only 1 *lb*; the scales went wrong. (4) We reject h_0; the thread did not break; it was only *observed* to break, but the professor who proposed h_1 & h_2 & h_3 was a well-known bourgeois liberal and his revolutionary laboratory assistants consistently *saw* his hypotheses refuted when in fact they were confirmed. (5) We reject h_3; the thread was not a 'thread', but a 'superthread', and 'superthreads' never break.[3] We could go on

[1] An experiment, for Duhem, can never *alone* condemn an isolated theory (such as the hard core of a research programme): for such 'condemnation' we *also* need 'common sense', 'sagacity', and, indeed, good metaphysical instinct which leads us towards (or *to*) 'a certain supremely eminent order'. (See the end of the *Appendix* of the second edition of his [1906].)

[2] Quine speaks of statements having 'varying distances from a sensory periphery', and thus more or less exposed to change. But both the sensory periphery and the metric are hard to define. According to Quine 'the considerations which guide [man] in warping his scientific heritage to fit his continuing sensory peripheries are, where rational, pragmatic' (Quine [1953]). But 'pragmatism' for Quine, as for James or LeRoy, is only psychological comfort; and I find it irrational to call this 'rational'.

[3] For such 'concept-narrowing defences' and 'concept-stretching refutations', cf. my [1963–64].

indefinitely. Indeed, there are infinitely many possibilities of how to replace—given sufficient imagination—any of the premisses (*in the deductive model*) by invoking a change in some *distant* part of our total knowledge (*outside the deductive model*) and thereby restore consistency.

Can we formulate this trivial observation by saying that '*each test is a challenge to the whole of our knowledge*'? I do not see any reason why not. The resistance of some falsificationists to this 'holistic dogma of the "global" character of all tests'[1] is due only to a semantic conflation of two different notions of 'test' (or 'challenge') which a recalcitrant experimental result presents to our knowledge.

The Popperian interpretation of a 'test' (or 'challenge') is that the result (O) contradicts ('challenges') a finite, well-specified conjunction of premisses (T): O & T cannot be true. But no protagonist of the Duhem–Quine argument would deny this point.

The Quinean interpretation of 'test' (or 'challenge') is that the *replacement* of O & T may invoke some change also outside O and T. The successor to O & T may be inconsistent with some H in some distant part of knowledge. But no Popperian would deny this point.

The conflation of the two notions of testing led to some misunderstandings and logical blunders. Some people felt intuitively that the *modus tollens* from refutation may 'hit' very distant premisses in our total knowledge and therefore were trapped in the idea that the '*ceteris paribus* clause' is a premiss which is joined *conjunctively* with the obvious premisses. But this 'hit' is achieved not by *modus tollens* but as a result of our subsequent replacement of our original deductive model.[2]

Thus 'Quine's weak thesis' trivially holds. But 'Quine's strong thesis' will be strenuously opposed, both by the naive and the sophisticated falsificationist.

The naive falsificationist insists that if we have an inconsistent set of scientific statements, we first must select from among them (1) a theory under test (to serve as a *nut*); then we must select (2) an accepted basic statement (to serve as a *hammer*) and the rest will be uncontested background knowledge (to provide an *anvil*). And in order to put teeth into this position, we must offer a method of 'hardening' the 'hammer' and the 'anvil' in order to enable us to crack the 'nut', and thus perform a 'negative

[1] Popper [1963], chapter 10, section xvi.

[2] The *locus classicus* of this confusion is Canfield's and Lehrer's wrongheaded criticism of Popper in their [1961]; Stegmüller followed them into the logical morass ([1966], p. 7). Coffa contributed to the clarification of the issue ([1968]).

Unfortunately, my own phraseology in this paper in places suggests that the '*ceteris paribus* clause' is an independent premiss in the theory under test. My attention was drawn to this easily repairable defect by Colin Howson.

crucial experiment'. But naive 'guessing' of this division is too arbitrary, it does not give us any serious hardening. (Grünbaum, on the other hand, swallowing his falsificationist pride, stoops down to accept help from the inductivist Salmon. He now applies Salmon's Reichenbachian theory of probability of hypotheses in order to show that, at least in some sense, the 'hammer' and the 'anvil' have high posterior probabilities and therefore are 'hard' enough for being used as a nutcracker.[1])

The sophisticated falsificationist allows *any* part of the body of science to be replaced *but* only on the condition that it is replaced in a 'progressive' way, so that the replacement successfully anticipates novel facts. In his rational reconstruction of falsification 'negative crucial experiments' play no role. He sees nothing wrong with a group of brilliant scientists conspiring to pack everything they can into their favourite research programme ('conceptual framework', if you wish) with a sacred hard core. As long as their genius—and luck—enables them to expand their programme '*progressively*', while sticking to its hard core, they are allowed to do it. And if a genius comes determined to *replace* ('progressively') a most uncontested and corroborated theory which he happens to dislike on philosophical, aesthetic or personal grounds, good luck to him. If two teams, pursuing rival research programmes, compete, the one with more creative talent is likely to succeed—unless God punishes them with an extreme lack of empirical success. The direction of science is determined primarily by human creative imagination and not by the universe of facts which surrounds us. Creative imagination is likely to find corroborating novel evidence even for the most 'absurd' programme, if the search has sufficient drive.[2] This look-out for *new confirming evidence* is perfectly permissible. Scientists dream up phantasies and then pursue a highly selective hunt for new facts which fit these phantasies. This process may be described as

[1] Grünbaum [1969]. He previously took a position which was one of radical dogmatic falsificationism and claimed that we *can* ascertain the falsity of scientific hypotheses (e.g. Grünbaum [1959*b*] and [1960]). His concrete case studies were thought-provoking for the philosopher and challenging for the physicist. But after criticisms from Feyerabend (cf. his [1959]), Laudan (cf. his [1965]), and others, he had to modify his position: such falsification cannot always be 'ascertained irrevocably': 'At least in some cases, we can ascertain the falsity of a component hypothesis to all scientific intents and purposes, although we cannot falsify it beyond any and all possibility of subsequent rehabilitation.'

[2] A typical such example is Newton's principle of gravitational attraction according to which bodies attract each other instantly from immense distances. Huyghens described this idea as 'absurd', Leibnitz as 'occult', and the best scientists of the age 'wondered how [Newton] could have given himself all the trouble of making such a number of investigations and difficult calculations that had no other foundation than this very principle' (cf. Koyré [1965], pp. 117–18). I had argued earlier that it is not so that theoretical progress is the merit of the theoretician but empirical success is *merely* a matter of luck. If the theoretician is *more* imaginative, it is likelier that his theoretical programme will achieve at least *some* empirical success. Cf. my [1968*a*], pp. 387–90.

'science creating its own universe' (as long as one remembers that 'creating' here is used in a provocative-idiosyncratic sense). A brilliant school of scholars (backed by a rich society to finance a few well-planned tests) might succeed in pushing any fantastic programme ahead, or, alternatively, if so inclined, in overthrowing any arbitrarily chosen pillar of 'established knowledge'.

The *dogmatic* falsificationist will throw up his hands in horror at this approach. He will see the spectre of Bellarmino's instrumentalism arising from the rubble under which Newtonian success of 'proven science' had buried it. He will accuse the sophisticated falsificationist of building arbitrary Procrustean pigeon hole systems and forcing the facts into them. He may even brand it as a revival of the unholy irrationalist alliance of James's crude pragmatism and of Bergson's voluntarism, triumphantly vanquished by Russell and Stebbing.[1] But our sophisticated falsificationism combines 'instrumentalism' (or 'conventionalism') with a strong empiricist requirement, which neither medieval 'saviours of phenomena' like Bellarmino, nor pragmatists like Quine and Bergsonians like Le Roy, had appreciated: the Leibnitz–Whewell–Popper requirement that *the—well planned—building of pigeon holes must proceed much faster than the recording of facts which are to be housed in them.* As long as this requirement is met, it does not matter whether we stress the 'instrumental' aspect of imaginative research programmes for finding novel facts and for making trustworthy predictions, or whether we stress the putative growing Popperian 'verisimilitude' (that is, the estimated difference between the truth-content and falsity-content) of their successive versions.[2] Sophisticated falsificationism thus combines the best elements of voluntarism, pragmatism and of the realist theories of empirical growth.

The sophisticated falsificationist sides neither with Galileo nor with Cardinal Bellarmino. He does not side with Galileo, for he claims that our basic theories may all be equally absurd and unverisimilar for the divine mind; and he does not side with Bellarmino, unless the Cardinal were to agree that scientific theories may yet lead, in the long run, to ever more true and ever fewer false consequences and, *in this strictly technical sense*, may have increasing 'verisimilitude'.[3]

[1] Cf. Russell [1914], Russell [1946] and Stebbing [1914]. Russell, a justificationist, despised conventionalism: 'As will has gone up in the scale, knowledge has gone down. This is the most notable change that has come over the temper of philosophy in our age. It was prepared by Rousseau and Kant...' ([1946], p. 787). Popper, of course, got some of his inspiration from Kant and Bergson. (Cf. his [1934], sections 2 and 4.)

[2] Cf. Popper [1963], chapter 10.

[3] '*Verisimilitude*' has two distinct meanings which must not be conflated. First, it may be used to mean intuitive truthlikeness of the theory; in this sense, in my view, all scientific theories created by the human mind are equally unverisimilar and 'occult'. Secondly, it

may be used to mean the set-theoretical difference between the true and false consequences of a theory which we can never know but certainly may guess. It was Popper who used 'verisimilitude' as a precise technical term to denote this difference ([1963], chapter 10). But his claim that this explication corresponds closely to *the* original meaning is mistaken and misleading. In the original prepopperian usage 'verisimilitude' could mean either *intuitive* truthlikeness or a naive proto-version of Popper's *empirical* truthlikeness. Popper gives interesting quotations for the latter ([1963], pp. 399 ff.) but none for the former. But Bellarmino might have agreed that Copernican theory had high 'verisimilitude' in Popper's technical sense but not that it had verisimilitude in the first, intuitive sense. Most 'instrumentalists' are 'realists' in the sense that they agree that the [Popperian] 'verisimilitude' of scientific theories is likely to be growing; but they are not 'realists' in the sense that they would agree that, for instance, the Einsteinian field approach is *intuitively* closer to the Blueprint of the Universe than the Newtonian action at a distance. *The 'aim of science' may then be increasing Popperian 'verisimilitude', but does not have to be also increasing classical verisimilitude.* The latter, as Popper himself said, is, unlike the former, a 'dangerously vague and metaphysical' idea ([1963], p. 231).

Popper's 'empirical verisimilitude' in a sense rehabilitates the idea of *cumulative growth* in science. But the driving force of cumulative growth in 'empirical verisimilitude' is revolutionary conflict in 'intuitive verisimilitude'.

When Popper was writing his 'Truth, rationality and the growth of knowledge', I had an uneasy feeling about his identification of the two concepts of verisimilitude. Indeed, it was I who asked him: 'Can we really speak about *better* correspondence? Are there such things as *degrees* of truth? Is it not dangerously misleading to talk as if Tarskian truth were located somewhere in a kind of metrical or at least topological space so that we can sensibly say of two theories—say an earlier theory t_1 and a later theory t_2, that t_2 has superseded t_1, or progressed beyond t_1, by approaching more closely to the truth than t_1?' (Popper [1963], p. 232). Popper rejected my vague misgivings. He felt—rightly—that he was proposing a very important new idea. But he was mistaken in believing that his new, technical conception of 'verisimilitude' completely absorbed the problems centred on the old *intuitive* 'verisimilitude'. Kuhn says: 'To say, for example, of a field theory that it "approaches more closely to the truth" than an older matter-and-force theory should mean, *unless words are being oddly used*, that the ultimate constituents of nature are more like fields than like matter and force' (*this volume, below*, p. 265; my italics). Indeed, Kuhn is right, except that words are *normally* 'oddly used'. I hope that this note may contribute to the clarification of the problem involved.

REFERENCES

Agassi [1959]: 'How are Facts Discovered?', *Impulse*, **3**, No. 10, pp. 2–4.

Agassi [1962]: 'The Confusion between Physics and Metaphysics in the Standard Histories of Sciences', in the *Proceedings of the Tenth International Congress of the History of Science*, 1964, **1**, pp. 231–8.

Agassi [1964]: 'Scientific Problems and Their Roots in Metaphysics', in Bunge (*ed.*): *The Critical Approach to Science and Philosophy*, 1964, pp. 189–211.

Agassi [1966]: 'Sensationalism', *Mind*, N.S. **75**, pp. 1–24.

Agassi [1968]: 'The Novelty of Popper's Philosophy of Science', *International Philosophical Quarterly*, **8**, pp. 442–63.

Agassi [1969]: 'Popper on Learning from Experience', in Rescher (*ed.*): *Studies in the Philosophy of Science*, 1969.

Ayer [1936]: *Language, Truth and Logic*, 1936; second edition 1946.

Bartley [1968]: 'Theories of Demarcation between Science and Metaphysics', in Lakatos and Musgrave (*eds*): *Problems in the Philosophy of Science*, 1968, pp. 40–64.

Becke and Sitte [1933]: 'Zur Theorie des β-Zerfalls', *Zeitschrift für Physik*, **86**, pp. 105–19.

Bernal [1965]: *Science in History*, third edition, 1965.

Bernstein [1961]: *A Comprehensible World: On Modern Science and its Origins*, 1961.
Bethe and Peierls [1934]: 'The "Neutrino"', *Nature*, **133**, p. 532.
Bohr [1913*a*]: 'On the Constitution of Atoms and Molecules', *Philosophical Magazine*, **26**, pp. 1–25, 476–502 and 857–75.
Bohr [1913*b*]: Letter to Rutherford, 6.3.1913; published in Bohr [1963], pp. xxxviii–ix.
Bohr [1913*c*]: 'The Spectra of Helium and Hydrogen', *Nature*, **92**, pp. 231–2.
Bohr [1922]: 'The Structure of the Atom', Nobel Lecture.
Bohr [1926]: Letter to *Nature*, **117**, p. 264.
Bohr [1930]: 'Chemistry and the Quantum Theory of Atomic Constitution', Faraday Lecture 1930, *Journal of the Chemical Society*, 1932/1, pp. 349–84.
Bohr [1933]: 'Light and Life', *Nature*, **131**, pp. 421–3 and 457–9.
Bohr [1936]: 'Conservation Laws in Quantum Theory', *Nature*, **138**, pp. 25–6.
Bohr [1949]: 'Discussion with Einstein on Epistemological Problems in Atomic Physics', in Schilpp (*ed.*): *Albert Einstein, Philosopher-Scientist*, 1949, **1**, pp. 201–41.
Bohr [1963]: *On the Constitution of Atoms and Molecules*, 1963.
Born [1948]: 'Max Karl Ernst Ludwig Planck', *Obituary Notices of Fellows of the Royal Society*, **6**, 161–80.
Born [1954]: 'The Statistical Interpretation of Quantum Mechanics', *Nobel Lecture* 1954.
Braithwaite [1938]: 'The Relevance of Psychology to Logic', *Aristotelian Society Supplementary Volumes*, **17**, pp. 19–41.
Braithwaite [1953]: *Scientific Explanation*, 1953.
Callendar [1914]: 'The Pressure of Radiation and Carnot's Principle,' *Nature*, **92**, p. 553.
Canfield and Lehrer [1961]: 'A Note on Prediction and Deduction', *Philosophy of Science*, 1961, **28**, pp. 204–8.
Carnap [1932–3]: 'Über Protokollsätze', *Erkenntnis*, **3**, pp. 215–28.
Carnap [1935]: Review of Popper's [1934], *Erkenntnis*, **5**, pp. 290–4.
Coffa [1968]: 'Deductive Predictions', *Philosophy of Science*, **35**, pp. 279–83.
Crookes [1886]: Presidential Address to the Chemistry Section of the British Association, *Report of British Association*, 1886, pp. 558–76.
Crookes [1888]: Report at the Annual General Meeting, *Journal of the Chemical Society*, **53**, pp. 487–504.
Davisson [1937]: 'The Discovery of Electron Waves', *Nobel Lecture*, 1937.
Dirac [1936]: 'Does Conservation of Energy Hold in Atomic Processes?', *Nature*, **137**, pp. 298–9.
Dirac [1951]: 'Is there an Aether?', *Nature*, **168**, pp. 906–7.
Dorling [1968]: 'Length Contraction and Clock Synchronisation: The Empirical Equivalence of the Einsteinian and Lorentzian Theories', *The British Journal for the Philosophy of Science*, **19**, pp. 67–9.
Dryer [1906]: *History of the Planetary Systems from Thales to Kepler*, 1906.
Duhem [1906]: *La Théorie Physique, Son Objet et Sa Structure*, 1905. English translation of the second (1914) edition: *The Aim and Structure of Physical Theory*, 1954.
Eccles [1964]: 'The Neurophysiological Basis of Experience', in Bunge (*ed.*): *The Critical Approach to Science and Philosophy*, 1964.
Ehrenfest [1911]: 'Welche Züge der Lichtquantenhypothese spielen in der Theorie der Wärmestrahlung eine wesentliche Rolle?', *Annalen der Physik*, **36**, pp. 91–118.
Ehrenfest [1913]: 'Zur Krise der Lichtäther-Hypothese', 1913.
Einstein [1909]: 'Über die Entwicklung unserer Anschauungen über das Wesen und die Konstitution der Strahlung', *Physikalische Zeitschrift*, **10**, pp. 817–26.
Einstein [1927]: 'Neue Experimente über den Einfluss der Erdbewegung auf die Lichtgeschwindigkeit relativ zur Erde', *Forschungen und Fortschritte*, **3**, p. 36.
Einstein [1928]: Letter to Schrödinger, 31.5.1928; published in K. Przibram (*ed.*): *Briefe Zur Wellenmechanik*, 1963.
Einstein [1931]: 'Gedenkworte auf Albert A. Michelson', *Zeitschrift für angewandte Chemie*, **44**, p. 658.

Einstein [1949]: 'Autobiographical Notes', in Schilpp (*ed.*): *Albert Einstein, Philosopher-Scientist*, **1**, pp. 2–95.

Ellis and Mott [1933]: 'Energy Relations in the β-Ray Type of Radioactive Disintegration', *Proceedings of the Royal Society of London*, Series A, **96**, pp. 502–11.

Ellis and Wooster [1927]: 'The average Energy of Disintegration of Radium E', *Proceedings of the Royal Society*, Series A, **117**, pp. 109–23.

Evans [1913]: 'The Spectra of Helium and Hydrogen', *Nature*, **92**, p. 5.

Fermi [1933]: 'Tentativo di una teoria dell emissione dei raggi "beta"', *Ricerci Scientifica*, **4(2)**, pp. 491–5.

Fermi [1934]: 'Versuch einer Theorie der β-Strahlen. I', *Zeitschrift für Physik*, **88**, pp. 161–77.

Feyerabend [1959]: 'Comments on Grünbaum's "Law and Convention in Physical Theory" ', in Feigl and Maxwell (*eds*): *Current Issues in the Philosophy of Science*, 1961, pp. 155–61.

Feyerabend [1965]: 'Reply to Criticism', in Cohen and Wartofsky (*eds*): *Boston Studies in the Philosophy of Science*, II, pp. 223–61.

Feyerabend [1968–9]: 'On a Recent Critique of Complementarity', *Philosophy of Science*, **35**, pp. 309–31 and **36**, pp. 82–105.

Feyerabend [1969]: 'Problems of Empiricism II', in Colodny (*ed.*): *The Nature and Function of Scientific Theory*, 1969.

Feyerabend [1970]: 'Against Method', *Minnesota Studies for the Philosophy of Science*, **4**, 1970.

Fowler [1912]: 'Observations of the Principal and Other Series of lines in the Spectrum of Hydrogen', *Monthly Notices of the Royal Astronomical Society*, **73**, pp. 62–71.

Fowler [1913*a*]: 'The Spectra of Helium and Hydrogen', *Nature*, **92**, p. 95.

Fowler [1913*b*]: 'The Spectra of Helium and Hydrogen', *Nature*, **92**, p. 232.

Fowler [1914]: 'Series Lines in Spark Spectra', *Proceedings of the Royal Society of London* (*A*), **90**, pp. 426–30.

Fresnel [1818]: 'Lettre à François Arago sur l'Influence du Mouvement Terrestre dans quelques Phénomènes Optiques', *Annales de Chimie et de Physique*, **9**, pp. 57 ff.

Galileo [1632]: *Dialogo dei Massimi Sistemi*, 1632.

Gamow [1966]: *Thirty Years that Shook Physics*, 1966.

Grünbaum [1959*a*]: 'The Falsifiability of the Lorentz-Fitzgerald Contraction Hypothesis', *British Journal for the Philosophy of Science*, **10**, pp. 48–50.

Grünbaum [1959*b*]: 'Law and Convention in Physical Theory', in Feigl and Maxwell (*eds*): *Current Issues in the Philosophy of Science*, 1961, pp. 40–155.

Grünbaum [1960]: 'The Duhemian Argument', *Philosophy of Science*, **11**, pp. 75–87.

Grünbaum [1969]: 'Can We Ascertain the Falsity of a Scientific Hypothesis?', *Studium Generale*, **22**, pp. 1061–93.

Heisenberg [1955]: 'The Development of the Interpretation of Quantum Theory', in Pauli (*ed.*): *Niels Bohr and the Development of Physics*, 1955.

Hempel [1937]: Review of Popper's [1934], *Deutsche Literaturzeitung*, 1937, pp. 309–14.

Hempel [1952]: 'Some Theses on Empirical Certainty', *The Review of Metaphysics*, **5**, pp. 620–1.

Henderson [1934]: 'The Upper Limits of the Continuous β-ray Spectra of Thorium C and C^{11}', *Proceedings of the Royal Society of London*, Series A, **147**, pp. 572–82.

Hevesy [1913]: 'Letter to Rutherford, 14.10.1913', quoted in Bohr [1963], p. XLII.

Hund [1961]: 'Göttingen, Copenhagen, Leipzig im Rückblick', in Bopp (*ed.*): *Werner Heisenberg und die Physik unserer Zeit*, Braunschweig 1961.

Jaffe [1960]: *Michelson and the Speed of Light*, 1960.

Jammer [1966]: *The Conceptual Development of Quantum Mechanics*, 1966.

Joffé [1911]: 'Zur Theorie der Strahlungserscheinungen', *Annalen der Physik*, **36**, pp. 534–52.

Juhos [1966]: 'Über die empirische Induktion', *Studium Generale*, **19**, pp. 259–72.

Keynes [1921]: *A Treatise on Probability*, 1921.

Koestler [1959]: *The Sleepwalkers*, 1959.

Konopinski and Uhlenbeck [1935]: 'On the Fermi theory of β-radioactivity', *Physical Review*, **48**, pp. 7–12.

Kramers [1923]: 'Das Korrespondenzprinzip und der Schalenbau des Atoms', *Die Naturwissenschaften*, **11**, pp. 550–9.

Kudar [1929–30]: 'Der wellenmechanische Charakter des β-Zerfalls, I–II–III', *Zeitschrift für Physik*, **57**, pp. 257–60, **60**, pp. 168–75 and 176–83.

Kuhn [1962]: *The Structure of Scientific Revolutions*, 1962.

Kuhn [1965]: 'Logic of Discovery or Psychology of Research', *this volume*, pp. 1–23.

Lakatos [1962]: 'Infinite Regress and the Foundations of Mathematics', *Aristotelian Society Supplementary Volume*, **36**, pp. 155–84.

Lakatos [1963–4]: 'Proofs and Refutations', *The British Journal for the Philosophy of Science*, **14**, pp. 1–25, 120–39, 221–43, 296–342.

Lakatos [1968*a*]: 'Changes in the Problem of Inductive Logic', in Lakatos (*ed.*): *The Problem of Inductive Logic*, 1968, pp. 315–417.

Lakatos [1968*b*]: 'Criticism and the Methodology of Scientific Research Programmes', in *Proceedings of the Aristotelian Society*, **69**, pp. 149–86.

Lakatos [1970]: *The Changing Logic of Scientific Discovery*, 1970.

Lakatos [1971]: *Proofs and Refutations and Other Essays in the Philosophy of Mathematics*, 1971.

Laplace [1796]: *Exposition du Système du Monde*, 1796.

Larmor [1904]: 'On the Ascertained Absence of Effects of Motion through the Aether, in Relation to the Constitution of Matter, and on the Fitzgerald-Lorentz Hypothesis', *Philosophical Magazine*, Series **6**, **7**, pp. 621–5.

Laudan [1965]: 'Grünbaum on "The Duhemian Argument" ', *Philosophy of Science*, **32**, pp. 295–9.

Leibnitz [1678]: Letter to Conring, 19.3.1678.

Le Roy [1899]: 'Science et Philosophie', *Revue de Métaphysique et de Morale*, **7**, pp. 375–425, 503–62, 706–31.

Le Roy [1901]: 'Un Positivisme Nouveau', *Revue de Métaphysique et de Morale*, **9**, pp. 138–53.

Lorentz [1886]: De l'Influence du Mouvement de la Terre sur les Phénomènes Lumineux', *Versl. Kon. Akad. Wetensch. Amsterdam*, **2**, pp. 297–358. Reprinted in Lorentz: *Collected Papers*, **4**, 1937, pp. 153–218.

Lorentz [1892*a*]: 'The Relative Motion of the Earth and the Ether', *Versl. Kon. Akad. Wetensch. Amsterdam*, **1**, pp. 74–7. Reprinted in Lorentz: *Collected Papers*, **4**, 1937, pp. 219–23.

Lorentz [1892*b*]: 'Stokes' Theory of Aberration', *Versl. Kon. Akad. Wetensch. Amsterdam*, **1**, pp. 97–103. Reprinted in Lorentz: *Collected Papers*, **4**, 1937, pp. 224–31.

Lorentz [1895]: *Versuch einer Theorie der electrischen und optischen Erscheinungen in bewegten Körpern*, 1895, § 89–92.

Lorentz [1897]: 'Concerning the Problem of the Dragging Along of the Ether by the Earth', *Versl. Kon. Akad. Wetensch. Amsterdam*, **6**, pp. 266–72. Reprinted in Lorentz: *Collected Papers*, **4**, 1937, pp. 237–44.

Lorentz [1923]: 'The Rotation of the Earth and its Influence on Optical Phenomena', *Nature*, **112**, pp. 103–4.

Lykken [1968]: 'Statistical Significance in Psychological Research', *Psychological Bulletin*, **70**, pp. 151–9.

McCulloch [1825]: *The Principles of Political Economy: With a Sketch of the Rise and Progress of the Science*, 1825.

MacLaurin [1748]: *Account of Sir Isaac Newton's Philosophical Discoveries*, 1748.

Margenau [1950]: *The Nature of Physical Reality*, 1950.

Marignac [1860]: 'Commentary on Stas' Researches on the Mutual Relations of Atomic Weights', reprinted in *Prout's Hypothesis*, Alembic Club Reprints, **20**, pp. 48–58.

Maxwell [1871]: *Theory of Heat*, 1871.

Medawar [1967]: *The Art of the Soluble*, 1967.

Medawar [1969]: *Induction and Intuition in Scientific Thought*, 1969.

Meehl [1967]: 'Theory Testing in Psychology and Physics: a Methodological Paradox', *Philosophy of Science*, **34**, pp. 103–115.

Meitner [1933]: 'Kernstruktur', in Geiger-Scheel (*eds.*): *Handbuch der Physik*, Zweite Auflage, **22/1**, pp. 118–52.

Meitner and Orthmann [1930]: 'Uber eine absolute Bestimmung der Energie der primären β-Strahlen von Radium E', *Zeitschrift für Physik*, **60**, pp. 143–55.

Michelson [1881]: 'The Relative Motion of the Earth and the Luminiferous Ether', *American Journal of Science*, Ser. 3, **22**, pp. 120–9.

Michelson [1891–2]: 'On the Application of Interference Methods to Spectroscopic Measurements, I–II', *Philosophical Magazine*, Ser. 3, **31**, pp. 338–46, and **34**, pp. 280–99.

Michelson [1897]: 'On the Relative Motion of the Earth and the Ether', *American Journal of Science*, Ser. 4, **3**, pp. 475–8.

Michelson and Gale [1925]: 'The Effect of the Earth's Rotation on the Velocity of Light', *Astrophysical Journal*, **61**, pp. 137–45.

Michelson and Morley [1887]: 'On the Relative Motion of the Earth and the Luminiferous Ether', *American Journal of Science*, Ser. 3, **34**, pp. 333–45.

Milhaud [1896]: 'La Science Rationnelle', *Revue de Métaphysique et de Morale*, **4**, pp. 280–302.

Mill [1843]: *A System of Logic, Ratiocinative and Inductive, Being a Connected View of the Principles of Evidence, and the Methods of Scientific Investigation*, 1843.

Miller [1925]: 'Ether-Drift Experiments at Mount Wilson', *Science*, **41**, pp. 617–21.

Morley and Miller [1904]: Letter to Kelvin, published in *Philosophical Magazine*, Ser. 6, **8**, pp. 753–4.

Moseley [1914]: 'Letter to Nature', *Nature*, **92**, p. 554.

Mott [1933]: "Wellenmechanik und Kernphysik', in Geiger and Scheel (*eds.*): *Handbuch der Physik*, Zweite Auflage, **24/1**, pp. 785–841.

Musgrave [1968]: 'On a Demarcation Dispute', in Lakatos and Musgrave (*eds.*): *Problems in the Philosophy of Science*, 1968, pp. 78–88.

Musgrave [1969*a*]: *Impersonal Knowledge*, Ph.D. Thesis, University of London, 1969.

Musgrave [1969*b*]: Review of Ziman's 'Public Knowledge: An Essay Concerning the Social Dimensions of Science', in *The British Journal for the Philosophy of Science*, **20**, pp. 92–4.

Nagel [1961]: *The Structure of Science*, 1961.

Nagel [1967]: 'What is True and False in Science: Medawar and the Anatomy of Research', *Encounter*, **29**, No. 3, pp. 68–70.

Nature [1913–14]: 'Physics at the British Association', *Nature*, **92**, pp. 305–9.

Neurath [1935]: 'Pseudorationalismus der Falsifikation', *Erkenntnis*, **5**, pp. 353–65.

Nicholson [1913]: 'A Possible Extension of the Spectrum of Hydrogen', *Monthly Notices of the Royal Astronomical Society*, **73**, pp. 382–5.

Pauli [1958]: 'Zur älteren und neueren Geschichte des Neutrinos', published in Pauli, *Aufsätze und Vorträge über Physik und Erkenntnistheorie*, 1961, pp. 156–80.

Pearce Williams [1968]: *Relativity Theory: Its Origins and Impact on Modern Thought*, 1968.

Peierls [1936]: 'Interpretation of Shankland's Experiment', *Nature*, **137**, p. 904.

Planck (1900*a*]: 'Über eine Verbesserung der Wienschen Spektralgleichung', *Verhandlungen der Deutschen Physikalischen Gesellschaft*, **2**, pp. 202–4; English translation in Ter Haar [1967].

Planck [1900*b*]: 'Zur Theorie des Gesetzes der Energieverteilung im Normalspektrum', *Verhandlungen der Deutschen Physikalischen Gesellschaft*, **2**, pp. 237–45; English translation in Ter Haar [1967].

Planck [1929]: 'Zwanzig Jahre Arbeit am Physikalischen Weltbild', *Physica*, **9**, pp. 193–222.

Planck [1947]: *Scientific Autobiography*, published posthumously in German in 1948, in English translation in 1950.

Poincaré [1891]: 'Les géométries non euclidiennes', *Revue des Sciences Pures et Appliquées*, **2**, pp. 769–74.

Poincaré [1902]: *La Science et l'Hypothèse*, 1902.

Polanyi [1958]: *Personal Knowledge, Towards a Post-critical Philosophy*, 1958.

Popkin [1968]: 'Scepticism, Theology and the Scientific Revolution in the Seventeenth Century', in Lakatos and Musgrave (*eds.*): *Problems in the Philosophy of Science*, 1968, pp. 1–28.

Popper [1933]: 'Ein Kriterium des empirischen Charakters theoretischer Systeme', *Erkenntnis*, **3**, pp. 426–7.

Popper [1934]: *Logik der Forschung*, 1935 (expanded English edition: Popper [1959*a*]).

Popper [1935]: 'Induktionslogik und Hypothesenwahrscheinlichkeit', *Erkenntnis*, **5**, pp. 170–2.

Popper [1940]: 'What is Dialectic?', *Mind*, N.S. **49**, pp. 403–26; reprinted in Popper [1963], pp. 312–35.

Popper [1945]: *The Open Society and its Enemies*, I–II, 1945.

Popper [1957]: 'The Aim of Science', *Ratio*, **1**, pp. 24–35.

Popper [1958]: 'Philosophy and Physics'; published in *Atti del XII Congresso Internazionale di Filosofia*, Vol. 2, 1960, pp. 363–74.

Popper [1959*a*]: *The Logic of Scientific Discovery*, 1959.

Popper [1959*b*]: 'Testability and "ad-Hocness" of the Contraction Hypothesis', *British Journal for the Philosophy of Science*, **10**, p. 50.

Popper [1963]: *Conjectures and Refutations*, 1963.

Popper [1965]: 'Normal Science and its Dangers', *this volume*, pp. 51–8.

Popper [1968*a*]: 'Epistemology without a Knowing Subject', in Rootselaar-Staal (*eds.*): *Proceedings of the Third International Congress for Logic, Methodology and Philosophy of Science*, Amsterdam, 1968, pp. 333–73.

Popper (1968*b*]: 'On the Theory of the Objective Mind', in *Proceedings of the XIV International Congress of Philosophy*, **1**, 1968, pp. 25–53.

Popper [1968*c*]: 'Remarks on the Problems of Demarcation and Rationality', in Lakatos and Musgrave (*eds.*): *Problems in the Philosophy of Science*, 1968, pp. 88–102.

Popper [1969]: 'A Realist View of Logic, Physics and History', in Yourgrau (*ed.*): *Logic, Physics and History*, 1969.

Power [1964]: *Introductory Quantum Electrodynamics*, 1964.

Prokhovnik [1967]: *The Logic of Special Relativity*, 1967.

Prout [1815]: 'On the Relation between the Specific Gravities of Bodies in their Gaseous State and the Weights of their Atoms', *Annals of Philosophy*, **6**, pp. 321–30; reprinted in *Prout's Hypothesis*, Alembic Club Reprints, **20**, 1932.

Quine [1953]: *From a Logical Point of View*, 1953.

Rabi [1961]: 'Atomic Structure', in G. M. Murphy and M. H. Shamos (*eds.*): *Recent Advances in Science*, 1956.

Reichenbach [1951]: *The Rise of Scientific Philosophy*, 1951.

Runge [1925]: 'Äther und Relativitätstheorie', *Die Naturwissenschaften*, **13**, p. 440.

Russell [1914]: *The Philosophy of Bergson*, 1914.

Russell [1943]: 'Reply to Critics', in Schilpp (*ed.*): *The Philosophy of Bertrand Russell*, 1943, pp. 681–741.

Russell [1946]: *History of Western Philosophy*, 1946.

Rutherford, Chadwick and Ellis [1930]: *Radiations from Radioactive Substances*, 1930.

Schlick [1934]: 'Über das Fundament der Erkenntnis', *Erkenntnis*, **4**, pp. 79–99; published in English in Ayer (*ed.*): *Logical Positivism*, 1959, pp. 209–27.

Schrödinger [1958]: 'Might perhaps Energy be merely a Statistical Concept?', *Il Nouvo Cimento*, **9**, pp. 162–70.

Shankland [1936]: 'An Apparent Failure of the Photon Theory of Scattering', *Physical Review*, **49**, pp. 8–13.

Shankland [1964]: 'Michelson-Morley Experiment', *American Journal of Physics*, **32**, pp. 16–35.

Soddy [1932]: *The Interpretation of the Atom*, 1932.

Sommerfeld [1916]: 'Zur Quantentheorie der Spektrallinien', *Annalen der Physik*, **51**, pp. 1–94 and 125–67.

Stebbing [1914]: *Pragmatism and French Voluntarism*, 1914.

Stegmüller [1966]: 'Explanation, Prediction, Scientific Systematization and Non-Explanatory Information', *Ratio*, **8**, pp. 1–24.

Stokes [1845]: 'On the Aberration of Light', *Philosophical Magazine*, Third Series, **27**, pp. 9–15.

Stokes [1846]: 'On Fresnel's Theory of the Aberration of Light', *Philosophical Magazine*, Third Series, **28**, pp. 76–81.

Synge [1952–4]: 'Effects of Acceleration in the Michelson-Morley Experiment', *The Scientific Proceedings of the Royal Dublin Society*, New Series, **26**, pp. 45–54.

Ter Haar [1967]: *The Old Quantum Theory*, 1967.

Thomson [1929]: 'On the Waves associated with β-rays, and the Relation between Free Electrons and their Waves', *Philosophical Magazine*, Seventh Series, **7**, pp. 405–17.

Toulmin [1967]: 'The Evolutionary Development of Natural Science', *American Scientist*, **55**, pp. 456–71.

Treiman [1959]: 'The Weak Interactions', *Scientific American*, **200**, pp. 72–84.

Truesdell [1960]: 'The Program toward Rediscovering the Rational Mechanics in the Age of Reason', *Archive of the History of Exact Sciences*, **1**, pp. 3–36.

Uhlenbeck and Goudsmit [1925]: 'Ersetzung der Hypothese vom unmechanischen Zwang durch eine Forderung bezüglich des inneren Verhaltens jedes einzelnen Electrons', *Die Naturwissenschaften*, **13**, pp. 953–4.

van der Waerden [1967]: *Sources of Quantum Mechanics*, 1967.

Watkins [1957]: 'Between Analytic and Empirical', *Philosophy*, **32**, pp. 112–31.

Watkins [1958]: 'Influential and Confirmable Metaphysics', *Mind*, N.S. **67**, pp. 344–65.

Watkins [1960]: 'When are Statements Empirical?', *British Journal for the Philosophy of Science*, **10**, pp. 287–308.

Watkins [1968]: 'Hume, Carnap and Popper', in Lakatos (*ed.*): *The Problem of Inductive Logic*, 1968, pp. 271–82.

Whewell [1837]: *History of the Inductive Sciences, from the Earliest to the Present Time.* Three volumes, 1837.

Whewell [1840]: *Philosophy of the Inductive Sciences, Founded upon their History.* Two volumes, 1840.

Whewell [1851]: 'On the Transformation of Hypotheses in the History of Science', *Cambridge Philosophical Transactions*, **9**, pp. 139–47.

Whewell [1858]: *Novum Organon Renovatum.* Being the second part of the philosophy of the inductive sciences. Third edition, 1858.

Whewell [1860]: *On the Philosophy of Discovery, Chapters Historical and Critical*, 1860.

Whittaker [1947]: *From Euclid to Eddington*, 1947.

Whittaker [1953]: *History of the Theories of Aether and Electricity*, Vol. II, 1953.

Wisdom [1963]: 'The Refutability of "Irrefutable" Laws', *The British Journal for the Philosophy of Science*, **13**, pp. 303–6.

Wu [1966]: 'Beta Decay', in *Rendiconti della Scuola Internazionale di Fisica*, "Enrico Fermi", *XXXII* Corso.

Wu and Moskowski [1966]: *Beta Decay*, 1966.

Consolations for the Specialist[1]

PAUL FEYERABEND
University of California, Berkeley

'I have been hanging people for years, but I have never had all this fuss before.' (Remark made by Edward 'Lofty' Milton, Rhodesia's part time executioner on the occasion of demonstrations against the death penalty.) 'He was'—says *Time* Magazine (15 March 1968)—'professionally incapable of understanding the commotion.'

1 INTRODUCTION

In the years 1960 and 1961 when Kuhn was a member of the philosophy department at the University of California in Berkeley I had the good fortune of being able to discuss with him various aspects of science. I have profited enormously from these discussions and I have looked at science in a new way ever since.[2] Yet while I thought I recognized Kuhn's *problems*; and while I tried to account for certain *aspects* of science to which he had drawn attention (the omnipresence of anomalies is one example); I was quite unable to agree with the *theory of science* which he himself proposed; and I was even less prepared to accept the general *ideology* which I thought formed the background of his thinking. This ideology, so it seemed to me, could only give comfort to the most narrowminded and the most conceited kind of specialism. It would tend to inhibit the advancement of knowledge. And it is bound to increase the anti-humanitarian tendencies

[1] An earlier version of this paper was read in Professor Popper's seminar at the London School of Economics (March 1967). I would like to thank Professor Popper for this opportunity as well as for his own detailed criticism. I am also grateful to Messrs Howson and Worrall for their valuable editorial and stylistic help.

[2] The criticism of some features of contemporary methodology which appears in my [1969] and [1970] is but one belated after-effect.

which are such a disquieting feature of much of post-Newtonian science.[1] On all these points my discussions with Kuhn remained inconclusive. More than once he interrupted a lengthy sermon of mine, pointing out that I had misunderstood him, or that our views are closer than I had made them appear. Now, looking back at our debates[2] as well as at the papers which Kuhn has published since his departure from Berkeley, I am not so sure that this was the case. And I am fortified in my behalf by the fact that almost every reader of Kuhn's *Structure of Scientific Revolutions* interprets him as I do, and that certain tendencies in modern sociology and modern psychology are the result of exactly this kind of interpretation. I hope that Kuhn will forgive me when therefore I once more raise the old issues and that he will not take it amiss when in my effort to be brief I do this in a somewhat blunt fashion.

2. AMBIGUITY OF PRESENTATION

Whenever I read Kuhn, I am troubled by the following question: are we here presented with *methodological prescriptions* which tell the scientist how to proceed; or are we given a *description*, void of any evaluative element, of those activities which are generally called 'scientific'? Kuhn's writings, it seems to me, do not lead to a straightforward answer. They are *ambiguous* in the sense that they are compatible with, and lend support to, both interpretations. Now this ambiguity (whose stylistic expression and mental impact has much in common with similar ambiguities in Hegel and in Wittgenstein) is not at all a side issue. It has had quite a definite effect on Kuhn's readers and has made them look at, and deal with their subject in a manner not altogether advantageous. More than one social scientist has pointed out to me that now at last he had learned how to turn his field into a 'science'—by which of course he meant that he had learned how to *improve* it. The recipe, according to these people, is to restrict criticism, to reduce the number of comprehensive theories to one, and to create a normal science that has this one theory as its paradigm.[3] Students must be prevented from speculating along different lines and the more restless colleagues must be made to conform and 'to do serious work'. *Is this what Kuhn wants to achieve?*[4] Is it his intention to provide a historico-

[1] Cf. my [1970].

[2] Some of which were carried out in the now defunct *Café Old Europe* on Telegraph Avenue and greatly amused the other customers by their friendly vehemence.

[3] See, e.g. Reagan [1967] p. 1385: He states: 'We [that is, we social scientists] are in what Kuhn might call a "pre-paradigm" stage of development in which consensus has yet to emerge on basic concepts and theoretical assumptions.'

[4] Neurophysiology, physiology, and certain parts of psychology are far ahead of contemporary physics in that they manage to make the discussion of fundamentals an essential part of even the most specific piece of research. Concepts are never completely stabilized

scientific justification for the ever growing need to identify with some group? Does he want every subject to imitate the monolithic character of, say, the quantum theory of 1930? Does he think that a discipline that has been constructed in this manner is in some ways better off? That it will lead to better, to more numerous, to more interesting results? Or is his following among sociologists an unintended side-effect of a work whose sole purpose is to report '*wie es wirklich gewesen*' without implying that the reported features are worthy of imitation? And if this is the sole purpose of the work, then why the constant misunderstanding, and why the ambiguous and occasionally highly moralizing style?

I venture to guess that the ambiguity is *intended* and that Kuhn wants to fully exploit its propagandistic potentialities. He wants on the one side to give solid, objective, historical support to value judgements which he just as many other people seem to regard as arbitrary and subjective. On the other side he wants to leave himself a safe second line of retreat: those who dislike the implied derivation of values from facts can always be told that no such derivation is made and that the presentation is purely descriptive. My first set of questions, therefore, is: why the ambiguity? How is it to be interpreted? What is Kuhn's attitude towards the kind of following I have described? Have they misread him? Or are they legitimate followers of a new vision of science?

3. PUZZLE SOLVING AS A CRITERION OF SCIENCE

Let us now disregard the problem of presentation and let us assume that Kuhn's aim is indeed to give but a *description* of certain influential historical events and institutions.

According to this interpretation it is the existence of a puzzle-solving tradition that *de facto* sets the sciences apart from other activities. It sets them apart in a 'far surer and more direct' way, in a manner that is 'at once . . . less equivocal and . . . more fundamental',[1] than do other and more recondite properties which they may also possess. But if the existence

but are left open and are elucidated now by the one, now by the other theory. There is no indication that progress is hampered by the more 'philosophical' attitude which, according to Kuhn, underlies such a procedure (cf. *this volume*, p. 6). (Thus the lack of clarity about the idea of perception has led to many interesting empirical investigations, some of them yielding quite unexpected and highly important results. Cf. Epstein [1967], especially pp. 6–18.) Quite the contrary, we find a greater awareness of the limits of our knowledge, of its connection with human nature, we find also a greater familiarity with the history of the subject and the ability not only to *record*, but to *actively use* past ideas for the advancement of contemporary problems. Must we not admit that all this contrasts most favourably with the humourless dedication and the constipated style of a 'normal' science?

[1] Cf. *this volume*, p. 7.

of a puzzle-solving tradition is so essential, if it is the occurrence of this property that unifies and characterizes a specific and well recognizable discipline; then I do not see how we shall be able to exclude say, Oxford philosophy, or, to take an even more extreme example, *organized crime* from our considerations.

For organized crime, so it would seem, is certainly puzzle-solving *par excellence*. Every statement which Kuhn makes about normal science remains true when we replace 'normal science' by 'organized crime'; and every statement he has written about the 'individual scientist' applies with equal force to, say, the individual safebreaker.

Organized crime certainly keeps foundational research to a minimum[1] although there are outstanding individuals, such as Dillinger, who introduce new and revolutionary ideas.[2] Knowing the rough outlines of the phenomena to be expected the professional safebreaker 'largely ceases to be an explorer . . . or at least an explorer of the unknown [after all, he is supposed to know all the existing types of safe]. Instead, he struggles to . . . concretize the known [i.e. to discover the idiosyncracies of the particular safe he is dealing with], designing much special-purpose apparatus and many special-purpose adaptations of theory for that task'.[3] According to Kuhn failure of achievement most certainly reflects 'on the competence of the [safebreaker] in the eyes of his professional compeers'[4] so that 'it is the individual [safebreaker] rather than current theory [of electromagnetism, for example] which is tested'[5]: 'only the practitioner is blamed, not his tools'[6]—and so we can continue step for step, down to the very last item on Kuhn's list. The situation is not improved by pointing to the existence of *revolutions*. First of all, because we are dealing with the thesis that it is normal science which is characterized by the activity of puzzle-solving. And secondly because there is no reason to believe that organized crime will fall behind in the mastery of major difficulties. Besides, if it is the *pressure* derived from the ever increasing number of anomalies that leads, first to a crisis, and then to a revolution, then the greater the pressure, the sooner the crisis must occur. Now the pressure exerted upon the members of a gang and their 'professional compeers' certainly can be expected to exceed the pressures upon a scientist—the latter hardly ever has to deal with the police. Wherever we look—the distinction we want to draw does not exist.

[1] Cf. Kuhn [1961*a*], p. 357.

[2] Dillinger considerably advanced the technique of the bank-holdup by staging dress rehearsals in life size models of the target-banks which he built at his farm. He thereby refuted Andrew Carnegie's 'Pioneering don't pay'.

[3] Kuhn [1961*a*], p. 363.

[4] *This volume*, p. 9; also cf. p. 7 and footnote 1 on p. 5.

[5] *This volume*, p. 5.

[6] *This volume*, p. 7; also cf. Kuhn [1962], p. 79.

This of course is no surprise. For Kuhn, as we interpret him now and as he himself very often wants to be interpreted, has failed to do one important thing. He has failed to discuss the *aim* of science. Every crook knows that apart from succeeding at his trade and being popular with his fellow crooks he wants one thing: money. He also knows that his normal criminal activity is going to give him just this. He knows that he will receive the more money and rise the faster on the professional ladder the better he is as a puzzle-solver and the better he fits into the criminal community. Money is his aim. What is the aim of the scientist? And, considering this aim, is normal science going to lead up to it? Or are perhaps scientists (and Oxford philosophers) less rational than crooks in that they 'are doing what they are doing' without regard to an aim?[1] These are the questions which arise if one wants to restrict oneself to the purely descriptive aspect of Kuhn's account.

4. FUNCTION OF NORMAL SCIENCE

In order to answer these questions we must now consider not only the *actual structure* of Kuhnian normal science, but also its *function*. Normal science, he says, is a *necessary presupposition of revolutions*.

According to this part of the argument the pedestrian activity associated with 'mature' science has far reaching effects both upon the *content* of our ideas, and upon their *substantiality*. This activity, this concern with 'tiny puzzles' leads to a close fit between theory and reality, and it also precipitates progress. It does so for various reasons. First of all the accepted paradigm gives the scientist a guide: 'As a glance at any Baconian natural history or a survey of the pre-paradigm development of any science will show, nature is vastly too complex to be explored even approximately at random'.[2] This point is not new. The attempt to create knowledge needs guidance, it cannot start from nothing. More specifically, it needs a theory, a point of view that allows the researcher to separate the relevant from the irrelevant, and that tells him in what areas research will be most profitable.

To this common idea Kuhn adds a specific twist of his own. He defends not only the *use* of theoretical assumptions, but the *exclusive choice* of one particular set of ideas, the monomaniac concern with only one single point of view. He defends such a procedure first, because it plays a role in actual science as he sees it. This is the description-recommendation ambiguity already dealt with. But he defends it also for a second reason that is somewhat more recondite as the preferences behind it are not made explicit. He defends it because he believes that its adoption will in the end lead to the

[1] 'I am doing what I am doing' was a favourite remark of Austin's.

[2] Kuhn [1961*a*], p. 363.

overthrow of the very same paradigm to which the scientists have restricted themselves in the first place. If even the most concerted effort to fit nature into its categories fails; if the very definite expectations created by these categories are disappointed again and again; then we are *forced* to look for something new. And we are forced to do this not just by an abstract discussion of possibilities which does not touch reality, but is rather guided by our own likes and dislikes[1]; we are forced to do it by procedures which have established a close contact with nature, and therefore, in the last resort, by nature itself. The debates of *pre-science* with their universal criticism and their uninhibited proliferation of ideas are 'often directed as much to the members of other schools as ... to nature'.[2] *Mature science*, especially in the quiet periods immediately before the storm, seems to address nature itself only and may therefore expect a definite *and objective* answer. In order to get such an answer we need more than a collection of facts assembled at random. But we need also more than an everlasting discussion of different ideologies. What is needed is the acceptance of *one* theory and the relentless attempt to fit nature into its pattern. This, I think, is the main reason why the rejection, by a mature science, of the uninhibited battle between alternatives would be defended by Kuhn not only as a *historical fact*, but also as a *reasonable move*. Is this defence acceptable?

5. THREE DIFFICULTIES OF FUNCTIONAL ARGUMENT

Kuhn's defence is acceptable *provided* revolutions are desirable and provided the particular way in which normal science leads to revolutions is desirable also.

Now I do not see how the desirability of revolutions can be established by Kuhn. Revolutions bring about a *change* of paradigm. But following Kuhn's account of this change, or 'gestalt-switch' as he calls it, it is impossible to say that they have led to something *better*. It is impossible to say this because pre- and post-revolutionary paradigms are frequently incommensurable.[3] This I would regard as the first difficulty of the functional argument if used in connection with the remainder of Kuhn's philosophy.

Secondly we have to examine what Lakatos has called the 'fine-structure' of the transition: normal science/revolution. This fine-structure may reveal elements we do not want to condone. Such elements would force us

[1] 'If any one offers conjectures about the truth of things from the mere possibility of hypothesis, then I do not see how any certainty can be determined in any science; for it is always possible to contrive hypotheses, one after another, which are found to lead to new difficulties' (Newton [1672]).

[2] Kuhn [1962], p. 13.

[3] Cf. *below*, section 9.

to consider different ways of bringing about a revolution. Thus it is quite imaginable that scientists abandon a paradigm out of frustration and not because they have arguments against it. (Killing the representatives of the *status quo* would be another way of breaking up a paradigm.[1]) How do scientists *actually* proceed? And how would we *want* them to proceed? An examination of these questions leads to a second difficulty for the functional argument.

In order to exhibit this difficulty as clearly as possible let us first consider the following *methodological problems*: Is it possible to give reasons for proceeding as Kuhn says normal science proceeds, that is, for trying to stick to a theory despite the existence of *prima facie* refuting evidence, of logical, and of mathematical counter arguments? And assuming it is possible to give such reasons—is it then possible to abandon the theory without violating them?

In what follows I shall call the advice to select from a number of theories the one that promises to lead to the most fruitful results, and to stick to this one theory even if the actual difficulties it encounters are considerable, the *principle of tenacity*.[2] The problem then is how this principle can be

[1] This is how *religious doctrines* or *political doctrines* were frequently replaced. The principle remains even today, though murder is no longer the accepted method. The reader should also consider Max Planck's remark that old theories disappear because their defenders die out.

[2] This formulation of the principle was suggested by an objection which Isaac Levi raised against an earlier version.

The principle of tenacity as formulated in the text should not be confused with Putnam's *rule of tenacity* (Putnam [1963], p. 772). For while Putnam's rule demands that a theory should be retained '*unless* it becomes inconsistent with the data' (his italics) tenacity as understood by Kuhn and by myself demands that it should be retained *even if there are data which are inconsistent with it*. This stronger version creates problems which do not appear in Putnam's methodology and which, I suggest, can be solved only if one is prepared to use a multiplicity of mutually inconsistent theories *at any time of the development of our knowledge*. It seems to me that neither Kuhn not Putnam is prepared to take this step. But while Kuhn sees the need for the use of alternatives (see below) Putnam demands that their number be always reduced either to one or to zero (*ibid.* pp. 770 ff.).

Lakatos differs from the account given in the text above in two respects. He distinguishes between *theories* and *research programmes*. And he applies tenacity to research programmes only.

Now while I admit that the distinction and the use he makes of it may increase clarity, I am still inclined to stick to my own and much more vague term 'theory' (for a partial explanation of this term, cf. my footnote 5 [1965*a*]) which covers both Lakatos's 'theories' and 'research programmes', to connect *it* with tenacity, and to *altogether eliminate* the more simple forms of refutation. One reason for this preference is given by Lakatos himself who has shown that even simple refutations involve a plurality of theories (see especially his paper in *this volume*, pp. 121 ff.). Another reason is my belief that progress can be brought about only by the active interaction of different 'theories' which of course assumes that the 'research programme'-component comes forth not only occasionally, *but is present all the time* (cf. also *below*, section 9).

defended, and how we can change our allegiance to paradigms in a manner that is either consistent with it, or perhaps even dictated by it. Remember that we are here dealing with a *methodological* problem and *not* with the question of how science *actually* proceeds. We are dealing with it because we hope that its discussion will sharpen our historical perception and will lead us to interesting historical discoveries.

Now the solution of the problem is quite straightforward. The principle of tenacity is reasonable because theories are capable of development, because they can be improved, and because they may eventually be able to accommodate the very same difficulties which in their original form they were quite incapable of explaining. Besides, it is not at all prudent to put too much trust in experimental results. Indeed, it would be a complete surprise and even a cause for suspicion, if all the available evidence should turn out to support a single theory, even if this theory should happen to be true. Different experimenters are liable to commit different errors and it usually needs considerable time before all experiments are brought to a common denominator.[1] To these arguments in favour of tenacity Professor Kuhn would add that a theory also provides *criteria* of excellence, of failure, of rationality, and that one must support it as long as possible, in order to keep the discourse rational as long as possible. The most important point is however this: it is hardly ever the case that theories are directly compared with 'the facts', or with 'the evidence'. What counts and what does not count as relevant evidence usually depends on the theory *as well* as on other subjects which may conveniently be called 'auxiliary sciences' ('touchstone theories' is Imre Lakatos's apt expression[2]). Such auxiliary sciences may function as additional premises in the derivation of testable statements. But they may also infect the observation language itself, providing the very concepts in terms of which experimental results are expressed. Thus a test of the Copernican view involves on the one hand assumptions concerning the terrestrial atmosphere, the effect of motion upon the object moved (dynamics); and on the other it also involves assumptions about the relation between sense experience and 'the world' (theories of cognition, theories of telescopic vision included).

The former assumptions function as premises while the latter determine which impressions are veridical and thus enable us not only to *evaluate*, but even to *constitute* our observations. Now there is no guarantee that a fundamental change in our cosmology, such as a change from a geostatic

[1] It took about twenty-five years before the disturbances of D. C. Miller's repetition of the Michelson–Morley experiment were accounted for in a satisfactory manner. H. A. Lorentz had given up in despair long before that time.

[2] Cf. his [1968*a*].

to a heliostatic point of view, will go hand in hand with an improvement of all the relevant auxiliary subjects. Quite the contrary: such a development is extremely unlikely. Who for example would expect the invention of Copernicanism and of the telescope to be at once followed by the appropriate physiological optics? Basic theories and auxiliary subjects are often 'out of phase'. As a result we obtain refuting instances which do not indicate that a new theory is doomed to failure, but only that it does not fit in at present with the rest of science. This being the case scientists must develop methods which permit them to retain their theories in the face of plain and unambiguously refuting facts, even if testable explanations for the clash are not immediately forthcoming. The principle of tenacity (which I call a 'principle' for mnemonic reasons only) is a first step in the construction of such methods.[1]

Having adopted tenacity we can no longer use recalcitrant facts for removing a theory, T, even if the facts should happen to be as plain and straight-forward as daylight itself. But we can use *other* theories, T', T'', T''', etc. which *accentuate* the difficulties of T while at the same time promising means for their solution. In this case elimination of T is urged by the principle of tenacity itself.[2] Hence, if change of paradigms is our aim, then we must be prepared to introduce and articulate alternatives to T or, as we shall express it (again for mnemonical reasons), we must be prepared to accept a *principle of proliferation*. Proceeding in accordance with such a principle is *one* method of precipitating revolutions. It is a *rational* method. Is it the method which science *actually* uses? Or do scientists stick to their paradigms to the bitter end until disgust, frustration and boredom makes it quite impossible for them to go on? What *does* happen at the end of a normal period? We see that our little methodological fairytale makes us indeed look at history with a sharpened vision.

I am sorry to say that I am quite dissatisfied with what Kuhn has to offer on this point. On the one side he steadfastly emphasizes the dogmatic,[3] authoritarian,[4] and narrowminded[5] features of normal science, the fact that it leads to a temporary 'closing of the mind',[6] that the scientist participating in it 'largely ceases to be an explorer . . . or at least an explorer of the unknown. Instead, he struggles to articulate and concretize the

[1] For details concerning the 'phase difference' between theories and the corresponding auxiliary sciences, cf. my [1969]. The idea already occurs in Lakatos's [1963–4]; it is a commonplace for Lenin and Trotsky (cf. my [1969]).

[2] This is of course not the whole story—but the present sketch suffices entirely for our purpose. Note that Kuhn's argument for tenacity (need for a rational background of argument) is not violated either as the better theory will of course also provide better standards of rationality and excellence.

[3] Kuhn [1961*a*], p. 349.

[4] *Ibid.* p. 393.

[5] *Ibid.* p. 350.

[6] *Ibid.* p. 393.

known . . .'[1] so that 'it is [almost always] the individual scientist rather than [the puzzle-solving tradition, or even some particular] current theory which is tested'.[2] 'Only the practitioner is blamed, not his tools.'[3] He realizes of course that a specific science such as physics may contain more than one puzzle-solving tradition, but he emphasizes their 'quasi-independence', asserting that each of them is 'guided by its own paradigms and pursuing its own problems'.[4] A single tradition therefore will be guided by a single paradigm only. This is one side of the story.

On the other side he points out that puzzle solving is replaced by more 'philosophical' arguments as soon as there exists a choice 'between competing theories'.[5]

Now if normal science is *de facto* as monolithic as Kuhn makes it out to be, then where do the competing theories come from? And if they *do* arise, then why should Kuhn take them seriously and allow them to bring about a change of the argumentative style, from 'scientific' (puzzle solving) to 'philosophical'?[6] I remember very well how Kuhn criticized Bohm for disturbing the uniformity of the contemporary quantum theory. Bohm's theory is *not* permitted to change the argumentative style. Einstein, whom Kuhn mentions in the above quotation, *is* permitted to do so, perhaps because his theory is now more firmly entrenched than Bohm's. Does this mean that proliferation is permitted as long as the competing alternatives are firmly entrenched? But pre-science which has exactly this feature is regarded as inferior to science. Besides, twentieth-century physics *does* contain a tradition which wants to isolate the general theory of relativity from the rest of physics, and restrict it to the very large. Why has Kuhn not supported *this* tradition which is in line with his view of the 'quasi-independence' of simultaneous paradigms? Conversely, if the existence of competing theories involves a change of argumentative style, must we not then doubt this alleged quasi-independence? I have been unable to find a satisfactory answer to these questions in Kuhn's writings.

Let us pursue the point a little further. Kuhn has not only *admitted* that multiplicity of theories changes the style of argumentation. He has also ascribed a definite *function* to such multiplicity. He has pointed out more than once,[7] in complete agreement with our brief methodological remarks, that refutations are impossible without the help of alternatives. Moreover,

[1] Kuhn [1961*a*], p. 363.

[2] *This volume*, p. 5.

[3] *This volume*, p. 7; also cf. Kuhn [1962], p. 79.

[4] Kuhn [1961*a*], p. 388.

[5] *This volume*, p. 7.

[6] 'Philosophical' in Kuhn's (and Popper's) sense and *not* in the sense of, say, contemporary linguistic philosophy.

[7] Cf. Kuhn [1961*b*] and also my acknowledgement in my [1962], p. 32.

he has described in some detail the magnifying effect which alternatives have upon anomalies and has explained how revolutions are brought about by such a magnification.[1] He has therefore said, in effect, that scientists create revolutions in accordance with our little methodological model and *not* by relentlessly pursuing one paradigm and suddenly giving up when the problems get too big.

All this leads now at once to difficulty number three, viz. the suspicion that normal or 'mature' science, as described by Kuhn, *is not even a historical fact.*

6. DOES NORMAL SCIENCE EXIST?

Let us recall what we have so far found to be asserted by Kuhn. First, it is asserted that theories *cannot* be refuted except with the help of alternatives. Secondly, it is asserted that proliferation also plays a *historical role* in the overthrow of paradigms. Paradigms *have been* overthrown because of the way in which alternatives have enlarged existing anomalies. Finally, Kuhn has pointed out that anomalies exist *at any point* of the history of a paradigm.[2] The idea that theories are blameless for decades and even centuries until a big refutation turns up and knocks them out—this idea, he asserts, is nothing but a myth. Now if this is true, then why should we not start proliferating *at once* and *never* allow a purely normal science to come into existence? And is it too much to be hoped that scientists thought likewise, and that normal periods, if they ever existed, cannot have lasted very long and cannot have extended over large fields either? A brief look at one example, viz. the last century, shows that this seems indeed to be the case.

In the second third of that century there existed at least three different and mutually incompatible paradigms. They were: (1) the *mechanical point of view* which found expression in astronomy, in the kinetic theory, in the various mechanical models for electrodynamics as well as in the biological sciences, especially in medicine (here the influence of Helmholtz was a decisive factor); (2) the point of view connected with the invention of an independent and phenomenological *theory of heat* which finally turned out to be inconsistent with mechanics; (3) the point of view implicit in Faraday's and Maxwell's *electrodynamics* which was developed, and freed from its mechanical concomitants, by Hertz.

[1] A minor disturbance, still accessible to treatment 'can be seen, from another viewpoint, as a counterinstance, and thus as a source of crisis' (Kuhn [1962], p. 79). 'Copernicus' astronomical proposal . . . *created* an increasing crisis for . . . the paradigm from which it had sprung' (*ibid.* p. 74, my italics), 'Paradigms are not corrigible by normal science *at all*' (*ibid.* p. 121, my italics).

[2] Kuhn [1962], pp. 80 ff. and p. 145.

Now these different paradigms were far from being 'quasi-independent'. Quite the contrary, it was their *active interaction* which brought about the downfall of classical physics. The troubles leading to the special theory of relativity could not have arisen without the tension that existed between Maxwell's theory on the one side and Newton's mechanics on the other (Einstein has described the situation in beautifully simple terms in his autobiography; Weyl has given an equally brief, though more technical account in *Raum, Zeit, Materie*; Poincaré exhibits this tension already in 1899, and then again in 1904, in his St Louis lecture). Nor was it possible to use the phenomenon of Brownian motion for a direct refutation of the second law of the phenomenological theory.[1] The kinetic theory had to be introduced from the very start. Here again Einstein, following Boltzmann, led the way. The investigations leading up to the discovery of the quantum of action, to mention still another example, brought together such different, incompatible, and occasionally even incommensurable disciplines as mechanics (kinetic theory as used in Wien's derivation of his law of radiation), thermodynamics (Boltzmann's principle of the equal distribution of energy over all degrees of freedom) and wave optics and they would have collapsed had the 'quasi-independence' of these subjects been respected by all scientists. Of course not everyone participated in the debate and the great majority may well have continued attending to their 'tiny puzzles'. However if we take seriously what Kuhn himself is teaching then it was not *this* activity that brought about progress, but the activity of the proliferating minority (and of those experimenters who attended to the problems of this minority, and to their strange predictions). And we may ask whether the majority does not continue solving the old puzzles right through the revolutions. But if this is true then Kuhn's account which *temporally separates* periods of proliferation and periods of monism altogether collapses.[2]

[1] Cf. my discussion in section VI of my [1965*b*].

[2] It might be objected that the puzzle-solving activity, though not *sufficient* for bringing about a revolution, is certainly *necessary* as it creates the material which eventually leads to trouble: puzzle solving is responsible for some conditions on which scientific progress depends. This objection is refuted by the Presocratics who progressed (their theories did not just *change*, they were also *improved*) without paying the slightest attention to puzzles. Of course, they did not produce the pattern: normal science—revolution—normal science—revolution, etc., in which professional stupidity is periodically replaced by philosophical outbursts only to return again at a 'higher level'. However there is no doubt that this is an advantage as it permits us to be open-minded all the time and not only in the middle of a catastrophe. Besides—is not 'normal science' full of 'facts' and 'puzzles' which belong, not to the current paradigm, *but to some earlier predecessors*? And is it not also the case that anomalous facts are often *introduced* by the critics of a paradigm, rather than *used by them* as a starting point for criticism? And if that is true, does it not follow that it is proliferation rather than the pattern normalcy-proliferation-normalcy that characterizes

7. A PLEA FOR HEDONISM

It seems, then, that the interplay between tenacity and proliferation which we described in our little methodological fairytale is also an essential feature of the actual development of science. It seems that it is not the puzzle-solving activity that is responsible for the growth of our knowledge but the active interplay of various tenaciously held views. Moreover, it is the invention of new ideas and the attempt to secure for them a worthy place in the competition that leads to the overthrow of old and familiar paradigms. Such inventing goes on all the time. Yet it is only during revolutions that the attention turns to it. This change of attention does not reflect any profound structural change (such as for example a transition from puzzle solving to philosophical speculation and testing of foundations). It is nothing but a change of interest and of publicity.

This is the picture of science that emerges from our brief analysis. Is it an attractive picture? Does it make the pursuit of science worthwhile? Is the presence of such a discipline, the fact that we have to live with it, study it, understand it, beneficial to us, or is it perhaps liable to corrupt our understanding and diminish our pleasure?

It is very difficult nowadays to approach such questions in the right spirit. What is worthwhile and what is not are to such a large extent determined by the existing institutions and forms of life that we hardly ever arrive at a proper evaluation of these institutions themselves.[1] The sciences especially are surrounded by an aura of excellence which checks any inquiry into their beneficial effect. Phrases such as 'search for the truth', or 'highest aim of mankind' are liberally used. Undoubtedly they ennoble their object, but they also remove it from the domain of critical discussion (Kuhn has gone one step further in this direction, conferring some dignity even on the most boring and most pedestrian part of the scientific enterprise: normal science). Yet why should a product of human ingenuity be allowed to put an end to the very same questions to which it owes its existence? Why should the existence of this product prevent us from asking the most important question of all, the question to what extent the happiness of individual human beings, and to what extent their freedom, has been increased? Progress has always been achieved by probing well-entrenched and well-founded forms of life with unpopular and unfounded values. This is how man gradually freed himself from fear and from the

---

science? So that Kuhn's position would be not only methodologically untenable (see the previous section) but also historically false?

[1] Modern analytic philosophers are trying to show that such evaluation is even *logically impossible*. In this they are but the followers of Hegel—except that they lack his knowledge, his comprehensiveness and his wit.

tyranny of unexamined systems. Our question therefore is: what values shall we choose to probe the sciences of today?

It seems to me that the happiness and the full development of an individual human being is now as ever the highest possible value. This value does not exclude the values which flow from institutionalized forms of life (truth; valour; self-negation; etc.). It rather encourages them *but only* to the extent to which they can contribute to the advance of some individual. What is excluded is the use of institutionalized values for the condemnation, or perhaps even the elimination, of those who prefer to arrange their lives in a different way. What is excluded is the attempt to 'educate' children in a manner that makes them lose their manifold talents so that they become restricted to a narrow domain of thought, action, emotion. Adopting this basic value we want a methodology and a set of institutions which enable us to lose as little as possible of what we are capable of doing and which force us as little as possible to deviate from our natural inclinations.

Now the brief methodological fairytale which we have sketched in section 6, says that a science that tries to develop our ideas and that uses rational means for the elimination of even the most fundamental conjectures must use a principle of tenacity together with a principle of proliferation. It must be allowed to *retain* ideas in the face of difficulties; and it must be allowed to introduce *new ideas* even if the popular views should appear to be fully justified and without blemish. We have also found that actual science, or at least the part of actual science that is responsible for change and for progress, is not very different from the ideal outlined in the fairytale. But this is a happy coincidence indeed! We are now in full agreement with our wishes as expressed above! Proliferation means that there is no need to suppress even the most outlandish product of the human brain. *Everyone may follow his inclinations* and science, conceived as a critical enterprise, will profit from such an activity. Tenacity: this means that one is encouraged not just to follow one's inclinations, but to develop them further, to raise them, with the help of criticism (which involves a comparison with the existing alternatives) to a higher level of articulation *and thereby to raise their defence to a higher level of consciousness.* The interplay between proliferation and tenacity also amounts to the continuation, on a new level, of the biological development of the species and it may even increase the tendency for useful *biological* mutations. It may be the only possible means of preventing our species from stagnation. This I regard as the final and the most important argument against a 'mature' science as described by Kuhn. Such an enterprise is not only ill-conceived and nonexistent; its defence is also incompatible with a humanitarian outlook.

8. AN ALTERNATIVE: THE LAKATOS MODEL OF SCIENTIFIC CHANGE

Let me now present in its entirety the picture of science which I think should replace Kuhn's account.

This picture is the synthesis of the following two discoveries. First, it contains Popper's discovery that science is advanced by a critical discussion of alternative views. Secondly, it contains Kuhn's discovery of the function of tenacity which he has expressed, mistakenly I think, by postulating tenacious *periods*. The synthesis consists in Lakatos's assertion (which is developed in his own comments on Kuhn) that proliferation and tenacity do not belong to *successive* periods of the history of science, but are always *copresent*.[1]

When speaking of 'discoveries' I do not mean to say that the ideas mentioned are entirely new, or that they now appear in a new form. Quite the contrary. Some of these ideas are as old as the hills. The idea that knowledge can be advanced by a struggle of alternative views and that it depends on proliferation was first put forth by the Presocratics (this has been emphasized by Popper himself), and it was developed into a general philosophy by Mill (especially in *On Liberty*). The idea that a struggle of alternatives is decisive for *science*, too, was introduced by Mach (*Erkenntnis und Irrtum*) and Boltzmann (see his *Populaerwissenschaftliche Vorlesungen*), mainly under the impact of Darwinism. The need for tenacity was emphasized by those dialectical materialists who objected to extreme 'idealistic' flights of fancy. And the synthesis, finally, is the very essence of dialectical materialism in the form in which it appears in the writings of Engels, Lenin, and Trotsky. Little of this is known to the 'analytic' or 'empiricist' philosophers of today who are still very much under the influence of the Vienna Circle. Considering this narrow, though quite 'modern' context we may therefore speak of genuine though quite belated, 'discoveries'.

According to Kuhn mature science is a *succession* of normal periods and of revolutions. Normal periods are monistic; scientists try to solve puzzles resulting from the attempt to see the world in terms of a single paradigm. Revolutions are pluralistic until a new paradigm emerges that gains sufficient support to serve as the basis for a new normal period.

This account leaves unanswered the problem how the transition from a normal period to a revolution is brought about. In section 6 we indicated

[1] Lakatos's analysis, I think, can be further improved by abandoning the distinction between theories and research programmes (cf. *above*, p. 203, footnote 2) and by allowing for incommensurability (jumps from quantity to quality in the language of dialectical materialism). Improved in this way it would be a truly dialectical account of the development of our knowledge.

how the transition could be achieved in a reasonable manner: one compares the central paradigm with alternative theories. Professor Kuhn seems to be of the same opinion. Moreover he points out that this is what actually happens. Proliferation sets in already *before* a revolution and is instrumental in bringing it about. But this means that the original account is faulty. Proliferation does not *start* with a revolution; it *precedes* it. A little imagination and a little more historical research then shows that proliferation not only *immediately precedes* revolutions, but that it is there *all the time*. Science as we know it is not a temporal succession of normal periods and of periods of proliferation; it is their *juxtaposition*.

Seen in this way the transition from pre-science to science does not *replace* the uninhibited proliferation and the universal criticism of the former by the puzzle-solving tradition of a normal science. It *supplements* it by this activity or, to express it even better, mature science *unites* two very different traditions which are often separate, the tradition of a pluralistic philosophical criticism and a more practical (and less humanitarian—see section 8) tradition which explores the potentialities of a given material (of a theory; of a piece of matter) without being deterred by the difficulties that might arise and without regard to alternative ways of thinking (and acting). We have learned from Professor Popper that the first tradition is closely connected with the cosmology of the Presocratics. The second tradition is best exemplified by the attitude of the members of a closed society towards their basic myth. Kuhn has conjectured that mature science consists in the *succession* of these two different patterns of thought and action. He is right in so far as he has noticed the normal, or conservative, or anti-humanitarian element. This is a genuine discovery. He is wrong as he has misrepresented the relation of this element to the more philosophical (i.e. critical) procedures. I suggest in accordance with Lakatos's model that the correct relation is one of *simultaneity* and *interaction*. I shall therefore speak of the normal *component* and the philosophical *component* of science and not of the normal *period* and the *period* of revolution.

It seems to me that such an account overcomes many difficulties, both logical and factual, which make Kuhn's point of view so fascinating but at the same time so unsatisfactory.[1] In considering it one should not be

[1] To take but one example, Kuhn writes (*this volume*, p. 6) that 'it is for the normal, not the extraordinary practice of science that professionals are trained; if they are nevertheless eminently successful in displacing and replacing the theories on which normal science depends, that is an oddity which must be explained'. It is certainly an oddity in Kuhn's account. In our account we only need to draw attention to the fact that revolutions are mostly made by members of the philosophical component who, while aware of the normal practice, are also able to think in a different way (in the case of Einstein the self-professed ability to escape from the normal training was essential for his freedom of thought and for his discoveries).

misled by the fact that the normal component almost always outweighs its philosophical part. For what we are investigating is not the size of a certain element of science, but its *function* (a single man can revolutionize an epoch). Nor must we be overly impressed by the fact that most scientists would ***regard*** the 'philosophical' component as lying outside science proper and that they could *support* this attitude by pointing to their own lack of philosophical acumen. For it is not *they* who carry out fundamental improvement but those who further the *active interaction* of the normal and the philosophical component (this interaction consists almost always in the criticism of what is well entrenched and unphilosophical by what is peripheral and philosophical). Now, granting all this, why is it that there seems to exist a definite fluctuation in the state of science? If science consists of the constant interaction of a normal and a philosophical part; if it is this interaction which advances it; then why do the revolutionary elements become visible only on such rare occasions? Is not this simple historical fact sufficient to support Kuhn's account over mine? Is it not typical philosophical sophistry to deny what is such an obvious historical fact?

I think that the answer to this question is obvious. The normal component is large and well entrenched. Hence, a change of the normal component is very noticeable. So is the resistance of the normal component to change. This resistance becomes especially strong and noticeable in periods where a change seems to be imminent. It is directed against the philosophical component and brings it into public consciousness. The younger generation, always eager for new things, seizes upon the new material and studies it avidly. Journalists, always on the lookout for headlines—the more absurd, the better—publicize the new discoveries (which are those elements of the philosophical component which most radically disagree with the current views while still possessing some plausibility and perhaps even some factual support). These are some reasons for the differences which we perceive. I do not think that one should look for anything more profound.

Now as regards the change of the normal component itself there is no reason to expect that it will follow a clearly recognizable and logical pattern. Kuhn like other philosophers before him (I am here mainly thinking of Hegel) assumes that a tremendous historical change must exhibit a logic of its own and that the change of an idea must be reasonable in the sense that there exists a link between the *fact* of change and the *content* of the idea changing. This is a plausible assumption as long as one is dealing with reasonable people: changes in the *philosophical component* most likely *can* be explained as the result of clear and unambiguous *arguments*. But to assume that people who habitually resist change; who frown at any criticism of things dear to them; and whose highest aim is to solve puzzles on a basis that is neither known nor understood; to assume that *such* people will change their allegiance in a reasonable fashion is carrying optimism and the quest for rationality too far. The normal elements, i.e. those elements which have the support of the majority, may change because the younger generation cannot be bothered to follow their elders; or because some public figure has changed his mind; or because some influential member of the establishment has died and has failed (perhaps because of his suspicious nature) to leave behind a strong and influential school, or because a powerful and non-scientific institution pushes thought in a definite direction.[1] Revolutions, then, are the outward manifestation of a change of the normal component that cannot be accounted for in any reasonable fashion. They are substance for anecdotes though they magnify and make visible the more rational elements of science, thus teaching us what science *could* be if there were more reasonable people around.

9. THE ROLE OF REASON IN SCIENCE

(1) So far I have *criticized* Kuhn from a point of view which is almost identical with that of Lakatos. (There are some slight differences, such as my reluctance to separate theories and research programmes,[2] but they will be disregarded. When speaking of 'theories' I always mean theories and/or research programmes.) I now want to *defend* Kuhn against Lakatos. More specifically, I want to argue that science both is, and should be, more irrational than Lakatos and Feyerabend$_1$ (the Popperian$_3$ author of

[1] It is plausible to assume that *one* of the causes for the transition to mature science with its various 'quasi-independent' traditions is to be sought in the decree of the Roman Catholic Church against the Copernican point of view. 'This must be taken into account by those who try to explain the special development of the many individual sciences and the absence of a conscious and secure philosophical background by regarding it as a peculiarity of seventeenth-century Italian culture.... Such an interpretation assumes ... that the condemnation of Galileo was but an *external* pressure which could not possibly have influenced the development of spiritual matters. However the Roman Judgement was regarded as a restriction of consciousness that could be broken only on pain of life and salvation.... The development of individual disciplines was allowed. Nobody was prevented from searching the heavens, from exploring physical phenomena, from thinking mathematically ... and from furthering the material culture by such a pursuit. Priests and religious orders, even the Jesuits who were responsible for Galileo's fate, diligently pursued these restricted tasks. But individual conscience as well as the omnipresent 'directeurs de conscience', the officials, the schools, the churches, the state watched carefully this simple fight for knowledge in order that no one might dare to use its results for philosophical speculation'. (Leonardo Olschki [1927], p. 400). *This is how 'mature science' came into being*, at least in the Roman countries. Cf. also chapter IX of Wohlwill's [1926] where the development after Galileo's death is sketched in some detail.

[2] Cf. *above*, p. 203, footnote 2.

the preceding sections of this paper and of 'Problems of Empiricism') are prepared to admit.[1]

This transition from criticism to defence does not mean that I have changed my mind. Nor can it be completely explained by my cynicism *vis-à-vis* the business of philosophy of science. It is rather connected with the nature of science itself, with its complexity, with the fact that it has different aspects, that it cannot be readily separated from the remainder of history, that it has always utilized and continues to utilize every talent and every folly of man. Contrary arguments bring out the different features it contains, they challenge us to make a decision, they challenge us to either *accept* this many-faced monster and be devoured by it, or else to *change* it in accordance with our wishes. Let us now see what can be said against the Lakatos model of scientific growth.

(2) Naive falsificationism judges (i.e. accepts, or condemns) a theory as soon as it is introduced into the discussion. Lakatos gives a theory time, he permits it to develop, he permits it to show its hidden strength, and he judges it only 'in the long run'. The 'critical standards' he employs provide for an interval of hesitation. They are applied 'with hindsight'.[2] They are applied *after* the occurrence of either 'progressive' or of 'degenerating' problem shifts.

Now it is easy to see that standards of this kind have practical force only if they are combined with a *time limit* (what looks like a degenerating problem shift may be the beginning of a much longer period of advance). But introduce the time limit and the argument against naive falsificationism reappears with only a minor modification (if you are permitted to wait, why not wait a little longer?) Thus the standards which Lakatos wants to defend are either *vacuous*—one does not know when to apply them—or they can be *criticized* on grounds very similar to those which led to them in the first place.

In these circumstances one can do one of the following two things. One can *stop* appealing to permanent standards which remain in force throughout history and govern every single period of scientific development and every transition from one period to another. Or one can retain such standards as a *verbal ornament*, as a memorial to happier times when it was still thought possible to run a complex and often catastrophic business like science by following a few simple and 'rational' rules. It seems that Lakatos wants to choose the second alternative.

[1] The indices are intended as an ironical criticism of Lakatos [1968*b*] where the practice of splitting a guy into three was first introduced. (Also cf. *this volume*, p. 181.) This practice has created a lot of confusion and has slowed down philosophers in their attempt to find the weak spots of critical rationalism.

[2] *This volume*, pp. 134, 158, and 173.

(3) Choosing the second alternative means abandoning permanent standards *in fact* though retaining them *in words*. *In fact*, Lakatos's position now seems to be identical with the position of Popper as summarized in a (because self-destructive) marvellous addendum of the fourth edition of the *Open Society*.[1] According to Popper we do not 'need any . . . definite frame of reference for our criticism', we may revise even the most fundamental rules and drop the most fundamental demands if the need for a different measure of excellence should arise.[2] Is such a position irrational? Does it imply that science is irrational? *Yes and no. Yes*—because there no longer exists a single set of rules that will guide us through all the twists and turns of the history of thought (science), either as participants, or as historians who want to reconstruct its course. One can of course *force* history into such a pattern, but the results will always be poorer and much less interesting than were the actual events. *No*—because each particular episode is rational in the sense that some of its features can be explained in terms of reasons which were either accepted at the same time as its occurrence, or invented in the course of its development. *Yes*—because even these logical reasons which change from age to age are never sufficient to explain *all* the important features of a particular episode. We must add accidents, prejudices, material conditions (such as the existence of a particular type of glass in one country and not in another), the vicissitudes of married life, oversight, superficiality, pride, and many other things in order to get a complete picture. *No*—because transported into the climate of the period under consideration and endowed with a lively and curious intelligence we might have had still more to say, we might have tried to overcome accidents, and to 'rationalize' even the most whimsical sequence of events. But—and now we come to a decisive point—how is the transition from certain standards to other standards to be achieved? More especially, what happens to our standards (as opposed to our theories) during a period of revolution? Are they changed in the Popperian manner, by a critical discussion of alternatives, or are there processes which defy a rational analysis? This is one of the questions raised by Kuhn. Let us see what answer we can give to it!

(4) That standards are not always adopted on the basis of argument has been emphasized by Popper himself. Children, he says, 'learn to imitate others . . . and so learn to look upon standards of behaviour as if they consisted of fixed, "given" rules . . . and such things as sympathy and imagination may play an important role in this development'.[3] Similar considerations apply to those grownups who want to continue learning and

[1] Popper [1961], p. 388.
[2] *Loc. cit.* p. 390.
[3] *Loc. cit.* p 390.

who are intent on expanding both their knowledge and their sensibility. We certainly cannot assume that what is possible in the case of children—to slide, on the smallest provocation, into entirely new reaction patterns—should be beyond the reach of adults and inaccessible to one of the most outstanding adult activities, science. Moreover, it is likely that catastrophic changes, frequent disappointment of expectations, crises in the development of our knowledge will change and, perhaps, multiply reaction patterns (including patterns of argumentation) just as an ecological crisis multiplies mutations. This may be an entirely *natural* process, like growing in size, and the only function of rational discourse may consist in increasing the mental tension that precedes *and causes* the behavioural outburst. Now—is this not exactly the kind of change we may expect at periods of scientific revolution? Does it not restrict the effectiveness of arguments (except as a causative agent leading to developments very different from what is demanded by their *content*)? Does not the occurrence of such a change show that science which, after all, is part of the evolution of man is not entirely rational and cannot be entirely rational? For if there are events, not necessarily arguments which *cause* us to adopt new standards, will it then not be up to the defenders of the status quo to provide, not just arguments, but also *contrary causes*? And if the old forms of argumentation turn out to be too weak a contrary cause, must they then not either give up, or resort to stronger and more 'irrational' means? (It is very difficult, and perhaps entirely impossible, to combat the effects of brainwashing by argument.) Even the most puritanical rationalist will then be forced to leave argument and to use, say, *propaganda* not because some of his arguments have ceased to be *valid*, but because the *psychological conditions* which enable him to effectively argue in this manner and thereby to influence others have disappeared. And what is the use of an argument that leaves people unmoved?

(5) Considering questions such as these a Popperian will reply that new standards may indeed be discovered, invented, accepted, imparted upon others in a very irrational manner, but that there always remains the possibility to criticize them *after* they have been adopted and that it is this possibility which keeps our knowledge rational. 'What, then, are we to trust?' asks Popper after a survey of possible sources for standards.[1] 'What are we to accept? The answer is: whatever we accept we should trust only tentatively, always remembering that we are in possession, at best, of partial truth (or rightness), and that we are bound to make at least some mistake or misjudgement somewhere—not only with respect to facts but also with respect to the adopted standards; secondly, we should trust (even

[1] *Loc. cit.* p. 391.

tentatively) our intuition only if it has been arrived at as the result of many attempts to use our imagination; of many mistakes, of many tests, of many doubts, and of searching criticism.'

Now this reference to tests and to criticism which is supposed to guarantee the rationality of science and, perhaps, of our entire life may be either to *well defined procedures* without which a criticism or test cannot be said to have taken place, or it may be purely *abstract* so that it is left to us to fill it now with this, and now with that concrete content. The first case has just been discussed. In the second case we have but a verbal ornament, just as Lakatos's defence of his own 'objective standards' turned out to be a verbal ornament. The questions of section 4 remain unanswered in either case.

(6) In a way even this situation has been described by Popper who says that 'rationalism is necessarily far from comprehensive or self-contained'.[1] But the question raised by Kuhn is not whether *there are* limits to our reason; the question is *where* these limits are *situated*. Are they outside the sciences so that science itself remains entirely rational, or are irrational changes an essential part of even the most rational enterprise that has been invented by man? Does the historical phenomenon 'science' contain ingredients which defy a rational analysis? Can the abstract aim to come closer to the truth be reached in an entirely rational manner, or is it perhaps inaccessible to those who decide to rely on argument only? These are the problems to which we must now address ourselves.

(7) Considering these further problems Popper and Lakatos reject 'mob psychology'[2] and assert the rational character of *all* science. According to Popper it is possible to arrive at a judgement as to which of two theories is closer to the truth, even if the theories should be separated by a catastrophic upheaval such as a scientific revolution. (A theory T is closer to the truth than another theory, T', if the class of the true consequences of T', the so-called truth content of T', exceeds the class of true consequences of T without an increase in the falsity content.) According to Lakatos the apparently unreasonable features of science occur only in the material world and in the world of (psychological) thought; they are absent from the 'world of ideas, [from] Plato's and Popper's "third world" '.[3] It is in this third world that the growth of knowledge takes place and that a rational judgement of all aspects of science becomes possible. It must be pointed out, however, that the scientist is unfortunately dealing with the world of matter and of (psychological) thought also and that the rules which create order in the third world may be entirely inappropriate for creating order

[1] Popper [1945], chapter 24.

[2] *This volume*, p. 178.

[3] *This volume*, p. 180.

in the brains of living human beings (unless these brains and their structural features are put into the third world, a point that does not become clear from Popper's account).[1] The numerous deviations from the straight path of rationality which we observe in actual science may well be *necessary* if we want to achieve progress with the brittle and unreliable material (instruments; brains; etc.) at our disposal.

However there is no need to pursue this objection further. There is no need to argue that real science may differ from its third world image *in precisely those respects* which make progress possible.[2] For the Popperian model of an approach to the truth breaks down even if we confine ourselves to ideas entirely. It breaks down because there are *incommensurable theories*.

(8) With the discussion of incommensurability, I come to a point of Kuhn's philosophy which I wholeheartedly accept. I am referring to his assertion that succeeding paradigms can be evaluated only with difficulty and that they may be altogether incomparable, at least as far as the more familiar standards of comparison are concerned (they may be readily comparable in other respects). I do not know who of us was the first to use the term 'incommensurable' in the sense that is at issue here. It occurs in Kuhn's '*Structure of Scientific Revolutions*' and in my essay 'Explanation, Reduction, and Empiricism' both of which appeared in 1962. I still remember marvelling at the pre-established harmony that made us not only defend similar ideas but use exactly the same words for expressing them. The coincidence is of course far from mysterious. I had read earlier drafts of Kuhn's book and had discussed their content with Kuhn. In these discussions we both agreed that new theories, while often better and more detailed than their predecessors were not always rich enough to deal with *all* the problems to which the predecessor had given a definite and precise answer. The growth of knowledge or, more specifically, the replacement of one comprehensive theory by another involves losses as well as gains. Kuhn was fond of comparing the scientific world view of the seventeenth century with the Aristotelian philosophy, while I used more recent examples such as the theory of relativity and the quantum theory. We also saw that it might be extremely difficult to compare successive theories in

[1] I am here referring to Popper [1968*a*] and Popper [1968*b*]. In the first paper birdnests are assigned to the 'Third World' (p. 341) and an interaction is assumed between them and the remaining worlds. They are assigned to the Third World *because of their function*. But then stones and rivers can be found in this third world, too, for a bird may sit on a stone, or take a bath in a river. As a matter of fact, everything that is noticed by some organism (and therefore plays a role in his *Umwelt*) will be found in the third world which will therefore contain the whole material world and all the mistakes mankind has made. It will also contain 'mob psychology'.

[2] Cf. my [1969].

the usual manner, that is, by an examination of consequence classes. The accepted scheme is as follows (*Fig. 1*): T is superseded by T'. T' explains why T fails where it does (in F); it also explains why T has been at least partly successful (in S); and makes additional predictions, (A). Now if this scheme is to work then there must be statements which follow (with, or without the help of definitions and/or correlation hypotheses) both from T and from T'. But there are cases which invite a comparative judgement without satisfying the conditions just stated. The relation between such theories is as shown in *Fig. 2*.[1] A judgement involving a comparison of content classes is now clearly impossible. For example, T' cannot be said to be either closer to, or farther from, the truth, than T.

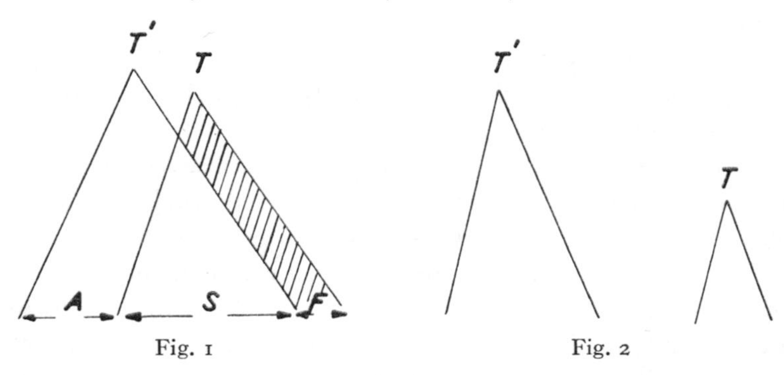

Fig. 1 Fig. 2

(9) As an example of two incommensurable theories let us briefly discuss classical celestial mechanics (CM) and the special theory of relativity (SR). To start with one should emphasize that the question 'are CM and SR incommensurable?' is not a complete question. Theories can be interpreted in different ways. They will be commensurable in some interpretations, incomparable in others. Instrumentalism, for example, makes commensurable all those theories which are related to the same observation language and are interpreted on its basis. A realist, on the other hand, wants to give a unified account, both of observable and of unobservable matters, and he will use the most abstract terms of whatever theory he is contemplating for that purpose. This is an entirely natural procedure. SR, so one would be inclined to say, does not just invite us to rethink *unobserved* length, mass, duration; it would seem to entail the relational character of *all* lengths, masses, durations, whether observed or unobserved, observable or unobservable. Now extending the concepts of a new theory T to all its consequences, observational reports included, may change the interpretation of these consequences to such an extent that they disappear from the consequence classes of earlier theories. These earlier theories will then all

[1] The area below T' should be imagined as lying either *in front of* the area below T, *or behind it*, so that there is no overlap.

become incommensurable with *T*. The relation between SR and CM is a case in point. The concept of length as used in SR and the concept of length as presupposed in CM are different concepts. Both are *relational* concepts, and very complex relational concepts at that (just consider determination of length in terms of the wave length of a specified spectral line). But relativistic length (or relativistic *shape*) involves an element that is absent from the classical concept and is in principle excluded from it.[1] It involves the *relative velocity* of the object concerned in some reference system. It is of course true that the relativistic scheme very often gives us *numbers* which are practically identical with the numbers we get from CM—but this does not make the *concepts* more similar. Even the case $c \to \infty$ (or $v \to \infty$) which gives *strictly identical* predictions cannot be used as an argument for showing that the concepts must coincide at least in this case: different magnitudes based on different concepts may give identical

[1] It is possible to base space time frames on this new element only and to avoid contamination by earlier modes of thought. All one has to do is to replace distances by light-times and to treat time intervals in the relativistic fashion, for example, by using the *k*-calculus. (Cf. chapter II of Synge [1964]. For the *k*-calculus, cf. Bondi [1967], pp. 29 ff., as well as Bohm [1965], chapter xxvi.) The resulting concepts (of distance, velocity, time, etc.) are a necessary part of relativity in the sense that all further ideas such as the idea of length as defined by the transport of rigid rods must be changed and adapted to them. They therefore suffice for explaining relativity.

Marzke and Wheeler [1963] have given a detailed account of the way in which the theory of relativity can be freed from external ingredients. They adopt the principle, ascribed by them to Bohr and Rosenfeld, 'that every proper theory should provide in and by itself its own means for defining the quantities with which it deals. According to this principle classical general relativity should admit to calibrations of space and time that are altogether free of any reference to the quantum of action [for atomic clocks, or minimal distances]' or to 'rigid rods' as described by, say, the non relativistic theory of elasticity (p. 48). They proceed to construct clocks and meters which use the properties of light and of inertial particle trajectories only (pp. 53–6). Equality of distances measured by such clocks and meters is intransitive in a classical universe, transitive in a relativistic universe. The results of distance measurements of this kind are invariant to translations in a relativistic universe, not so invariant in a classical universe. Two different events are always separated by a finite distance in a relativistic universe, they are not always so separated in a classical universe. The unity of measurement in the relativistic universe is the interval between the two effective equinoxes of 1900 and it can be compared with any interval (spatial or temporal) in an invariant way. No such comparison is possible in the classical case (p. 62). 'The number 3.10^8 never shows itself. The importance of lightrays and the lightcone in the intrinsic geometry of physics comes more directly to the surface. The true function of the speed of light is no longer confused with the trivial task of relating two separate units of interval, the meter and the second, of purely historical and accidental origin' (p. 56). General relativity theory, then, can be shown to 'provide its own means of defining intervals of space and time' (p. 62) and the intervals so defined are incommensurable with classical intervals.

Space forbids to argue this interesting case in detail but it is hoped that those who are turned on by the problem of incommensurability will use Marzke and Wheeler as a basis for concrete discussion.

values on their respective scales without ceasing to be different magnitudes (the same remark applies to the attempt to identify classical mass with relative *rest* mass).[1] This conceptual disparity, if taken seriously, infects even the most 'ordinary' situations: the relativistic concept of a certain shape, such as a table, or of a certain temporal sequence, such as my saying 'yes', will differ from the corresponding classical concept also. It is therefore vain to expect that sufficiently long derivations may eventually return us to the older ideas.[2] The consequence classes of SR and CM are related as in Fig. 2. A comparison of content and a judgement of verisimilitude cannot be made.[3]

(10) In what follows I shall discuss a few objections which have been raised, not against this *particular* analysis of the relation between SR and CM, but against the very *possibility*, or *desirability* of incommensurable theories (almost all objections against incommensurability are of this general kind). They express methodological ideas which we must criticize if we want to increase our freedom *vis-à-vis* the sciences.

One of the most popular objections proceeds from the version of realism that I just described in (9). 'A realist', we said, 'wants to give a unified account, both of observable and of unobservable matters, and he will use the most abstract terms of whatever theory he is contemplating for that purpose'. He will use such terms in order to either *give* meaning to observation sentences, or else to *replace* their customary interpretation (for example, he will use the ideas of SR in order to replace the customary CM-interpretation of everyday statements about shapes, temporal sequences, and so on). As against this it is pointed out that theoretical terms receive their interpretation by being connected either with a pre-existing observation language, or with another theory that has already been connected with such an observation language and that they are devoid of content without such a connection. Thus Carnap asserts[4] that 'there is no independent interpretation for L_T [the language in terms of which a certain theory, or a certain world view, is formulated]. The system T [consisting of the axioms of the theory and the rules of derivation] is itself an uninterpreted postulate

[1] For this point and further arguments, cf. Eddington [1924], p. 33.

[2] This takes care of an objection which John Watkins has raised on various occasions.

[3] For further details, especially concerning the concept of mass, the function of 'bridge laws' or 'correspondence rules', and the two-language model, cf. section IV of my [1965*b*]. It is clear that, given the situation described in the text, we cannot derive classical mechanics from relativity, not even approximately (for example, we cannot derive the classical law of mass conservation from a corresponding relativistic law). The possibility to connect the formulae of the two disciplines in a manner that might satisfy a pure mathematician (or an instrumentalist) is however not excluded. For an analogous situation in the case of quantum mechanics cf. section 3 of my [1968–9]. Cf. also section 2 of the same article for more general considerations.

[4] Cf. Carnap [1956], p. 47.

system. [Its] terms obtain only an indirect and incomplete interpretation by the fact that some of them are connected by the [correspondence rules] C with observational terms'. Now, if theoretical terms have no 'independent interpretation' then they cannot be used for correcting the interpretation of the observation statements which is the one and only source of their meaning. It follows that realism as described by us is an impossible doctrine.

The guiding idea behind this objection is that new and abstract languages cannot be introduced in a direct way but must be first connected with an already existing, and presumably stable, observational idiom.[1]

This guiding idea is refuted at once by pointing to the way in which children learn to speak and in which anthropologists and linguists learn the unknown language of a newly discovered tribe.

The first example is instructive for other reasons also, for incommensurability plays an important role in the early months of human development. As has been suggested by Piaget and his school,[2] the child's perception develops through various stages before it reaches its relatively stable adult form. In one stage objects seem to behave very much like afterimages[3]—and they are treated as such: the child follows the object with his eyes until it disappears and he does not make the slightest attempt to recover it even if this would require a minimal physical (or intellectual) effort, an effort moreover, that is already within the child's reach. There is not even a tendency to search—and this is quite appropriate, 'conceptually' speaking. For it would indeed be nonsensical to 'look for' an afterimage. Its 'concept' does not provide for such an operation.

The arrival of the concept, and of the perceptual image, of material objects changes the situation quite dramatically. There occurs a drastic reorientation of behavioural patterns and, so one may conjecture, of thought. Afterimages or things somewhat like them still exist, but they are now difficult to find and must be discovered by special methods (the earlier visual world therefore *literally disappears*). Such methods proceed from a new conceptual scheme (afterimages occur in *humans*, not in the outer physical world, and are tied to them) and cannot lead back to the exact

[1] An even more conservative principle is sometimes used when discussing the possibility of languages with a logic different from our own. Thus Stroud, in his [1968], *discussing*, and not just stating the principle, says that 'any allegedly new possibility must be capable of being fitted into, or understood in terms of, our present conceptual or linguistic apparatus' from which it follows (172) that 'any "alternative" is either something we already understand and can make sense of, or it is no alternative at all'. What is overlooked is that an initially ununderstood alternative may be *learned* in the way in which one learns a new and unfamiliar language, not by *translation*, but by *living* with the members of the community where the language is spoken.

[2] As an example the reader is invited to consult Piaget [1954].

[3] Piaget [1954], pp. 5 ff.

phenomena of the previous stage (these phenomena should therefore be called by a different name, such as 'pseudo-afterimages'). Neither afterimages, nor pseudo-afterimages are given a special position in the new world. For example, they are not treated as *evidence* on which the new notion of a material object is supposed to rest. Nor can they be used to *explain* this notion: afterimages arise *together with it* and are absent from the mind of those who do not yet recognize material objects; and pseudo-afterimages *disappear* as soon as such recognition takes place. It is to be admitted that every stage possesses a kind of observational 'basis' to which one pays special attention and from which one receives a multitude of suggestions. However this basis (*1*) *changes* from stage to stage; and (*2*) it is *part* of the conceptual apparatus of a given stage, *not* its one and only source of interpretation.

Considering developments such as these we may suspect that the family of concepts centering upon 'material object' and the family of concepts centering upon 'pseudo-afterimages' are incommensurable in precisely the sense that is at issue here. Is it reasonable to expect that conceptual changes of this kind occur only in childhood? Should we welcome the fact—if it is a fact—that an adult is stuck with a stable perceptual world and an accompanying stable conceptual system which he can modify in many ways but whose general outlines have forever become immobilized? Or is it not more realistic to assume that fundamental changes, entailing incommensurability, are still possible, and that they should be encouraged lest we remain forever excluded from what might be a higher stage of knowledge and of consciousness? Besides, the question of the mobility of the adult stage is at any rate an empirical question which must be attacked by *research* and cannot be settled by methodological *fiat*. An attempt to break through the boundaries of a given conceptual system and to escape the range of 'Popperian spectacles'[1] is an essential part of such research.[2]

(11) Looking now at the second element of the refutation—anthropological field work—we see that what is anathema here (and for very good

[1] Cf. Lakatos's paper, *this volume*, p. 179, footnote 1.

[2] For the condition of research formulated in the last sentence, cf. section 8 of my [1965*a*]. For the role of observation cf. section 7 of the same article. For the application of Piaget's work to physics and, more especially, to the theory of relativity, cf. the appendix of Bohm [1965]. Bohm and Schumacher have also carried out an analysis of the different informal structures which underlie our theories. One of the main results of their work is that Bohr and Einstein argued from incommensurable points of view. Seen in this way the case of Einstein, Podolski and Rosen cannot refute the Copenhagen interpretation, and it cannot be refuted by it. The situation is rather that we have two theories, one permitting us to formulate the Einstein–Podolski–Rosen thought-experiment, the other not providing the machinery necessary for such a formulation so that we must find independent means of deciding which one to adopt. For further comments on this problem, cf. section 9 of my [1968–9].

reasons) is still a fundamental principle for the contemporary representatives of the philosophy of the Vienna Circle. According to Carnap, Feigl, Nagel, and others the terms of a theory receive their interpretation, in an indirect fashion, by being related to a different conceptual system which is either an older theory, or an observation language.[1] Older theories, or observation languages are adopted not because of their theoretical excellence (they cannot possibly be: the older theories are usually refuted). They are adopted because they are 'used by a certain language community as a means of communication'.[2] According to this method, the phrase 'having much larger relativistic mass than . . . ' is partially interpreted by first connecting it with some *prerelativistic terms* (classical terms; commonsense terms) which are 'commonly understood' (presumably as the result of previous teaching in connection with crude weighing methods). This is even worse than the once quite popular demand to clarify doubtful points by translating them into Latin. For while Latin was chosen because of its precision and clarity and also because it was conceptually richer than the slowly evolving vulgar idioms, the choice of an observation language or of an older theory as a basis for interpretation is due to the fact that they are 'antecedently understood', it is due to their *popularity*. Besides, if prerelativistic terms which are pretty far removed from reality—especially in view of the fact that they come from an incorrect theory—can be taught ostensively, for example, with the help of crude weighing methods (and we must assume that they can be so taught, or the whole scheme collapses) then why should we not introduce the relativistic terms *directly*, and *without* assistance from the terms of some other idiom? Finally, it is but plain commonsense that the teaching, or the learning, of new and unknown languages must not be contaminated by external material. Linguists remind us that a perfect translation is never possible, even if we use complex contextual definitions. This is one of the reasons for the importance of *field work* where new languages are learned *from scratch* and for the rejection, as inadequate, of any account that relies on (complete, or partial) translation. *Yet just what is anathema in linguistics is now taken for granted by logical empiricists*, a mythical 'observation language' replacing the English of the translators. Let us commence field work in this domain also and let us study the language of new theories not in the definition factories of the double language model, but in the company of those metaphysicians, experimenters, theoreticians, playwrights, courtesans, who have constructed new world views! This finishes our discussion of the guiding principle of the first objection against realism and the possibility of incommensurable theories.

[1] For what follows, cf. also my [1966].

[2] Carnap [1956], p. 40. Cf. also Hempel [1966], pp. 74 ff.

(12) Next I shall deal with a mixed bag of asides which have never been presented in a systematic fashion and which can be disposed of in a few words.

To start with, there is the suspicion that observations which are interpreted in terms of a new theory can no longer be used to refute that theory. The suspicion is allayed by pointing out that the predictions of a theory depend on its postulates, the associated grammatical rules *as well as* on initial conditions, while the meaning of the primitive notions depends on the postulates (and the associated grammatical rules) only: it is possible to refute a theory by an experience that is entirely interpreted in its terms.

Another point that is often made is that there exist *crucial experiments* which refute one or two allegedly incommensurable theories and confirm the other, for example: the Michelson–Morley experiment, the variation of the mass of elementary particles, the transversal Doppler effect refute CM and confirm SR. The answer to this problem is not difficult either: adopting the point of view of relativity we find that the experiments *which of course will now be described in relativistic terms*, using the relativistic notions of length, duration, speed, and so on,[1] are relevant to the theory and we shall also find that they support the theory. Adopting CM (with, or without an aether) we again find that the experiments (which are now described in the very different terms of classical physics, roughly in the manner in which Lorentz described them) are relevant, but we also find that they *undermine* (the conjunction of classical electrodynamics and of) CM. Why should it be necessary to possess terminology that allows us to say that it is the *same* experiment which confirms one theory and refutes the other? But did we not ourselves use such terminology? Well, for one thing it should be easy, though somewhat laborious, to express what was just said *without* asserting identity. Secondly, the identification is of course not contrary to our thesis, for we are now not *using* the terms of either relativity, or of classical physics, as is done in a test, but are *referring to* them and their relation to the physical world. The language in which *this* discourse is carried out can be classical, or relativistic, or ordinary. It is no good insisting that scientists act as if the situation were much less complicated. If they act that way, then they are either instrumentalists (see above, section 9) or mistaken: many scientists are nowadays interested in *formulae* while we are discussing *interpretations*. It is also possible that being well acquainted with both CM and SR they change back and forth between these theories with such speed that they seem to remain within a single domain of discourse.

(13) It is also said that in admitting incommensurability into science we

[1] For examples of such descriptions cf. Synge [1964].

can no longer decide whether a new view explains what it is supposed to explain or whether it does not wander off into different fields. For example, we would not know whether a newly invented physical theory is still dealing with problems of space and time or whether its author has not by mistake made a biological assertion. But there is no need to possess such knowledge. For once the fact of incommensurability has been admitted the question which underlies the objection does not arise (conceptual progress often makes it impossible to ask certain questions; thus we can no longer ask for the absolute velocity of an object—at least as long as we take relatively seriously). Yet is this not a serious loss for science? Not at all! Progress was made by the very same 'wandering off into different fields' whose undecidability now so greatly exercises the critic: Aristotle saw the world as a super*organism*, that is, as a *biological* entity, while one essential element of the new science of Descartes, Galileo, and of their followers in medicine and in biology is its exclusively *mechanistic* outlook. Are such developments to be forbidden? And if they are not, then what is left of the complaint?

A closely connected objection starts from the notion of *explanation*, or *reduction*, and emphasizes that this notion presupposes continuity of concepts (other notions could be used for starting exactly the same kind of argument). Now to take our above example, relativity is supposed to explain the valid parts of classical physics, hence it cannot be incommensurable with it! The reply is again obvious. Why should the relativist be concerned with the fate of classical mechanics except as part of a historical exercise? There is only *one* task we can legitimately demand of a theory and it is that it should give us a correct account of the *world*. What have the principles of explanation got to do with this demand? Is it not reasonable to assume that a point of view such as the point of view of classical mechanics that has been found wanting in various respects cannot have entirely adequate concepts, and is it not equally reasonable to try replacing its concepts by those of a more successful cosmology? Besides, why should the notion of explanation be burdened by the demand for conceptual continuity? This notion has been found to be too narrow before (demand of derivability) and it had to be widened so as to include partial and statistical connections. Nothing prevents us from widening it still further to admit, say, 'explanation by equivocation'.

(14) Incommensurable theories, then, can be *refuted* by reference to their own respective kinds of experience (in the absence of commensurable alternatives these refutations are quite weak, however).[1] Their *content* cannot be compared. Nor is it possible to make a judgement of *verisimilitude*

[1] For this point cf. section 1 of my [1965*a*], as well as my [1965*b*].

except within the confines of a particular theory. None of the methods which Popper wants to use for rationalizing science can be applied and the one that can be applied, refutation, is greatly reduced in strength. What remains are aesthetic judgements, judgements of taste, and our own subjective wishes. Does this mean that we are ending up in subjectivism? Does this mean that science has become arbitrary, that it has become one element of the general relativism which Popper wants to attack? Let us see.

To start with, it seems to me that an enterprise whose human character can be seen by all is preferable to one that looks 'objective', and impervious to human actions and wishes.[1] The sciences, after all, are our own creation, including all the severe standards they seem to impose upon us. It is good to be constantly reminded of this fact. It is good to be constantly reminded of the fact that science as we know it today is not inescapable and that we may construct a world in which it plays no role whatever (such a world, I venture to suggest, would be more pleasant than the world we live in today). What better reminder is there than the realization that the choice between theories which are sufficiently general to provide us with a comprehensive world view and which are empirically disconnected may become a matter of taste? That the choice of our basic cosmology may become a matter of taste?

Secondly, matters of taste are not completely beyond the reach of argument. Poems, for example, can be compared in grammar, sound structure, imagery, rhythm, and can be evaluated on such a basis (cf. Ezra Pound on progress in poetry).[2] Even the most elusive mood can be analysed, *and must be* analysed if the purpose is to present it in a manner that can either be enjoyed, or that increases the emotional (cognitive, perceptual) inventory of the reader. Every poet who is not completely irrational compares, improves, argues until he finds the correct formulation of what he wants to say.[3] Would it not be marvellous if this process played a role in the sciences also?

Finally, there are more pedestrian ways of explaining the same matter which may be somewhat less repulsive to the ears of a professional philosopher of science. We may consider the *length* of derivations leading from

[1] For this problem of 'alienation' cf. Marx [1844*a*] and [1844*b*].

[2] Popper has repeatedly asserted, both in his lectures, and in his writings that while there is progress in the sciences there is no progress in the arts. He bases his assertion on the belief that the content of succeeding theories can be compared and that a judgement of verisimilitude can be made. The refutation of this belief eliminates an important difference (and perhaps the *only* important difference) between science and the arts and makes it possible to speak of styles and preferences in the first, and of progress in the second.

[3] Cf. Brecht [1964], p. 119. In my lectures on the theory of knowledge I usually present and discuss the thesis that finding a new theory for given facts is like finding a new production for a well-known play. For painting, cf. also Gombrich [1960].

the principles of a theory to its observation language, and we may also draw attention to the number of *approximations* made in the course of the derivation (all derivations must be standardized for this purpose so that an unambiguous judgement of length can be made; this standardization concerns the *form* of the derivation, it does not concern the *content* of the concepts used). Smaller length and smaller number of approximations would seem to be preferable. It is not easy to see how this requirement can be made compatible with the demand for simplicity and generality which, so it seems, would tend to increase both parameters. However that may be—there are many ways open to us once the fact of incommensurability is understood, and taken seriously.

(15) I started by pointing out that scientific method, as softened up by Lakatos, is but an ornament which makes us forget that a position of 'anything goes' has in fact been adopted. I then considered the argument that the method of problemshifts, while perhaps useless in the first world might still give a correct account of what goes on in the third world and that it might permit us to view the whole 'third world' through 'Popperian spectacles'. The reply was that there is trouble in the third world also and that the attempt to judge cosmologies by their content may have to be given up. Such a development, far from being undesirable, changes science from a stern and demanding mistress into an attractive and yielding courtesan who tries to anticipate every wish of her lover. Of course, it is up to us to choose either a dragon or a pussy cat for our company. I do not think I need to explain my own preferences.

REFERENCES

Bohm [1965]: *The Special Theory of Relativity*, 1965.

Bondi [1967]: *Assumption and Myth in Physical Theory*, 1967.

Brecht [1964]: 'Über das Zerpflücken von Gedichten', in *Über Lyrik*, 1964.

Carnap [1956]: 'The Methodological Character of Theoretical Concepts', in Feigl and Scriven (*eds.*): *Minnesota Studies in the Philosophy of Science*, **1**, pp. 38–76.

Eddington [1924]: *The Mathematical Theory of Relativity*, 1924.

Epstein [1967]: *Varieties of Perceptual Learning*, 1967.

Feyerabend [1962]: 'Explanation, Reduction and Empiricism', in Feigl-Maxwell (*eds.*): *Minnesota Studies in the Philosophy of Science*, **3**, pp. 28–97.

Feyerabend [1965*a*]: 'Reply to Criticism', in Cohen and Wartofsky (*eds.*): *Boston Studies in the Philosophy of Science*, **2**, pp. 223–61.

Feyerabend [1965*b*]: 'Problems of Empiricism', in Colodny (*ed.*): *Beyond the Edge of Certainty*, pp. 145–260.

Feyerabend [1966]: Review of Nagel's 'Structure of Science', *The British Journal for the Philosophy of Science*, **17**, pp. 237–49.

Feyerabend [1968–9]: 'On a Recent Critique of Complementarity', *Philosophy of Science*, **35**, pp. 309–31 and **36**, pp. 82–105.

Feyerabend [1969]: 'Problems of Empiricism, part 2', in Colodny (*ed.*): *The Nature and Function of Scientific Theory*, 1969.

Feyerabend [1970*a*]: 'Classical Empiricism', in Butts (*ed.*): *The Methodological Heritage of Newton*, 1970.

Feyerabend [1970*b*]: 'Against Method', *Minnesota Studies in the Philosophy of Science*, **4**.

Gombrich [1960]: *Art and Illusion*, 1960.

Hempel [1966]: *Philosophy of Natural Science*, 1966.

Kuhn [1961*a*]: 'The Function of Dogma in Scientific Research', in Crombie (*ed.*): *Scientific Change*, 1963, pp. 347–69 and 386–95.

Kuhn [1961*b*]: 'Measurement in Modern Physical Science', *Isis*, **52**, pp. 161–93.

Kuhn [1962]: *The Structure of Scientific Revolutions*, 1962.

Lakatos [1963–4]: 'Proofs and Refutations', *The British Journal for the Philosophy of Science*, **14**, pp. 1–25, 120–39, 221–43 and 296–342.

Lakatos [1968*a*]: 'Changes in the Problem of Inductive Logic', in Lakatos (*ed.*): *The Problem of Inductive Logic*, pp. 315–417.

Lakatos [1968*b*]: 'Criticism and the Methodology of Scientific Research Programmes', in *Proceedings of the Aristotelian Society*, **69**, pp. 149–86.

Marx [1844*a*]: *Nationalökonomie und Philosophie*, 1932.

Marx [1844*b*]: 'Zur Kritik der Hegelschen Rechtsphilosophie,' *Deutsch-Französische Jahrbücher*, 1844.

Marzke and Wheeler [1963]: 'Gravitation and Geometry I: the geometry of space-time and geometrodynamical standard meter', in Chiu and Hoffmann (*eds.*): *Gravitation and Relativity*, pp. 40–64.

Newton [1672]: Letter to Pardies, 10.6.1672, in Turnbull (*ed.*): *The Correspondence of Isaac Newton*, **1**, 1959, pp. 163–71.

Olschki [1927]: *Geschichte der neusprachlichen wissenschaftlichen Literatur*, **3**, *Galilei und seine Zeit*, 1927.

Piaget [1954]: *The Construction of Reality in the Child*, 1954.

Popper [1945]: *The Open Society and its Enemies*, I–II, 1945.

Popper [1961]: 'Fact, Standards, and Truth: a further criticism of relativism', *Addendum* 1 in the fourth edition of Popper [1945], vol. II. pp. 369–96, 1962.

Popper [1968*a*]: 'Epistemology without a Knowing Subject', in Rootselaar-Staal (*eds.*): *Proceedings of the Third International Congress for Logic, Methodology and Philosophy of Science*, pp. 333–73.

Popper [1968*b*]: 'On the Theory of the Objective Mind', in *Proceedings of the XIV International Congress of Philosophy*, **1**, pp. 25–53.

Putnam [1963]: ' "Degree of Confirmation" and Inductive Logic', in Schilpp (*ed.*): *The Philosophy of Rudolf Carnap*, pp. 761–83.

Reagan [1967]: 'Basic and Applied Research: A Meaningful Distinction?', *Science*, **155**, pp. 1383–86.

Stroud [1968]: 'Conventionalism and the Indeterminacy of Translation', *Synthese*, **18**, pp. 82–96.

Synge [1964]: 'Introduction to General Relativity', in de Witt and de Witt (*eds.*): *Relativity, Groups and Topology*, 1964.

Wohlwill [1926]: *Galileo und sein Kampf für die Kopernikanische Lehre*, **2**, 1926.

Reflections on my Critics[1]

THOMAS S. KUHN
Princeton University

1. *Introduction.*
2. *Methodology: the role of history and sociology.*
3. *Normal Science: its nature and functions.*
4. *Normal Science: its retrieval from history.*
5. *Irrationality and Theory-Choice.*
6. *Incommensurability and Paradigms.*

1. INTRODUCTION

It is now four years since Professor Watkins and I exchanged mutually impenetrable views at the International Colloquium in the Philosophy of Science held at Bedford College, London. Rereading our contributions together with those that have since accreted to them, I am tempted to posit the existence of two Thomas Kuhns. $Kuhn_1$ is the author of this essay and of an earlier piece in this volume. He also published in 1962 a book called *The Structure of Scientific Revolutions*, the one which he and Miss Masterman discuss above. $Kuhn_2$ is the author of another book with the same title. It is the one here cited repeatedly by Sir Karl Popper as well as by Professors Feyerabend, Lakatos, Toulmin, and Watkins. That both books bear the same title cannot be altogether accidental, for the views they present often overlap and are, in any case, expressed in the same words. But their central concerns are, I conclude, usually very different. As reported by his critics (his original has unfortunately been unavailable to me), $Kuhn_2$ seems on occasion to make points that subvert essential aspects of the position outlined by his namesake.

Lacking the wit to extend this introductory fantasy, I will instead explain why I have embarked upon it. Much in this volume testifies to what I described above as the gestalt-switch that divides readers of my *Scientific Revolutions* into two groups. Together with that book, this collection of essays therefore provides an extended example of what I have elsewhere called partial or incomplete communication—the

[1] Though my battle with a publication deadline allowed them almost no time for it, my colleagues C. G. Hempel and R. E. Grandy both managed to read my first manuscript and offer useful suggestions for its improvement, conceptual and stylistic. I am most grateful to them, but they should not be blamed for my views.

talking-through-each-other that regularly characterizes discourse between participants in incommensurable points of view.

Such communication breakdown is important and needs much study. Unlike Paul Feyerabend (at least as I and others are reading him), I do not believe that it is ever total or beyond recourse. Where he talks of incommensurability *tout court*, I have regularly spoken also of partial communication, and I believe it can be improved upon to whatever extent circumstances may demand and patience permit, a point to be elaborated below. But neither do I believe, as Sir Karl does, that the sense in which 'we are prisoners caught in the framework of our theories; our expectations; our past experiences; our language' is merely 'Pickwickian'. Nor do I suppose that 'we can break out of our framework at any time . . . [into] a better and roomier one . . . [from which] we can at any moment break out . . . again.'[1] If that possibility were routinely available, there ought to be no very special difficulties about stepping into someone else's framework in order to evaluate it. My critics' attempts to step into mine suggest, however, that changes of framework, of theory, of language, or of paradigm pose deeper problems of both principle and practice than the preceding quotations recognize. These problems are not simply those of ordinary discourse, nor will they be resolved by quite the same techniques. If they could be, or if changes of framework were normal, occurring at will and at any moment, they would not be comparable, in Sir Karl's phrase, to 'the culture clash[es] which [have] stimulated some of the greatest intellectual revolutions.'[2] The very possibility of that comparison is what makes them so very important.

One especially interesting aspect of this volume is, then, that it provides a developed example of a minor culture clash, of the severe communication difficulties which characterize such clashes, and of the linguistic techniques deployed in the attempt to end them. Read as an example, it could be an object for study and analysis, providing concrete information concerning a type of developmental episode about which we know very little. For some readers, I suspect, the recurrent failure of these essays to intersect on intellectual issues will provide this book's greatest interest. Indeed, because those failures illustrate a phenomenon at the heart of my own point of view, the book has that interest for me. I am, however, too much a participant, too deeply involved, to provide the analysis which the breakdown of communication warrants. Instead, though I remain convinced that their fire is frequently misplaced and that it often obscures the deeper differences between Sir Karl's views and my own, I must here speak primarily to the points raised by my present critics.

[1] *This volume*, p. 56.

[2] *This volume*, p. 57.

Those points, excepting for the moment the ones raised in Miss Masterman's stimulating paper, fall into three coherent categories, each of which illustrates what I have just called the failure of our discussion to intersect on issues. The first, for purposes of my discussion, is the perceived difference in our methods: logic versus history and social psychology; normative versus descriptive. These, as I shall shortly try to show, are odd contrasts with which to discriminate among the contributors to this volume. All of us, unlike the members of what has until recently been the main movement in philosophy of science, do historical research and rely both on it and on observation of contemporary scientists in developing our viewpoints. In those viewpoints, furthermore, the descriptive and the normative are inextricably mixed. Though we may differ in our standards and surely differ about some matters of substance, we are scarcely to be distinguished by our methods. The title of my earlier paper, 'Logic of Discovery or Psychology of Research?' was not chosen to suggest what Sir Karl *ought* to do but rather to describe *what he does*. When Lakatos writes, 'But Kuhn's conceptual framework... is socio-psychological: mine is normative',[1] I can only think that he is employing a sleight of hand to reserve the philosophical mantle for himself. Surely Feyerabend is right in claiming that my work repeatedly makes normative claims. Equally surely, though the point will require more discussion, Lakatos's position is social-psychological in its repeated reliance on decisions governed not by logical rules but by the mature sensibility of the trained scientist. If I differ from Lakatos (or Sir Karl, Feyerabend, Toulmin, or Watkins), it is with respect to substance rather than method.

As to substance, our most apparent difference is about normal science, the topic to which I shall turn immediately after discussing method. A disproportionate part of this volume is devoted to normal science, and it calls forth some of the oddest rhetoric: normal science does not exist *and* is uninteresting. On this issue we do disagree, but not, I think, either consequentially or in the ways my critics suppose. When I take it up, I shall deal in part with the real difficulties in retrieving normal scientific traditions from history, but my first and more central point will be a logical one. The existence of normal science is a corollary of the existence of revolutions, a point implicit in Sir Karl's paper and explicit in Lakatos's. If it did not exist (or if it were non-essential, dispensable for science), then revolutions would be in jeopardy also. But, about the latter, I and my critics (excepting Toulmin) agree. Revolutions through criticism demand normal science no less than revolutions through crisis. Inevitably, the term 'cross-purposes' better catches the nature of our discourse than 'disagreement.'

[1] *This volume*, p. 177.

Discussion of normal science raises the third set of issues about which criticism has here clustered: the nature of the change from one normal-scientific tradition to another and of the techniques by which the resulting conflicts are resolved. My critics respond to my views on this subject with charges of irrationality, relativism, and the defence of mob rule. These are all labels which I categorically reject, even when they are used in my defence by Feyerabend. To say that, in matters of theory-choice, the force of logic and observation cannot in principle be compelling is neither to discard logic and observation nor to suggest that there are not good reasons for favouring one theory over another. To say that trained scientists are, in such matters, the highest court of appeal is neither to defend mob rule nor to suggest that scientists could have decided to accept any theory at all. In this area, too, my critics and I differ, but our points of difference have yet to be seen for what they are.

These three sets of issues—method, normal science, and mob rule—are the ones which bulk largest in this volume and, for that reason, in my response. But my reply cannot close without going one step beyond them to consider the problem of paradigms to which Miss Masterman's essay is devoted. I concur in her judgement that the term 'paradigm' points to the central philosophical aspect of my book but that its treatment there is badly confused. No aspect of my viewpoint has evolved more since the book was written, and her paper has helped in that development. Though my present position differs from hers in many details, we approach the prolem in the same spirit including a common conviction of the relevance of the philosophy of language and of metaphor.

I shall not here be able to deal at all fully with the problems presented by my initial treatment of paradigms, but two considerations necessitate my touching upon them. Even brief discussion should permit the isolation of two quite different ways in which the term is deployed in my book and thus eliminate a constellation of confusions which has handicapped me as well as my critics. The resulting clarification will, in addition, permit me to suggest what I take to be the root of my single most fundamental difference from Sir Karl.

He and his followers share with more traditional philosophers of science the assumption that the problem of theory-choice can be resolved by techniques which are semantically neutral. The observational consequences of both theories are first stated in a shared basic vocabulary (not necessarily complete or permanent). Some comparative measure of their truth/falsity count then provides the basis for a choice between them. For Sir Karl and his school, no less than for Carnap and Reichenbach, canons of rationality thus derive exclusively from those of logical and linguistic syntax. Paul

Feyerabend provides the exception which proves that rule. Denying the existence of a vocabulary adequate to neutral observation reports, he at once concludes to the intrinsic irrationality of theory-choice.

That conclusion is surely Pickwickian. No process essential to scientific development can be labelled 'irrational' without vast violence to the term. It is therefore fortunate that the conclusion is unnecessary. One can deny, as Feyerabend and I do, the existence of an observation language shared in its entirety by two theories and still hope to preserve good reasons for choosing between them. To achieve that goal, however, philosophers of science will need to follow other contemporary philosophers in examining, to a previously unprecedented depth, the manner in which language fits the world, asking how terms attach to nature, how those attachments are learned, and how they are transmitted from one generation to another by the members of a language community. Because paradigms, in one of the two separable senses of the term, are fundamental to my own attempts to answer questions of that sort, they must also find a place in this essay.

2. METHODOLOGY: THE ROLE OF HISTORY AND SOCIOLOGY

Doubts about the appropriateness of my methods to my conclusions unite many of the essays in this volume. History and social-psychology are not, my critics claim, a proper basis for philosophical conclusions. Their reservations are not, however, all of a piece. I shall therefore consider *seriatim* the somewhat different forms they take in the essays by Sir Karl, Watkins, Feyerabend, and Lakatos.

Sir Karl concludes his paper by pointing out that to him 'the idea of turning for enlightenment concerning the aims of science, and its possible progress, to sociology or psychology (or ... to the history of science) is surprising and disappointing. ... how,' he asks, 'can the regress to these often spurious sciences help us in this particular difficulty?'[1] I am puzzled to know what these remarks intend, for in this area I think there are no differences between Sir Karl and myself. If he means that the generalizations which constitute received theories in sociology and psychology (and history?) are weak reeds from which to weave a philosophy of science, I could not agree more heartily. My work relies on them no more than his. If, on the other hand, he is challenging the relevance to philosophy of science of the sorts of observations collected by historians and sociologists, I wonder how his own work is to be understood. His writings are crowded with historical examples and with generalizations about scientific behaviour, some of them discussed in my earlier essay. He does write on

[1] *This volume*, pp. 57–8.

historical themes, and he cites those papers in his central philosophical works. A consistent interest in historical problems and a willingness to engage in original historical research distinguishes the men he has trained from the members of any other current school in philosophy of science. On these points I am an unrepentant Popperian.

John Watkins voices a different sort of doubt. Early in his paper he writes that 'methodology . . . is concerned with science at its best, or with science as it should be conducted, rather than with hack science,'[1] a point with which, at least in a more careful formulation, I fully agree. Later he argues that what I have called normal science is hack science, and he then asks why I am so 'concerned to up-value Normal Science and down-value Extraordinary Science?'[2] In so far as that question is about normal science in particular, I reserve my response until later (at which point I shall attempt also to unravel Watkins's extraordinary distortion of my position). But Watkins seems also to be asking a more general question, one that relates closely to an issue raised by Feyerabend. Both grant, at least for the sake of their argument, that scientists do behave as I have said they do (I shall later consider their qualifications of that concession). Why should the philosopher or methodologist, they then ask, take the facts seriously? He is, after all, concerned not with a full description of science but with the discovery of the essentials of the enterprise, i.e., with rational reconstruction. By what right and what criteria does the historian-observer or sociologist-observer tell the philosopher which facts of scientific life he must include in his reconstruction, which he may ignore?

To avoid lengthy disquisitions on the philosophy of history and of sociology, I restrict myself to a personal response. I am no less concerned with rational reconstruction, with the discovery of essentials, than are philosophers of science. My objective, too, is an understanding of science, of the reasons for its special efficacy, of the cognitive status of its theories. But unlike most philosophers of science, I began as an historian of science, examining closely the facts of scientific life. Having discovered in the process that much scientific behaviour, including that of the very greatest scientists, persistently violated accepted methodological canons, I had to ask why those failures to conform did not seem at all to inhibit the success of the enterprise. When I later discovered that an altered view of the nature of science transformed what had previously seemed aberrant behaviour into an essential part of an explanation for science's success, the discovery was a source of confidence in that new explanation. My criterion for emphasizing any particular aspect of scientific behaviour is therefore not simply that it occurs, nor merely that it occurs frequently, but rather that it fits a

[1] *This volume*, p. 27. [2] *This volume*, p. 31.

theory of scientific knowledge. Conversely, my confidence in that theory derives from its ability to make coherent sense of many facts which, on an older view, had been either aberrant or irrelevant. Readers will observe a circularity in the argument, but it is not vicious, and its presence does not at all distinguish my view from those of my present critics. Here, too, I am behaving as they do.

That my criteria for discriminating between the essential and non-essential elements of observed scientific behaviour are to a significant extent theoretical provides also an answer to what Feyerabend calls the ambiguity of my presentation. Are Kuhn's remarks about scientific development, he asks, to be read as descriptions or prescriptions?[1] The answer, of course, is that they should be read in both ways at once. If I have a theory of how and why science works, it must necessarily have implications for the way in which scientists should behave if their enterprise is to flourish. The structure of my argument is simple and, I think, unexceptionable: scientists behave in the following ways; those modes of behaviour have (here theory enters) the following essential functions; in the absence of an alternate mode *that would serve similiar functions*, scientists should behave essentially as they do if their concern is to improve scientific knowledge.

Note that nothing in that argument sets the value of science itself, and that Feyerabend's 'plea for hedonism' is correspondingly irrelevant.[2] Partly because they have misconstrued my prescription (a point to which I shall return), both Sir Karl and Feyerabend find menace in the enterprise I have described. It is 'liable to corrupt our understanding and diminish our pleasure' (Feyerabend); it is 'a danger . . . indeed to our civilization' (Sir Karl).[3] I am not led to that evaluation nor are many of my readers, but nothing in my argument depends on its being wrong. To explain why an enterprise works is not to approve or disapprove it.

Lakatos's paper raises a fourth problem about method, and it is the most fundamental of all. I have already confessed my inability to understand what he means when he says things like, 'Kuhn's conceptual framework . . . is socio-psychological: mine is normative'. If I ask, however, not what he intends, but why he finds this sort of rhetoric appropriate, an important point emerges, one that is almost explicit in the first paragraph of his section 4. Some of the principles deployed in my explanation of science are irreducibly sociological, at least at this time. In particular, confronted with the problem of theory-choice, the structure of my response runs roughly as follows: take a *group* of the ablest available people with the

[1] *This volume*, p. 198. For a far deeper and more careful examination of some contexts in which the descriptive and normative merge, see Cavell [1969].

[2] *This volume*, p. 209.

[3] *This volume*, pp. 209 and 53.

most appropriate motivation; train them in some science and in the specialties relevant to the choice at hand; imbue them with the value system, the ideology, current in their discipline (and to a great extent in other scientific fields as well); and, finally, *let them make the choice.* If that technique does not account for scientific development as we know it, then no other will. There can be no set of rules of choice adequate to dictate desired *individual* behaviour in the concrete cases that scientists will meet in the course of their careers. Whatever scientific progress may be, we must account for it by examining the nature of the scientific group, discovering what it values, what it tolerates, and what it disdains.

That position is intrinsically sociological and, as such, a major retreat from the canons of explanation licensed by the traditions which Lakatos labels justificationism and falsificationism, both dogmatic and naive. I shall later specify it further and defend it. But my present concern is simply with its structure, which both Lakatos and Sir Karl find unacceptable in principle. My question is, why should they? Both repeatedly use arguments of the same structure themselves.

Sir Karl does not, it is true, do so all the time. That part of his writing which seeks an algorithm for verisimilitude would, if successful, eliminate all need for recourse to group values, to judgements made by minds prepared in a particular way. But, as I pointed out at the end of my previous essay, there are many passages throughout Sir Karl's writings which can only be read as descriptions of the values and attitudes which scientists must possess if, when the chips are down, they are to succeed in advancing their enterprise. Lakatos's sophisticated falsificationism goes even further. In all but a few respects, only two of them essential, his position is now very close to my own. Among the respects in which we agree, though he has not yet seen it, is our common use of explanatory principles that are ultimately sociological or ideological in structure.

Lakatos's sophisticated falsificationism isolates a number of issues about which scientists employing the method must make decisions, individually or collectively. (I distrust the term 'decision' in this context since it implies conscious deliberation on each issue prior to the assumption of a research stance. For the moment, however, I shall use it. Until the last section of this paper very little will depend upon the distinction between making a decision and finding oneself in the position that would have resulted from making it.) Scientists must, for example, *decide* which statements to make 'unfalsifiable by *fiat*' and which not.[1] Or, dealing with a probabilistic theory, they must *decide* on a probability threshold below which statistical evidence will be held ' "inconsistent" ' with that theory.[2] Above all, viewing theories

[1] *This volume*, p. 106.

[2] *This volume*, p. 109.

as research programmes to be evaluated over time, scientists must *decide* whether a given programme at a given time is 'progressive' (whence scientific) or 'degenerative' (whence pseudo-scientific).[1] If the first, it is to be pursued; if the latter, rejected.

Notice now that a call for decisions like these may be read in two ways. It may be taken to name or describe decision points for which procedures applicable in concrete cases must still be supplied. On this reading Lakatos has yet to tell us how scientists are to select the particular statements that are to be unfalsifiable by their *fiat*; he must also still specify criteria which can be used at the time to distinguish a degenerative from a progressive research programme; and so on. Otherwise, he has told us nothing at all. Alternatively, his remarks about the need for particular decisions may be read as already complete descriptions (at least in form—their particular content may be preliminary) of directives, or maxims which the scientist is required to follow. On this interpretation, the third decision directive would read: 'As a scientist, you may not refrain from deciding whether your research programme is progressive or degenerative, and you must take the consequences of your decision, abandoning the programme in one case, pursuing it in the other.' Correspondingly, the second directive would read: 'Working with a probabilistic theory, you must constantly ask yourself whether the result of some particular experiment is not so improbable as to be inconsistent with your theory, and you must, as a scientist, also answer.' Finally, the first directive would read: 'As a scientist, you will have to take risks, choosing certain statements as the basis for your work and ignoring, at least until your research programme has developed, all actual and potential attacks upon them.'

The second reading is, of course, far weaker than the first. It demands the same decisions, but it neither supplies nor promises to supply rules which would dictate their outcomes. Instead, it assimilates these decisions to judgements of value (a subject about which I shall have more to say) rather than to measurements or computations, say, of weight. Nevertheless, conceived merely as imperatives which commit the scientist to making certain sorts of decisions, these directives are strong enough to affect scientific development profoundly. A group whose members felt no obligations to wrestle with such decisions (but which instead emphasized others, or none at all) would behave in notably different ways, and their discipline would change accordingly. Though Lakatos's discussion of his decision-directives is often equivocal, I believe that it is just this second sort of efficacy upon which his methodology depends. Certainly he does little to specify algorithms by which the decisions he

[1] *This volume*, pp. 118 ff.

demands are to be made, and the tenor of his discussion of naive and dogmatic falsificationism suggests that he no longer thinks such specification possible. In that case, however, his decision-imperatives are, in form though not always in content, identical to my own. They specify ideological commitments which scientists must share if their enterprise is to succeed. They are therefore irreducibly sociological in the same sense and to the same extent as my explanatory principles.

Under these circumstances I am not sure what Lakatos is criticizing or what, in this area, he thinks we disagree about. A strange footnote late in his paper may, however, provide a clue[1]:

'There are *two kinds of psychologistic philosophies of science*. According to one kind there can be no philosophy of science: only a psychology of individual scientists. According to the other kind there is a psychology of the "scientific," "ideal," or "normal" mind: this turns philosophy of science into a psychology of this ideal mind. . . . Kuhn does not seem to have noticed this distinction.'

If I understand him correctly, Lakatos identifies the first kind of psychologistic philosophy of science with me, the second with himself. But he is misunderstanding me. We are not nearly so far apart as his description would suggest, and, where we do differ, his literal position would demand a renunciation of our common goal.

Part of what Lakatos is rejecting is explanations that demand recourse to the factors which individuate particular scientists ('the psychology of the individual scientist' versus 'the psychology of the . . . "normal" mind'). But that does not separate us. My recourse has been exclusively to social psychology (I prefer 'sociology'), a field quite different from individual psychology reiterated *n* times. Correspondingly, my unit for purposes of explanation is the normal (i.e. non-pathological) scientific group, account being taken of the fact that its members differ but not of what makes any given individual unique. In addition, Lakatos would like to reject those characteristics of even normal scientific minds which make them the minds of human beings. Apparently he sees no other way to retain the methodology of an ideal science in explaining the observed success of actual science. But his way will not do if he hopes to explain an enterprise practiced by people. There are no ideal minds, and the 'psychology of this ideal mind' is therefore unavailable as a basis for explanation. Nor is Lakatos's manner of introducing the ideal needed to achieve what he aims at. Shared ideals affect behaviour without making those who hold them ideal. The type of question I ask has therefore been: how will a particular constellation of beliefs, values, and imperatives affect group behaviour? My explanations follow from the answer. I am not sure Lakatos

[1] *This volume*, p. 180, footnote 3.

means anything else, but, if he does not, there is nothing in this area for us to disagree about.

Having misconstrued the sociological base of my position, Lakatos and my other critics inevitably fail to note a special feature which follows from taking the normal group rather than the normal mind as unit. Given a shared algorithm adequate, let us say, to individual choice between competing theories or to the identification of severe anomaly, all members of a scientific group will reach the same decision. That would be the case even if the algorithm were probabilistic, for all those who used it would evaluate the evidence in the same way. The effects of a shared ideology, however, are less uniform, for its mode of application is of a different sort. Given a group all the members of which are committed to choosing between alternative theories and also to considering such values as accuracy, simplicity, scope, and so on while making their choice, the concrete decisions of individual members in individual cases will nevertheless vary. Group behaviour will be affected decisively by the shared commitments, but individual choice will be a function also of personality, education, and the prior pattern of professional research. (These variables *are* the province of individual psychology.) To many of my critics this variability seems a weakness of my position. When considering the problems of crisis and of theory-choice I shall want, however, to argue that it is instead a strength. If a decision must be made under circumstances in which even the most deliberate and considered judgement may be wrong, it may be vitally important that different individuals decide in different ways. How else could the group as a whole hedge its bets?[1]

3. NORMAL SCIENCE: ITS NATURE AND FUNCTIONS

As to methods, then, the ones I employ are not significantly different from those of my Popperian critics. Applying those methods, we, of course, draw somewhat different conclusions, but even they are not so far apart as several of my critics believe. In particular, all of us excepting Toulmin share the conviction that the central episodes in scientific advance—those which make the game worth playing and the play worth studying—are revolutions. Watkins is constructing an opponent from his own straw when he describes me as having 'down-valued' scientific revolutions, taken a 'philosophical dislike' to them, or suggested that they 'can hardly be called science at all.'[2] Discovering the puzzling nature of revolutions was

[1] If human motivation were not at issue, the same effect could be achieved by first computing a probability and then *assigning* a certain fraction of the profession to each of the competing theories, the exact fraction to depend on the result of the probabilistic computation. Somehow that alternative makes my point by *reductio ad absurdum*.

[2] *This volume*, pp. 31, 32 and 29.

what drew me to history and philosophy of science in the first place. Almost everything I have written since deals with them, a fact which Watkins points out and then ignores.

If, however, we agree about this much, we cannot altogether disagree about normal science, the aspect of my work which most disturbs my present critics. By their nature revolutions cannot be the whole of science: something different must necessarily go on in between. Sir Karl sets up the point admirably. Underlining what I have always recognized as one of our principle areas of agreement, he stresses that 'scientists *necessarily* develop their ideas within a definite theoretical framework'.[1] For him, as for me, furthermore, revolutions demand such frameworks, since they always involve the rejection and replacement of a framework or of some of its integral parts. Since the science which I call normal is precisely research within a framework, it can only be the opposite side of a coin the face of which is revolutions. No wonder Sir Karl has been 'dimly aware of the distinction' between normal science and revolutions.[2] It follows from his premises.

Something else follows as well. If frameworks are necessary to scientists, if to break with one is inevitably to break into another—points which Sir Karl embraces explicitly—then the hold of a framework on a scientist's mind may not be accounted for *merely* as the result of his having 'been badly taught, . . . a victim of indoctrination'.[3] Nor may it, as Watkins supposes, be explained *entirely* by reference to the prevalence of third-rate minds, fit only for 'plodding, uncritical' work.[4] Those things do exist, and most of them do damage. Nevertheless, if frameworks are the prerequisite of research, their grip on the mind is not merely 'Pickwickian', nor can it be quite right to say that, 'if we try, we can break out of our framework at any time'.[5] To be simultaneously essential and freely dispensible is very nearly a contradiction in terms. My critics become incoherent when they embrace it.

None of that is said in an effort to show that my critics really agree with me, if only they knew it. They do not! Rather I am trying, by eliminating irrelevancies, to discover what we disagree about. I have so far argued that Sir Karl's phrase 'revolutions in permanence' does not, any more than 'square-circle', describe a phenomenon that could exist. Frameworks must be lived with and explored before they can be broken. But that does not imply that scientists ought not aim at perpetual framework-breaking, however unobtainable that goal. 'Revolutions in permanence' could name an

[1] *This volume*, p. 51, italics added. Unless explicitly noted all italic passages in the quotations in this paper are in the originals.
[2] *This volume*, p. 52.
[3] *This volume*, p. 53.
[4] *This volume*, p. 32.
[5] *This volume*, p. 56

important ideological imperative. If Sir Karl and I disagree at all about normal science, it is over this point. He and his group argue that the scientist should try at all times to be a critic and a proliferator of alternate theories. I urge the desirability of an alternate strategy which reserves such behaviour for special occasions.

That disagreement, being restricted to research strategy, is already narrower than the one my critics have envisaged. To see what is at stake it must be narrowed further. Everything that has been said so far, though phrased for science and scientists, applies equally to a number of other fields. My methodological prescription is, however, directed exclusively to the sciences and, among them, to those fields which display the special developmental pattern known as progress. Sir Karl neatly catches the distinction I have in mind. At the start of his paper he writes: ' "A scientist engaged in a piece of research . . . can go at once to the heart of . . . an organized structure . . . [and of] a generally accepted problem-situation . . . [leaving] it to others to fit his contribution into the framework of scientific knowledge." . . . the philosopher', he continues, 'finds himself in a different position.'[1] Nevertheless, having pointed to the difference, Sir Karl thereafter ignores it, recommending the same strategy to both scientists and philosophers. In the process he misses the consequences for research design of the special detail and precision with which, as he says, the framework of a mature science informs its practitioners what to do. In the absence of that detailed guidance, Sir Karl's critical strategy seems to me the very best available. It will not induce the special developmental pattern which characterizes, say, physics, but neither will any other methodological prescription. Given a framework which does provide such guidance, however, then I do intend my methodological recommendations to apply.

Consider for a moment the evolution of philosophy or of the arts since the end of the Renaissance. These are fields often contrasted with the established sciences as ones which do not progress. That contrast cannot be due to the absence of revolutions or of an intervening mode of normal practice. On the contrary, long before the similar structure of scientific development was noticed, historians portrayed these fields as developing through a succession of traditions punctuated by revolutionary alterations of artistic style and taste or of philosophical viewpoint and goal. Nor can the contrast be due to the absence from philosophy and the arts of a Popperian methodology. As Miss Masterman observes for philosophy,[2]

[1] *This volume*, p. 51. Readers who know my [1962*a*] will recognize how closely Sir Karl's phrase 'leaving it to others to fit his contribution into the framework of scientific knowledge' catches the essential implications of my description of normal science.

[2] *This volume*, pp. 69 ff.

these are just the fields in which it is best exemplified, in which practitioners do find current tradition stifling, do struggle to break with it, and do regularly seek a style or a philosophical viewpoint of their own. In the arts, in particular, the work of men who do not succeed in innovation is described as 'derivative', a term of derogation significantly absent from scientific discourse which does, on the other hand, repeatedly refer to 'fads'. In none of these fields, whether arts or philosophy, does the practitioner who fails to alter traditional practice have significant impact on the discipline's development.[1] These are, in short, fields to which Sir Karl's method is essential because without constant criticism and the proliferation of new modes of practice there would be no revolutions. Substituting my own methodology for Sir Karl's would induce stagnation for exactly the reasons my critics underscore. In no obvious sense, however, does his methodology produce progress. The relation of pre- to post-revolutionary practice in these fields is not what we have learned to expect from the developed sciences.

My critics will suggest that the reasons for that difference are obvious. Fields like philosophy and the arts do not claim to be sciences, nor do they satisfy Sir Karl's demarcation criterion. They do not, that is, generate results which can in principle be tested through a point-by-point comparison with nature. But that argument seems to me mistaken. Without satisfying Sir Karl's criterion these fields could not be sciences, but they could nevertheless progress as the sciences do. In antiquity and during the Renaissance, the arts rather than the sciences provided the accepted paradigms of progress.[2] Few philosophers find reasons of principle why their field should not move steadily ahead, though many bemoan its failure to do so. In any case, there are many fields—I shall call them proto-sciences—in which practice does generate testable conclusions but which nonetheless resemble philosophy and the arts rather than the established sciences in their developmental patterns. I think, for example, of fields like chemistry and electricity before the mid-eighteenth century, of the study of heredity and phylogeny before the mid-nineteenth, or of many of the social sciences today. In these fields, too, though they satisfy Sir Karl's demarcation criterion, incessant criticism and continual striving for a fresh start are primary forces, and need to be. No more than in philosophy and the arts, however, do they result in clear-cut progress.

I conclude, in short, that the proto-sciences, like the arts and philosophy, lack some element which, in the mature sciences, permits the

[1] For a fuller discussion of differences between scientific and artistic communities and between the corresponding developmental patterns, see my [1969].

[2] Gombrich [1960], pp. 11 ff.

more obvious forms of progress. It is not, however, anything that a methodological prescription can provide. Unlike my present critics, Lakatos at this point included, I claim no therapy to assist the transformation of a proto-science to a science, nor do I suppose that anything of the sort is to be had. If, as Feyerabend suggests, some social scientists take from me the view that they can improve the status of their field by first legislating agreement on fundamentals and then turning to puzzle solving, they are badly misconstruing my point.[1] A sentence I once used when discussing the special efficacy of mathematical theories applies equally here: 'As in individual development, so in the scientific group, maturity comes most surely to those who know how to wait.'[2] Fortunately, though no prescription will force it, the transition to maturity does come to many fields, and it is well worth waiting and struggling to attain. Each of the currently established sciences has emerged from a previously more speculative branch of natural philosophy, medicine, or the crafts at some relatively well-defined period in the past. Other fields will surely experience the same transition in the future. Only after it occurs does progress become an obvious characteristic of a field. And only then do those prescriptions of mine which my critics decry come into play.

About the nature of that change I have written at length in my *Scientific Revolutions* and more briefly when discussing demarcation criteria in my earlier contribution to this volume. Here I shall be content with an abstract descriptive summary. Confine attention first to fields which aim to explain in detail some range of natural phenomena. (If, as my critics point out, my further description fits theology and bank-robbery as well, no problems are thereby created.) Such a field first gains maturity when provided with theory and technique which satisfy the four following conditions. First is Sir Karl's demarcation criterion without which no field is potentially a science: for some range of natural phenomena concrete predictions must emerge from the practice of the field. Second, for some interesting sub-class of phenomena, whatever passes for predictive success must be consistently achieved. (Ptolemaic astronomy always predicted planetary position within widely recognized limits of error. The companion astrological tradition could not, excepting for the tides and the average menstrual cycle, specify in advance which prediction would succeed, which fail.) Third, predictive techniques must have roots in a theory which, however metaphysical, simultaneously justifies them, explains their limited success, and suggests means for their improvement in both precision and

[1] *This volume*, p. 198. Note, however, that the passage Feyerabend quotes in footnote 3 does not say at all what he reports.

[2] See p. 190 of my [1962*b*].

scope. Finally, the improvement of predictive technique must be a challenging task, demanding on occasions the very highest measure of talent and devotion.

These conditions are, of course, tantamount to the description of a good scientific theory. But once hope for a therapeutic prescription is abandoned, there is no reason to expect anything less. My claim has been—it is my single genuine disagreement with Sir Karl about normal science—that with such a theory in hand the time for steady criticism and theory proliferation has passed. Scientists for the first time have an alternative which is not merely aping what has gone before. They can instead apply their talents to the puzzles which lie in what Lakatos now calls the 'protective belt'. One of their objectives then is to extend the range and precision of existing experiment and theory as well as to improve the match between them. Another is to eliminate conflicts both between the different theories employed in their work and between the ways in which a single theory is used in different applications. (Watkins is right, I now think, in charging that my book gives too small a role to these inter-and intra-theoretic puzzles, but Lakatos's attempt to reduce science to mathematics, leaving no significant role to experiment, goes vastly too far. He could not, for example, be more mistaken about the irrelevance of the Balmer formula to the development of Bohr's atom model.[1]) These puzzles and others like them constitute the main activity of normal science. Though I cannot argue the point again, they are not, *pace* Watkins, for hacks, nor do they, *pace* Sir Karl, resemble the problems of applied science and engineering. Of course the men fascinated by them are a special breed, but so are philosophers or artists.

Even given a theory which permits normal science, however, scientists need not engage the puzzles it supplies. They could instead behave as practitioners of the proto-sciences must; they could, that is, seek potential weak spots, of which there are always large numbers, and endeavour to erect alternate theories around them. Most of my present critics believe they should do so. I disagree but exclusively on strategic grounds. Feyerabend mispresents me in a way I particularly regret when he reports, for example, that I 'criticized Bohm for disturbing the uniformity of the contemporary quantum theory'.[2] My record as a trouble maker should be hard to reconcile with that report. In fact, I confessed to Feyerabend that I shared Bohm's discontent but thought his exclusive attention to it almost

[1] *This volume*, p. 147, for the remarks on the Balmer formula. This attitude towards the role of experiment is found throughout much of Lakatos's paper. For the actual role of the Balmer formula in Bohr's work, see the paper cited in footnote 3, p. 256 *below*.

[2] *This volume*, p. 206. An implicit answer to the contrast Feyerabend draws between my attitudes towards Bohm and Einstein as critics will be found *below*, on pp. 257ff.

certain to fail. No one, I suggested, was likely to resolve the paradoxes of the quantum theory until he could relate them to some concrete technical puzzle of current physics. In the developed sciences, unlike philosophy, it is technical puzzles that provide the usual occasion and often the concrete materials for revolution. Their availability together with the information and signals they provide account in large part for the special nature of scientific progress. Because they can ordinarily take current theory for granted, exploiting rather than criticizing it, the practitioners of mature sciences are freed to explore nature to an esoteric depth and detail otherwise unimaginable. Because that exploration will ultimately isolate severe trouble spots, they can be confident that the pursuit of normal science will inform them when and where they can most usefully become Popperian critics. Even in the developed sciences, there is an essential role for Sir Karl's methodology. It is the strategy appropriate to those occasions when something goes wrong with normal science, when the discipline encounters crisis.

I have discussed those points at great length elsewhere and shall not elaborate them here. Let me instead conclude this section by returning to the generalization with which it began. Despite the energy and space which my critics have devoted to it, I do not think the position just outlined departs very greatly from Sir Karl's. On this set of questions our differences are over nuances. I hold that in the developed sciences occasions for criticism need not, and by most practitioners ought not, deliberately be sought. When they are found, a decent restraint is the appropriate first response. Sir Karl, though he sees the need to defend a theory when first attacked, gives more emphasis than I to the purposeful search for weak points. There is not a great deal to choose between us.

Why is it, then, that my present critics see our crucial differences here? One reason I have already suggested: their sense—which I do not share but which is in any case irrelevant—that my strategic prescription violates a higher morality. A second reason, which I shall discuss in the next section, is their apparent inability to see in historical examples the detailed functions of the breakdown of normal science in setting the stage for revolutions. Lakatos's case histories are in this respect particularly interesting, for he describes clearly the transition from the progressive to the degenerative phase of a research programme (the transition from normal science to crisis) and then appears to deny the critical importance of what results. With a third reason, however, I must deal at this point. It emerges from a criticism voiced by Watkins, which, however, in the present context serves a purpose he by no means intends.

'By contrast with the relatively sharp idea of testability,' Watkins writes, 'the notion of [normal science's] "ceasing adequately to support a puzzle-solving tradition" is essentially vague.'[1] With the charge of vagueness I agree, but it is a mistake to suppose that it differentiates my position from Sir Karl's. What is precise about Sir Karl's position is, as Watkins also points out, the idea of testability in principle. On that much I rely too, for no theory that was not *in principle* testable could function or cease to function adequately when applied to scientific puzzle solving. I do, despite Watkins's strange failure to see it, take Sir Karl's notion of the asymmetry of falsification and confirmation very seriously indeed. What is vague, however, about my position is the actual criteria (if that is what is called for) to be applied when deciding whether a particular failure in puzzle-solving is or is not to be attributed to fundamental theory and thus to become an occasion for deep concern. That decision is, however, identical in kind with the decision whether or not the result of a particular test actually falsifies a particular theory, and on that subject Sir Karl is necessarily as vague as I. To drive a wedge between us on this issue, Watkins transfers the sharpness of testability-in-principle to the shady area of testability-in-practice without even hinting how the transfer is to be effected. It is not an unprecedented mistake, and it regularly makes Sir Karl's methodology appear more a logic, less an ideology, than it is.

Besides, reverting to a point made at the end of the last section, one may legitimately ask whether what Watkins calls vagueness is a disadvantage. All scientists must be taught—it is a vital element in their ideology—to be alert for and responsible to theory-breakdown, whether it be described as severe anomaly or falsification. In addition, they must be supplied with examples of what their theories can, with sufficient care and skill, be expected to do. Given only that much, they will, of course, often reach different judgements in concrete cases, one man seeing a cause of crisis where another sees only evidence of limited talent for research. But they do reach judgements, and their lack of unanimity may then be what saves their profession. Most judgements that a theory has ceased adequately to support a puzzle-solving tradition prove to be wrong. If everyone agreed in such judgements, no one would be left to show how existing theory could account for the apparent anomaly as it usually does. If, on the other hand, no one were willing to take the risk and then seek an alternate theory, there would be none of the revolutionary transformations on which scientific development depends. As Watkins says, 'there must be a critical level at which a tolerable turns into an intolerable amount of anomaly'.[2] But that level ought not be the same for everyone, nor need any individual specify his own tolerance level in advance. He need only be certain that he has one

[1] *This volume*, p. 30.

[2] *This volume*, p. 30.

and aware of some sorts of discrepancies which would drive him towards it.

4. NORMAL SCIENCE: ITS RETRIEVAL FROM HISTORY

I have so far argued that, if there are revolutions, then there must be normal science. One may, however, legitimately ask whether either exists. Toulmin has done so, and my Popperian critics have difficulties in retrieving from history a significant normal science upon the existence of which that of revolutions depends. Toulmin's questions are of particular value, for a response to them will require me to confront some genuine difficulties presented by my *Scientific Revolutions* and to modify my original presentation accordingly. Unfortunately, however, those difficulties are not the ones Toulmin sees. Before they can be isolated, the dust he has imported must be swept away.

Though there have been important changes in my position during the seven years since my book was published, the retreat from a concern with macro- to a concentration on micro-revolutions is not among them. Part of that retreat Toulmin finds by contrasting a paper *read* in 1961 with a book *published* in 1962.[1] The paper was, however, both written and published after the book, and its first footnote specifies the relationship which Toulmin inverts. Other evidence of retreat Toulmin retrieves from a comparison of the book with the manuscript of my first essay in this volume.[2] But no one else has, to my knowledge, even noticed the differences which he underlines, and the book is in any case quite explicit about the centrality of the concern which Toulmin finds only in my more recent work. Among the revolutions discussed in the body of the book are, for example, discoveries like those of X-rays and of the planet Uranus. 'Admittedly', the preface states, 'the extension [of the term "revolution" to episodes like these] strains customary usage. Nevertheless, I shall continue to speak even of discoveries as revolutionary, because it is just the possibility of relating their structure to that of, say, the Copernican revolution that makes the extended conception seem to me so important.'[3] My concern, in short, has never been with scientific revolutions as 'something that tended to happen in a given branch of science only once every two hundred years or so'.[4] Rather it has been throughout what Toulmin now takes it to have become: a little studied type of conceptual change which occurs frequently in

[1] *This volume*, p. 39 ff.

[2] See also Toulmin [1967], especially p. 471, footnote 8. The publication of this biographical canard in advance of the article on which it claims to be based has given me much trouble.

[3] Cf. my [1962*a*], pp. 7 f. On p. 6 the possibility of extending the conception to micro-revolutions is described as 'a fundamental thesis' of the book.

[4] *This volume*, p. 44.

science and is fundamental to its advance.

To that concern Toulmin's geological analogy is entirely appropriate, but not in the way he uses it. He emphasizes the aspect of the uniformitarian-catastrophist debate which dealt with the possibility of attributing catastrophes to natural causes, and he suggests that once that issue had been resolved ' "catastrophes" became *uniform* and law-governed just like any other geological and palaeontological phenomena'.[1] But his insertion of the term 'uniform' is gratuitous. Besides the issue of natural causes, the debate had a second central aspect: the question whether catastrophes existed, whether a major role in geological evolution should be attributed to phenomena like earthquakes and volcanic action which acted more suddenly and destructively than erosion and sedimentary deposition. This part of the debate the uniformitarians lost. When it was over, geologists recognized two sorts of geological change, no less distinct because both due to natural causes; one acted gradually and uniformly, the other suddenly and catastrophically. Even today we do not treat tidal waves as special cases of erosion.

Correspondingly, my claim has been, not that revolutions were inscrutable unit events, but that in science as in geology there are two sorts of change. One of them, normal science, is the generally cumulative process by which the accepted beliefs of a scientific community are fleshed out, articulated, and extended. It is what scientists are trained to do, and the main tradition in English-speaking philosophy of science derives from the examination of the exemplary works in which that training is embodied. Unfortunately, as indicated in my previous essay, proponents of that philosophical tradition generally choose their examples from changes of another sort which are then tailored to fit. The result is a failure to recognize the prevalence of changes in which conceptual commitments fundamental to the practice of some scientific specialty must be jettisoned and replaced. Of course, as Toulmin says, the two sorts of change interpenetrate: revolutions are no more total in science than in other aspects of life, but recognizing continuity through revolutions has not led historians or anyone else to abandon the notion. It was a weakness of my *Scientific Revolutions* that it could only name, not analyse, the phenomenon it repeatedly referred to as 'partial communication'. But partial communication was never, as Toulmin would have it, 'complete [mutual] incomprehension'.[2] It named a problem to be worked on, not elevated to inscrutability. Unless we can learn more about it (I shall offer some hints in the next section), we shall continue to mistake the nature of scientific progress and thus perhaps of knowledge. Nothing in Toulmin's essay begins to

[1] *This volume*, p. 43; my italics.
[2] *This volume*, p. 43

persuade me that we shall succeed if we continue to treat all scientific change as one.

The fundamental challenge of his paper, however, remains. Can we distinguish mere articulations and extensions of shared belief from changes which involve reconstruction? The answer in extreme cases is obviously 'Yes'. Bohr's theory of the hydrogen spectrum was revolutionary as Sommerfeld's theory of the hydrogen fine-structure was not; Copernican astronomical theory was revolutionary but the caloric theory of adiabatic compression was not. These examples are, however, too extreme to be fully informative: there are too many differences between the theories contrasted, and the revolutionary changes affected too many people. Fortunately, however, we are not restricted to them: Ampère's theory of the electric circuit was revolutionary (at least among French electricians), because it severed electric-current and electrostatic effects which had previously been conceptually united. Ohm's Law was again revolutionary, and was resisted accordingly, because it demanded a reintegration of concepts previously applied separately to current and charge.[1] On the other hand, the Joule–Lenz law relating the heat generated in a wire to the resistance and current was a product of normal science, for both the qualitative effects and the concepts required for quantification were in hand. Again, at a level which is not so obviously theoretical, Lavoisier's discovery of oxygen (though perhaps not Scheele's and surely not Priestley's) was revolutionary, for it was inseparable from a new theory of combustion and acidity. The discovery of neon, however, was not, for helium had supplied both the notion of an inert gas and the needed column of the periodic table.

One may question, however, how far and how universally this process of discrimination can be pressed. I am repeatedly asked whether such-and-such a development was 'normal or revolutionary', and I usually have to answer that I do not know. Nothing depends upon my, or anyone else's, being able to respond in every conceivable case, but much depends on the discrimination's being applicable to a far larger number of cases than have been supplied so far. Part of the difficulty in answering is that the discrimination of normal from revolutionary episodes demands close historical study, and few parts of the history of science have received it. One must know not simply the name of the change, but the nature and structure of group commitments before and after it occurred. Often, to determine these, one must also know the manner in which the change was received when first proposed. (There is no area in which I am more deeply conscious of the need for additional historical research, though I dissent from the

[1] On these topics, see Brown [1969] and Schagrin [1963].

conclusions Pearce Williams draws from that need and doubt that the results of investigation will draw Sir Karl and me closer.) My difficulty, however, has a deeper aspect. Though much depends upon more research, the investigations required are not simply of the sort indicated above. Furthermore, the structure of the argument in my *Scientific Revolutions* somewhat obscures the nature of what is missing. If I were rewriting the book now I would significantly change its organization.

The gist of the problem is that to answer the question 'normal or revolutionary?' one must first ask, 'for whom?' Sometimes the answer is easy: Copernican astronomy was a revolution for everyone; oxygen was a revolution for chemists but not for, say, mathematical astronomers unless, like Laplace, they were interested in chemical and thermal subjects too. For the latter group oxygen was simply another gas, and its discovery was merely an increment to their knowledge; nothing essential to them as astronomers had to be changed in the discovery's assimilation. It is not, however, usually possible to identify groups which share cognitive commitments simply by naming a scientific subject matter—astronomy, chemistry, mathematics, or the like. That is, however, what I have just done here and did earlier in my book. Some scientific subjects, for example the study of heat, have belonged to different scientific communities at different times, sometimes to several at once without becoming the special province of any. In addition, though scientists are much more nearly unanimous in their commitments than practitioners of, say, philosophy and the arts, there are such things as schools in science, communities which approach the same subject from very different points of view. French electricians in the first decades of the nineteenth century were members of a school which included almost none of the British electricians of the day, and so on. If I were writing my book again now, I would therefore begin by discussing the community structure of science, and I would not rely exclusively on shared subject matter in doing so. Community structure is a topic about which we have very little information at present, but it has recently become a major concern for sociologists, and historians are now increasingly concerned with it as well.[1]

The research problems involved are by no means trivial. Historians of science who engage in them must cease to rely exclusively on the techniques of the intellectual historian and use those of the social and cultural historian as well. Even though work has scarcely begun, there is every reason to expect it to succeed, particularly for the developed sciences, those which have severed their historical roots in the philosophical or

[1] A somewhat more detailed discussion of this reorganization together with some preliminary bibliography is included in my [1970].

medical communities. What one would then have would be a roster of the different specialists' groups through which science was advanced at various periods of time. The analytic unit would be the practitioners of a given specialty, men bound together by common elements in their education and apprenticeship, aware of each other's work, and characterized by the relative fullness of their professional communication and the relative unanimity of their professional judgement. In the mature sciences the members of such communities would ordinarily see themselves and be seen by others as the men exclusively responsible for a given subject matter and a given set of goals, including the training of their successors. Research would, however, disclose the existence of rival schools as well. Typical communities, at least on the contemporary scientific scene, may consist of a hundred members, sometimes significantly fewer. Individuals, particularly the ablest, may belong to several such groups, either simultaneously or in succession, and they will change or at least adjust their thinking caps as they go from one to another.

Groups like these should, I suggest, be regarded as the units which produce scientific knowledge. They could not, of course, function without individuals as members, but the very idea of scientific knowledge as a private product presents the same intrinsic problems as the notion of a private language, a parallel to which I shall return. Neither knowledge nor language remains the same when conceived as something an individual can possess and develop alone. It is, therefore, with respect to groups like these that the question 'normal or revolutionary?' should be asked. Many episodes will then be revolutionary for no communities, many others for only a single small group, still others for several communities together, a few for all of science. Posed in that way, the question will, I believe, have answers as precise as my distinction requires. One reason for thinking so I shall illustrate in a moment by applying this approach to some of the concrete cases used by my critics to raise doubts about the existence and role of normal science. First, however, I must point out one aspect of my present position which, far more clearly than normal science, represents a deep divide between my viewpoint and Sir Karl's.

The programme just outlined makes even clearer than it has been before the sociological base of my position. More important, it highlights what has perhaps not been clear before, the extent to which I regard scientific knowledge as intrinsically a product of a congeries of specialists' communities. Sir Karl sees 'a great danger in . . . specialization', and the context in which he provides this evaluation suggests that the danger is the same one he sees in normal science.[1] But with respect to the former, at least, the

[1] *This volume*, p. 53.

battle has clearly been lost from the start. Not that one might not wish for good reasons to oppose specialization and even succeed in doing so, but that the effort would necessarily be to oppose science as well. Whenever Sir Karl contrasts science with philosophy, as he does at the start of his paper, or physics with sociology, psychology, and history, as he does at the end, he is contrasting an esoteric, isolated, and largely self-contained discipline with one that still aims to communicate with and persuade an audience larger than their own profession. (Science is not the only activity the practitioners of which can be grouped into communities, but it is the only one in which each community is its own exclusive audience and judge.[1]) The contrast is not a new one, characteristic, say, of Big Science and the contemporary scene. Mathematics and astronomy were esoteric subjects in antiquity; mechanics became so after Galileo and Newton; electricity after Coulomb and Poisson; and so on until economics today. For the most part that transition to a closed specialists' group was part of the transition to maturity that I discussed above when considering the emergence of puzzle solving. It is hard to believe that it is a dispensable characteristic. Perhaps science could again become like philosophy, as Sir Karl wishes, but I suspect that he would then admire it less.

To conclude this part of my discussion, I turn to some concrete cases by means of which my critics illustrate their difficulties in finding normal science and its functions in history, taking up first a problem raised by Sir Karl and Watkins. Both point out that nothing like a consensus over fundamentals 'emerged during the long history of the theory of *matter*: here from the pre-Socratics to the present day there has been an unending *debate* between continuous and discontinuous concepts of matter, between various atomic theories on the one hand, and ether, wave and field theories on the other'.[2] Feyerabend makes a very similar point for the second half of the nineteenth century by contrasting the mechanical, phenomenological and field-theoretic approaches to problems of physics.[3] With all of their descriptions of what went on I agree. But the term 'theories of matter' does not, at least until the last thirty years, even differentiate the concerns of science from those of philosophy, much less single out a community or small group of communities responsible for and expert in the subject.

I am not suggesting that scientists do not have and use theories of matter, nor that their work is unaffected by such theories, nor that their research results have no role in the theories of matter held by others. But

[1] See my [1969].

[2] *This volume*, pp. 34 ff, and 54–5. As Watkins notes, Dudley Shapere has made a similar point in his [1964] in connection with the role of atomism in chemistry in the first half of the nineteenth century. I deal with that case immediately below.

[3] *This volume*, p. 207.

until this century theories of matter have been a tool for scientists rather than a subject matter. That different specialties have chosen different tools and sometimes criticized each others' choices does not mean that they have not each been practising normal science. The frequently heard generalization that, before the advent of wave mechanics, physicists and chemists deployed characteristic and irreconcilable theories of matter is too simplistic (partly because it can equally well be said about different chemical specialties even today). But the very possibility of such a generalization suggests the way in which the issue raised by Watkins and Sir Karl must be approached. For that matter, the practitioners of a given community or school need not always share a theory of matter. Chemistry during the first half of the nineteenth century is a case in point. Though many of its fundamental tools—constant proportion, multiple proportion, combining weights, and so on—had been developed and become common property through Dalton's atomic theory, the men who used them could, after the event, adopt widely varying attitudes about the nature and even the existence of atoms. Their discipline, or at least many parts of it, did not depend upon a shared model for matter.

Even where they admit the existence of normal science, my critics regularly have difficulty discovering crisis and its role. Watkins provides an example, and its resolution follows at once from the sort of analysis deployed above. Kepler's Laws, Watkins reminds us, were incompatible with Newton's planetary theory, but astronomers had not previously been dissatisfied with them. Newton's revolutionary treatment of planetary motions was not, Watkins therefore asserts, preceeded by astronomical crisis. But why should it have been? In the first place, the transition from Keplerian to Newtonian orbits need not have been (I lack the evidence to be certain) a revolution *for astronomers*. Most of them followed Kepler and explained the shape of the planetary orbits in mechanical rather than geometrical terms. (Their explanation did not, that is, make use of the ellipse's 'geometric perfection', if any, or of some other characteristic of which the orbit was deprived by Newtonian perturbations.) Though the transition from circle to ellipse had been part of a revolution for them, a minor adjustment of mechanism would account, as it did with Newton, for departure from ellipticity. More important, Newton's adjustment of Keplerian orbits was a by-product of his work in mechanics, a field to which the community of mathematical astronomers made passing reference in their prefaces but which thereafter played only the most global role in their work. In mechanics, however, where Newton did induce a revolution, there had been a widely recognized crisis since the acceptance of Copernicanism. Watkins's counter-example is the best sort of grist for my mill.

I turn finally to one of Lakatos's extended case histories, that of the Bohr research programme, for it illustrates what most puzzles me about his often admirable paper and suggests how deep even residual Popperianism can be. Though his terminology is different, his analytic apparatus is as close to mine as need be: hard core, work in the protective belt, and degenerative phase are close parallels for my paradigms, normal science, and crisis. Yet in important ways Lakatos fails to see how these shared notions function even when applying them to what is for me an ideal case. Let me illustrate some of the things he could have seen and might have said. My version, like his or like any other bit of historical narrative, will be a rational reconstruction. But I shall not ask my readers to apply 'tons of salt' nor add footnotes pointing out that what is said in my text is false.[1]

Consider Lakatos's account of the origin of the Bohr atom. 'The background problem', he writes, 'was the riddle of how Rutherford atoms . . . can remain stable; for, according to the well-corroborated Maxwell–Lorentz theory of electromagnetism they should collapse.'[2] That is a genuine Popperian problem (not a Kuhnian puzzle) arising from the conflict between two increasingly well-established parts of physics. It had, in addition, been available for some time as a potential focal point for criticism. It did not originate with Rutherford's model in 1911; radiative instability was equally a difficulty for most older atom-models, including both Thomson's and Nagaoka's. Furthermore, it is the problem which Bohr (in some sense) solved in his famous three-part paper of 1913, thereby inaugurating a revolution. No wonder Lakatos would like it to be the 'background problem' for the research programme that produced the revolution, but it emphatically is not.[3]

Instead, the background was an entirely normal puzzle. Bohr set out to improve the physical approximations in a paper by C. G. Darwin on the energy lost by charged particles passing through matter. In the process he

[1] *This volume*, pp. 138, 140 and 146, and elsewhere. One may reasonably ask about the evidential force of examples that call for this sort of qualification (and is 'qualification' quite the right word?). I shall, however, in another context be very grateful for these 'case histories' of Lakatos's. More clearly, because more explicitly, than any other examples I know, they illustrate the differences between the way philosophers and historians usually do history. The problem is not that philosophers are likely to make errors—Lakatos knows the facts better than many historians who have written on these subjects, and historians do make egregious errors. But a historian would not *include in his narrative* a factual report which he *knew* to be false. If he had done so, he would be so sensitive to the offence that he could not conceivably compose a footnote calling attention to it. Both groups are scrupulous, but they differ in what they are scrupulous about. I have discussed some differences of this sort, in my unpublished Isenberg Lecture, 'The Relations between History and Philosophy of Science', read in March 1968.

[2] *This volume*, p. 141.

[3] For what follows, see Heilbron and Kuhn [1969].

made what was to him the surprising discovery that the Rutherford atom, unlike other current models, was mechanically unstable and that a Planck-like *ad hoc* device for stabilizing it provided a promising explanation of the periodicities in Mendeleev's table, something else for which he had not been looking. At that point his model still had no excited states, nor was Bohr yet concerned to apply it to atomic spectra. Those steps followed, however, as he attempted to reconcile his model with the apparently incompatible one developed by J. W. Nicholson and, in the process, encountered Balmer's formula. Like much of the research that produces revolutions, Bohr's biggest achievements in 1913 were products, therefore, of a research programme directed to goals very different from those obtained. Though he could not have stabilized the Rutherford model by quantization if unaware of the crisis which Planck's work had introduced to physics, his own work illustrates with particular clarity the revolutionary efficacy of normal research puzzles.

Examine, finally, the concluding portion of Lakatos's case history, the degenerative phase of the old quantum theory. Most of the story he tells well, and I shall simply point it up. From 1900 on it was increasingly widely recognized among physicists that Planck's quantum had introduced a fundamental inconsistency into physics. At first many of them tried to eliminate it, but, after 1911 and particularly after the invention of Bohr's atom, those critical efforts were increasingly abandoned. Einstein was, for more than a decade, the only physicist of note who continued to direct his energies towards the search for a consistent physics. Others learned to live with inconsistency and tried instead to solve technical puzzles with the tools at hand. Particularly in the areas of atomic spectra, atomic structure, and specific heats, their achievements were unprecedented. Though the inconsistency of physical theory was widely acknowledged, physicists could nevertheless exploit it and by doing so made fundamental discoveries at an extraordinary rate between 1913 and 1921. Quite suddenly, however, beginning in 1922, these very successes were seen to have isolated three obdurate problems—the helium model, the anomalous Zeeman effect, and optical dispersion—which could not, physicists were increasingly convinced, be resolved by anything quite like existing technique. As a result, many of them changed their research stance, proliferating more and wilder versions of the old quantum theory than before, designing and testing each attempt against the three recognized trouble spots.

It is this last phase, 1922 and after, which Lakatos calls the degenerative stage of Bohr's programme. For me it is a case book example of crisis, clearly documented in publications, correspondence, and anecdote. We see it in very nearly the same way. Lakatos might therefore have told the rest of the story. To those who were experiencing this crisis, two of the three problems which had provoked it proved immensely informative, dispersion and the anomalous Zeeman effect. By a series of connected steps too complex to be outlined here, their pursuit led first to the adoption in Copenhagen of an atom model in which so-called virtual oscillators coupled discrete quantum states, then to a formula for quantum-theoretical dispersion, and finally to matrix mechanics which terminated the crisis barely three years after it had begun. For that first formulation of quantum mechanics, the degenerative phase of the old quantum theory provided both occasion and much detailed technical substance. History of science, to my knowledge, offers no equally clear, detailed, and cogent example of the creative functions of normal science and crisis.

Lakatos, however, ignores this chapter and jumps instead to wave mechanics, the second and at first quite different formulation of a new quantum theory. First, he describes the degenerative phase of the old quantum theory as filled with 'ever more sterile inconsistencies and ever more *ad hoc* hypotheses' ('*ad hoc*' and 'inconsistencies' are right; 'sterile' could not be more wrong; not only did these hypotheses lead to matrix mechanics but also to electron spin). Then, he produces the crisis-resolving innovation like a magician pulling a rabbit from a hat: 'A rival research programme soon appeared: wave mechanics . . . [which] soon caught up with, vanquished and replaced Bohr's programme. De Broglie's paper came at a time when Bohr's programme was degenerating. *But this was mere coincidence.* One wonders what would have happened if de Broglie had published his paper in 1914 instead of 1924.'[1]

To the closing rhetorical question, the answer is clear: nothing at all. Both de Broglie's paper and the route from it to the Schrödinger wave equation depend in detail on developments which occurred after 1914: on work by Einstein and by Schrödinger himself as well as on the discovery of the Compton effect in 1922.[2] Even if that point could not be documented in detail, however, is not coincidence strained beyond recognition when used to explain the simultaneous emergence of two independent and at first quite different theories, both capable of resolving a crisis that had been visible for only three years?

Let me be scrupulous. Though Lakatos entirely misses the essential creative functions of the crisis of the old quantum theory, he is not altogether wrong about its relevance to the invention of wave mechanics. The wave equation was not a response to the crisis which began in 1922 but to the one which dates from Planck's work in 1900 and on which most physicists had turned their backs after 1911. If Einstein had not tenaciously refused to

[1] *This volume*, p. 154; my italics. [2] See Klein [1964] and Raman and Forman [1969].

set aside his deep dissatisfaction with the fundamental inconsistencies of the old quantum theory (and if he had not been able to attach that discontent to the concrete technical puzzles of electromagnetic fluctuation phenomena—something for which he found no equivalent after 1925), the wave equation would not have emerged when and as it did. The research route which leads to it is not the same as the route to matrix mechanics.

But neither are the two independent, nor is the simultaneity of their termination due merely to coincidence. Among the several research episodes which tie them together is, for example, Compton's convincing demonstration in 1922 of the particulate properties of light, the by-product of a very high-class piece of normal research on X-ray scattering. Before physicists could consider the idea of matter waves, they had first to take the idea of the photon seriously, and this few of them had done before 1922. De Broglie's work started as photon theory, its main thrust being to reconcile Planck's radiation law with the particulate structure of light; matter waves entered along the way. De Broglie himself may not have needed Compton's discovery in order to take the photon seriously, but his audience, French and foreign, certainly did. Though wave mechanics in no sense follows from the Compton effect, there are historical ties between the two. On the road to matrix mechanics the role of the Compton effect is even clearer. The first use of the virtual oscillator model in Copenhagen was to show how that effect could be explained *without* recourse to Einstein's photon, a concept that Bohr had been notoriously reluctant to accept. The same model was next applied to dispersion and the clues to matrix mechanics found. The Compton effect is therefore one bridge across the gap which Lakatos hides under 'coincidence'.

Having provided elsewhere many other examples of the significant roles of normal science and crisis, I shall not multiply them further here. For lack of additional research I could not, in any case, provide enough. When completed, that research need not bear me out, but what has been done so far surely fails to support my critics. They must look further for counter-examples.

5. IRRATIONALITY AND THEORY-CHOICE

I consider now one last set of concerns voiced by my present critics, in this case one they share with a number of other philosophers. It arises mainly from my description of the procedures by which scientists choose between competing theories, and it results in charges which cluster about such terms as 'irrationality', 'mob rule', and 'relativism'. In this section I aim to eliminate misunderstandings for which my own past rhetoric is doubtless partially responsible. In my concluding section, which follows, I shall touch upon some deeper issues raised by the problem of theory-choice. At that point the terms 'paradigm' and 'incommensurability', which I have so far almost entirely avoided, will necessarily re-enter the discussion.

In my *Scientific Revolutions* normal science is at one point described as 'a strenuous and devoted attempt to force nature into the conceptual boxes supplied by professional education.'[1] Later, discussing the problems which surround the choice between competing sets of boxes, theories, or paradigms, I described them as[2]:

> about techniques of persuasion, or about argument and counter argument in a situation in which . . . neither proof nor error is at issue. The transfer of allegiance from paradigm to paradigm is a conversion experience that cannot be forced. Lifelong resistance . . . is not a violation of scientific standards but an index to the nature of scientific research itself. . . . Though the historian can always find men—Priestley, for instance—who were unreasonable to resist for as long as they did, he will not find a point at which resistance becomes illogical or unscientific. At most he may wish to say that the man who continues to resist after his whole profession has been converted has *ipso facto* ceased to be a scientist.

Not surprisingly (though I have myself been very much surprised), passages like these are in some quarters read as implying that, in the developed sciences, might makes right. Members of a scientific community can, I am held to have claimed, believe anything they please if only they will first decide what they agree about and then enforce it both on their colleagues and on nature. The factors which determine what they do choose to believe are fundamentally irrational, matters of accident and personal taste. Neither logic nor observation nor good reason is implicated in theory-choice. Whatever scientific truth may be, it is through-and-through relativistic.

These are all damaging misinterpretations, whatever my responsibility may be for making them possible. Though their elimination will still leave a deep divide between my critics and me, it is prerequisite even to discovering our disagreement. Before treating them individually, however, one general remark should be helpful. The sorts of misinterpretations just outlined are voiced only by philosophers, a group already familiar with the points at which I aim in passages like the above. Unlike readers to whom the point is less familiar, they sometimes suppose that I intend more than I do. What I mean to be saying, however, is only the following.

In a debate over choice of theory, neither party has access to an argument which resembles a proof in logic or formal mathematics. In the latter, both premises and rules of inference are stipulated in advance. If there

[1] Cf. my [1962*a*], p. 5. [2] *Op. cit.* p. 151.

is disagreement about conclusions, the parties to the debate can retrace their steps one by one, checking each against prior stipulation. At the end of that process, one or the other must concede that at an isolable point in the argument he has made a mistake, violated or misapplied a previously accepted rule. After that concession he has no recourse, and his opponent's proof is then compelling. Only if the two discover instead that they differ about the meaning or applicability of a stipulated rule, that their prior agreement does not provide a sufficient basis for proof, does the ensuing debate resemble what inevitably occurs in science.

Nothing about this relatively familiar thesis should suggest that scientists do not *use* logic (and mathematics) in their arguments, including those which aim to persuade a colleague to renounce a favoured theory and embrace another. I am dumbfounded by Sir Karl's attempt to convict me of self-contradiction because I employ logical arguments myself.[1] What might better be said is that I do not expect that, merely because my arguments are logical, they will be compelling. Sir Karl underscores my point, not his, when he describes them as logical but mistaken, and then makes no attempt to isolate the mistake or to display its logical character. What he means is that, though my arguments are logical, he disagrees with my conclusion. Our disagreement must be about premises or the manner in which they are to be applied, a situation which is standard among scientists debating theory-choice. When it occurs, their recourse is to persuasion as a prelude to the possibility of proof.

To name persuasion as the scientist's recourse is not to suggest that there are not many good reasons for choosing one theory rather than another.[2] It is emphatically *not* my view that 'adoption of a new scientific theory is an intuitive or mystical affair, a matter for psychological description rather than logical or methodological codification'.[3] On the contrary, the chapter of my *Scientific Revolutions* from which the preceding quotation was abstracted explicitly denies 'that new paradigms triumph ultimately through some mystical aesthetic', and the pages which precede that denial contain a preliminary codification of good reasons for theory choice.[4] These are, furthermore, reasons of exactly the kind standard in philosophy of science: accuracy, scope, simplicity, fruitfulness, and the like. It is vitally important that scientists be taught to value these characteristics and that they be provided with examples that illustrate them in practice. If they did not hold values like these, their disciplines would

[1] *This volume*, pp. 55 and 57.

[2] For one version of the view that Kuhn insists that 'the decisions of a scientific group to adopt a new paradigm cannot be based on good reasons of any kind, factual or otherwise', see Shapere [1966], especially p. 67.

[3] Cf. Scheffler [1967], p. 18.

[4] Cf. my [1962*a*], p. 157.

develop very differently. Note, for example, that the periods in which the history of art was a history of progress were also the periods in which the artist's aim was accuracy of representation. With the abandonment of that value, the developmental pattern changed drastically though very significant development continued.[1]

What I am denying then is neither the existence of good reasons nor that these reasons are of the sort usually described. I am, however, insisting that such reasons constitute values to be used in making choices rather than rules of choice. Scientists who share them may nevertheless make different choices in the same concrete situation. Two factors are deeply involved. First, in many concrete situations, different values, though all constitutive of good reasons, dictate different conclusions, different choices. In such cases of value-conflict (e.g. one theory is simpler but the other is more accurate) the relative weight placed on different values by different individuals can play a decisive role in individual choice. More important, though scientists share these values and must continue to do so if science is to survive, they do not all apply them in the same way. Simplicity, scope, fruitfulness, and even accuracy can be judged quite differently (which is not to say they may be judged arbitrarily) by different people. Again, they may differ in their conclusions without violating any accepted rule.

That variability of judgement may, as I suggested above in connection with the recognition of crises, even be essential to scientific advance. The choice of a theory, which is, as Lakatos says, equally the choice of a research programme, involves major risks, particularly in its early stages. Some scientists must, by virtue of a value system differing in its applicability from the average, choose it early, or it will not be developed to the point of general persuasiveness. The choices dictated by these atypical value systems are, however, generally wrong. If all members of the community applied values in the same high-risk way, the group's enterprise would cease. This last point, I think, Lakatos misses, and with it the essential role of individual variability in what is only belatedly the unanimous decision of the group. As Feyerabend also emphasizes, to give these decisions a '*historical character*' or to suggest that they are made only '*with hindsight*' deprives them of their function.[2] The scientific community cannot wait for history, though some individual members do. The needed results are instead achieved by distributing the risk that must be taken among the group's members.

Does anything in this argument suggest the appropriateness of phrases

[1] Gombrich, [1960], pp. 11 f.

[2] *This volume*, pp. 120 and 215 ff.

like decision by 'mob psychology'?[1] I think not. On the contrary, one characteristic of a mob is its rejection of values which its members ordinarily share. Done by scientists, the result should be the end of their science, and the Lysenko case suggests that it would be. My argument, however, goes even further, for it emphasizes that, unlike most disciplines, the responsibility for applying shared scientific values, must be left to the specialists' group.[2] It may not even be extended to all scientists, much less to all educated laymen, much less to the mob. If the specialists' group behaves as a mob, renouncing its normal values, then science is already past saving.

By the same token, no part of the argument here or in my book implies that scientists may choose any theory they like so long as they agree in their choice and thereafter enforce it.[3] Most of the puzzles of normal science are directly presented by nature, and all involve nature indirectly. Though different solutions have been received as valid at different times, nature cannot be forced into an arbitrary set of conceptual boxes. On the contrary, the history of proto-science shows that normal science is possible only with very special boxes, and the history of developed science shows that nature will not indefinitely be confined in any set which scientists have constructed so far. If I sometimes say that any choice made by scientists on the basis of their past experience and in conformity with their traditional values is *ipso facto* valid science for its time, I am only underscoring a tautology. Decisions made in other ways or decisions that could not be made in this way provide no basis for science and would not be scientific.

The charges of irrationality and relativism remain. To the first, however, I have already spoken, for I have discussed the issues, excepting incommensurability, from which it seems to arise. I am not sanguine in this matter, however, for I have not previously and do not now understand quite what my critics mean when they employ terms like 'irrational' and 'irrationality' to characterize my views. These labels seem to me mere shibboleths, barriers to a joint enterprise whether conversation or research. My difficulties in understanding are, however, even clearer and more acute when these terms are used not to criticize my position but in its defence. Obviously there is much in the last part of Feyerabend's paper with

[1] *This volume*, pp. 140, footnote 3, and 178.

[2] Cf. my [1962*a*], p. 167.

[3] Some sense of my surprise and chagrin over this and related ways of reading my book may be generated by the following anecdote. During a meeting I was talking to a usually far-distant friend and colleague whom I knew, from a published review, to be enthusiastic about my book. She turned to me and said, 'Well, Tom, it seems to me that your biggest problem now is showing in what sense science can be empirical'. My jaw dropped and still sags slightly. I have total visual recall of that scene and of no other since de Gaulle's entry into Paris in 1944.

which I agree, but to describe the argument as a defence of irrationality in science seems to me not only absurd but vaguely obscene. I would describe it, together with my own, as an attempt to show that existing theories of rationality are not quite right and that we must readjust or change them to explain why science works as it does. To suppose, instead, that we possess criteria of rationality which are independent of our understanding of the essentials of the scientific process is to open the door to cloud-cuckoo land.

An answer to the charge of relativism must be more complex than those which precede, for the charge arises from more than misunderstanding. In one sense of the term I may be a relativist; in a more essential one I am not. What I can hope to do here is separate the two. It must already be clear that my view of scientific development is fundamentally evolutionary. Imagine, therefore, an evolutionary tree representing the development of the scientific specialties from their common origin in, say, primitive natural philosophy. Imagine, in addition, a line drawn up that tree from the base of the trunk to the tip of some limb without doubling back on itself. Any two theories found along this line are related to each other by descent. Now consider two such theories, each chosen from a point not too near its origin. I believe it would be easy to design a set of criteria—including maximum accuracy of predictions, degree of specialization, number (but not scope) of concrete problem solutions—which would enable any observer involved with neither theory to tell which was the older, which the descendant. For me, therefore, scientific development is, like biological evolution, unidirectional and irreversible. One scientific theory is not as good as another for doing what scientists normally do. In that sense I am not a relativist.

But there are reasons why I get called one, and they relate to the contexts in which I am wary about applying the label 'truth'. In the present context, its intra-theoretic uses seem to me unproblematic. Members of a given scientific community will generally agree which consequences of a shared theory sustain the test of experiment and are therefore true, which are false as theory is currently applied, and which are as yet untested. Dealing with the comparison of theories designed to cover the same range of natural phenomena, I am more cautious. If they are historical theories, like those considered above, I can join Sir Karl in saying that each was believed to be true in its time but was later abandoned as false. In addition, I can say that the later theory was the better of the two as a tool for the practice of normal science, and I can hope to add enough about the senses in which it was better to account for the main developmental characteristics of the sciences. Being able to go that far, I do not myself feel that I am a relativist. Nevertheless, there is another step, or kind of step, which many philosophers of science wish to take and which I refuse. They wish, that is, to compare theories as representations of nature, as statements about 'what is really out there'. Granting that neither theory of a historical pair is true, they nonetheless seek a sense in which the later is a better approximation to the truth. I believe nothing of that sort can be found. On the other hand, I no longer feel that anything is lost, least of all the ability to explain scientific progress, by taking this position.

What I am rejecting will be clarified by reference to Sir Karl's paper and to his other writings. He has proposed a criterion of verisimilitude which permits him to write that 'a later theory . . . t_2 has superseded t_1 . . . by approaching more closely to the truth than t_1'. Also, when discussing a succession of frameworks, he speaks of each later member of the series as 'better *and roomier*' than its predecessors; and he implies that the limit of the series, at least if carried to infinity, is ' "absolute" or "objective" truth, in Tarski's sense'.[1] Those positions present, however, two problems, about the first of which I am uncertain of Sir Karl's position. To say, for example, of a field theory that it 'approach[es] more closely to the truth' than an older matter-and-force theory should mean, unless words are being oddly used, that the ultimate constituents of nature are more like fields than like matter and force. But in this ontological context it is far from clear how the phrase 'more like' is to be applied. Comparison of historical theories gives no sense that their ontologies are approaching a limit: in some fundamental ways Einstein's general relativity resembles Aristotle's physics more than Newton's. In any case, the evidence from which conclusions about an ontological limit are to be drawn is the comparison not of whole theories but of their empirical consequences. That is a major leap, particularly in the face of a theorem that any finite set of consequences of a given theory can be derived from another incompatible one.

The other difficulty is highlighted by Sir Karl's reference to Tarski and is more fundamental. The semantic conception of truth is regularly epitomized in the example: 'Snow is white' is true if and only if snow is white. To apply that conception in the comparison of two theories, one must therefore suppose that their proponents agree about technical equivalents of such matters of fact as whether snow is white. If that supposition were exclusively about objective observation of nature, it would present no insuperable problems, but it involves as well the assumption that the objective observers in question understand 'snow is white' in the same way, a matter which may not be obvious if the sentence reads 'elements combine in constant proportion by weight'. Sir Karl takes it for granted that the

[1] Popper [1963], chapter 10, particularly p. 232; and *this volume*, p. 56; my italics.

proponents of competing theories do share a neutral language adequate to the comparision of such observation reports. I am about to argue that they do not. If I am right, then 'truth' may, like 'proof', be a term with only intra-theoretic applications. Until this problem of a neutral observation language is resolved, confusion will only be perpetuated by those who point out (as Watkins does when responding to my closely parallel remarks about 'mistakes'[1]) that the term is regularly used as though the transfer from intra- to inter-theoretic contexts made no difference.

6. INCOMMENSURABILITY AND PARADIGMS

At last we arrive at the central constellation of issues which separate me from most of my critics. I regret the length of the journey to this point but accept only partial responsibility for the brush that has had to be cleared from the path. Unfortunately, the necessity of relegating these issues to my concluding section results in a relatively cursory and dogmatic treatment. I can hope only to isolate some aspects of my viewpoint which my critics have generally missed or dismissed and to provide motives for further reading and discussion.

The point-by-point comparison of two successive theories demands a language into which at least the empirical consequences of both can be translated without loss or change. That such a language lies ready to hand has been widely assumed since at least the seventeenth century when philosophers took the neutrality of pure sensation-reports for granted and sought a 'universal character' which would display all languages for expressing them as one. Ideally the primitive vocabulary of such a language would consist of pure sense-datum terms plus syntactic connectives. Philosophers have now abandoned hope of achieving any such ideal, but many of them continue to assume that theories can be compared by recourse to a basic vocabulary consisting entirely of words which are attached to nature in ways that are unproblematic and, to the extent necessary, independent of theory. That is the vocabulary in which Sir Karl's basic statements are framed. He requires it in order to compare the verisimilitude of alternate theories or to show that one is 'roomier' than (or includes) its predecessor. Feyerabend and I have argued at length that no such vocabulary is available. In the transition from one theory to the next words change their meanings or conditions of applicability in subtle ways.[2]

[1] *This volume*, p. 26, footnote 3.

[2] In his [1964], Shapere criticizes, in part quite properly, the way I discuss meaning-change in my book. In the process he challenges me to specify the 'cash difference' between a change in meaning and an alteration in the application of a term. Need I say that, in the present state of the theory of meaning, there is none. The identical point can be made using either term.

Though most of the same signs are used before and after a revolution—e.g. force, mass, element, compound, cell—the ways in which some of them attach to nature has somehow changed. Successive theories are thus, we say, incommensurable.

Our choice of the term 'incommensurable' has bothered a number of readers. Though it does not mean 'incomparable' in the field from which it was borrowed, critics have regularly insisted that we cannot mean it literally since men who hold different theories do communicate and sometimes change each others' views.[1] More important, critics often slide from the observed existence of such communication, which I have underscored myself, to the conclusion that it can present no essential problems. Toulmin seems content to admit 'conceptual incongruities' and then go on as before.[2] Lakatos inserts parenthetically the phrase 'or from semantical reinterpretations' when telling us how to compare successive theories and thereafter treats the comparison as purely logical.[3] Sir Karl exorcises the difficulty in a way that has particular interest: 'It is just a dogma—a dangerous dogma—that the different frameworks are like mutually untranslatable languages. The fact is that even totally different languages (like English and Hopi, or Chinese) are not untranslatable, and that there are many Hopis or Chinese who have learnt to master English very well.'[4]

I accept the utility, indeed the importance, of the linguistic parallel, and shall therefore dwell for a bit upon it. Presumably Sir Karl accepts it too since he uses it. If he does, the dogma to which he objects is not that frameworks are like languages but that languages are untranslatable. But no one ever believed they were! What people have believed, and what makes the parallel important, is that the difficulties of learning a second language are different from and far less problematic than the difficulties of translation. Though one must know two languages in order to translate at all, and though translation can then always be managed up to a point, it can present grave difficulties to even the most adept bilingual. He must find the best available compromises between incompatible objectives. Nuances must be preserved but not at the price of sentences so long that communication breaks down. Literalness is desirable but not if it demands introducing too many foreign words which must be separately discussed in a glossary or appendix. People deeply committed both to accuracy and to felicity of expression find translation painful, and some cannot do it at all.

[1] See, for example, *this volume*, pp. 43–4.

[2] *This volume*, p. 44.

[3] *This volume*, p. 118. Perhaps only because of its excessive brevity, Lakatos's other reference to this problem on p. 179, note 1, is equally little helpful.

[4] *This volume*, p. 56.

Translation, in short, always involves compromises which alter communication. The translator must decide what alterations are acceptable. To do that he needs to know what aspects of the original it is most important to preserve and also something about the prior education and experience of those who will read his work. Not surprisingly, therefore, it is today a deep and open question what a perfect translation would be and how nearly an actual translation can approach the ideal. Quine has recently concluded 'that rival systems of analytic hypotheses [for the preparation of translations] can conform to all speech dispositions within each of the languages concerned and yet dictate, in countless cases, utterly disparate translation . . . Two such translations might even be patently contrary in truth value.'[1] One need not go that far to recognize that reference to translation only isolates but does not resolve the problems which have led Feyerabend and me to talk of incommensurability. To me at least, what the existence of translations suggests is that recourse is available to scientists who hold incommensurable theories. That recourse need not, however, be to full restatement in a neutral language of even the theories' consequences. The problem of theory-comparison remains.

Why is translation, whether between theories or languages, so difficult? Because, as has often been remarked, languages cut up the world in different ways, and we have no access to a neutral sub-linguistic means of reporting. Quine points out that, though the linguist engaged in radical translation can readily discover that his native informant utters 'Gavagai' because he has seen a rabbit, it is more difficult to discover how 'Gavagai' should be translated. Should the linguist render it as 'rabbit', 'rabbit-kind', 'rabbit-part', 'rabbit-occurrence', or by some other phrase he may not even have thought to formulate? I extend the example by supposing that, in the community under examination, rabbits change colour, length of hair, characteristic gait, and so on during the rainy season, and that their appearance then elicits the term 'Bavagai'. Should 'Bavagai' be translated 'wet rabbit', 'shaggy rabbit', limping rabbit', all of these together, or should the linguist conclude that the native community has not recognized that 'Bavagai' and 'Gavagai' refer to the same animal? Evidence relevant to a choice among these alternatives will emerge from further investigation, and the result will be a reasonable analytic hypothesis with implication for the translation of other terms as well. But it will be only a hypothesis (none of the alternatives considered above need be right); the result of any error may be later difficulties in communication; when it occurs, it will be far from clear whether the problem is with translation and, if so, where the root difficulty lies.

[1] Quine [1960], pp. 73 ff.

These examples suggest that a translation manual inevitably embodies a theory, which offers the same sorts of reward, but also is prone to the same hazards, as other theories. To me they also suggest that the class of translators includes both the historian of science and the scientist trying to communicate with a colleague who embraces a different theory.[1] (Note, however, that the motives and correlated sensitivities of the scientists and historian are very different, which accounts for many systematic differences in their results.) They often have the inestimable advantage that the signs used in the two languages are identical or nearly so, that most of them function the same way in both languages, and that, where function has changed, there are nevertheless informative reasons for retaining the same sign. But those advantages bring with them penalties illustrated in both scientific discourse and history of science. They make it excessively easy to ignore functional changes that would be apparent if they had been accompanied by a change of sign.

The parallel between the task of the historian and the linguist highlights an aspect of translation with which Quine does not deal (he need not) and that has made trouble for linguists.[2] Teaching Aristotelian physics to students, I regularly point out that matter (in the *Physics*, not the *Metaphysics*), just because of its omnipresence and qualitative neutrality, is a physically dispensable concept. What populates the Aristotelian universe, accounting for both its diversity and regularity, is immaterial 'natures' or 'essences'; the appropriate parallel for the contemporary periodic table is not the four Aristotelian elements, but the quadrangle of four fundamental forms. Similarly, when teaching the development of Dalton's atomic theory, I point out that it implied a new view of chemical combination with the result that the line separating the referents of the terms 'mixture' and 'compound' shifted; alloys were compounds before Dalton, mixtures after.[3] Those remarks are part and parcel of my attempt to translate older theories into modern terms, and my students characteristically read source materials, though already rendered into English, differently after I have

[1] A number of these ideas about translation were developed in my Princeton seminar. I cannot now distinguish my contributions from those of the students and colleagues who attended. A paper by Tyler Burge was, however, particularly helpful.

[2] See particularly Nida [1964]. I am much indebted to Sarah Kuhn for calling this paper to my attention.

[3] This example makes particularly clear the inadequacy of Scheffler's suggestion that the problems raised by Feyerabend and me vanish if one substitutes sameness-of-reference for sameness-of-meaning (Scheffler [1967], chapter 3). Whatever the reference of 'compound' may be, in this example it changes. But, as the following discussion will indicate, sameness-of-reference is no more free of difficulty than sameness-of-meaning in any of the applications that concern me and Feyerabend. Is the referent of 'rabbit' the same as that of 'rabbit-kind' or of 'rabbit-occurrence'? Consider the criteria of individuation and of self-identity which fit each of the terms.

made them than they did before. By the same token, a good translation manual, particularly for the language of another region and culture, should include or be accompanied by discursive paragraphs explaining how native speakers view the world, what sorts of ontological categories they deploy. Part of learning to translate a language or a theory is learning to describe the world with which the language or theory functions.

Having introduced translation to illustrate the illumination that can be had by regarding scientific communities as language communities, I now leave it for a time in order to examine a particularly important aspect of the parallelism. In learning either a science or a language, vocabulary is generally acquired together with at least a minimal battery of generalizations which exhibit it applied to nature. In neither case, however, do the generalizations embody more than a fraction of the knowledge of nature which has been acquired in the learning process. Much of it is embodied instead in the mechanism, whatever it may be, which is used to attach terms to nature.[1] Both natural and scientific language are designed to describe the world as it is, not any conceivable world. The former, it is true, adapts to the unexpected occurrence more easily than the latter, but often at the price of long sentences and dubious syntax. Things which cannot *readily* be said in a language are things that its speakers do not expect to have occasion to say. If we forget this or underestimate its importance, that is probably because its converse does not hold. We can readily describe many things (unicorns, for example) which we do not expect to see.

How, then, do we acquire the knowledge of nature that is built into language? For the most part by the same techniques and at the same time as we acquire language itself, whether everyday or scientific. Parts of the process are well known. The definitions in a dictionary tell us something about what words mean and simultaneously inform us of the objects and situations about which we may need to read or speak. About some of these words we learn more, and about others everything we know, by encountering them in a variety of sentences. Under those circumstances, as Carnap has shown, we acquire laws of nature together with a knowledge of meanings. Given a verbal definition of two tests, each definitive, for the presence of an electric charge, we learn both about the term 'charge' and also that a body which passes one test will also pass the other. These procedures for language-nature learning are, however, purely linguistic. They relate words to other words and thus can function only if we already possess some vocabulary acquired by a non-verbal or incompletely verbal process.

[1] For an extended example, see my [1964]. A more analytic discussion will be found in my [1970].

Presumably that part of learning is by ostension or some elaboration of it, the direct matching of whole words or phrases to nature. If Sir Karl and I have a fundamental philosophic dispute, it is about the relevance of this last mode of language-nature learning to philosophy of science. Though he knows that many words needed by scientists, particularly for the formulation of basic sentences, are learned by a process not fully linguistic, he treats those terms and the knowledge acquired with them as unproblematic, at least in the context of theory-choice. I believe he misses a central point, the one which led me to introduce the notion of paradigms in my *Scientific Revolutions*.

When I speak of knowledge embedded in terms and phrases learned by some non-linguistic process like ostension, I am making the same point that my book aimed to make by repeated reference to the role of paradigms as concrete problem solutions, the exemplary objects of an ostension. When I speak of that knowledge as consequential for science and for theory-construction, I am identifying what Miss Masterman underscores about paradigms by saying that they 'can function when the theory is not there'.[1] These ties are not, however, likely to be apparent to anyone who has taken the notion of paradigm less seriously than Miss Masterman, for, as she quite properly emphasizes, I have used the term in a number of different ways. To discover what is presently the issue, I must briefly digress to unravel confusions, in this case ones that are entirely of my own making.

In Section 4, above, I remarked that a new version of my *Scientific Revolutions* would open with a discussion of community structure. Having isolated an individual specialists' group, I would next ask what its members shared that enabled them to solve puzzles and that accounted for their relative unanimity in problem-choice and in the evaluation of problem-solutions. One answer which my book licences to that question is 'a paradigm' or 'a set of paradigms'. (This is Miss Masterman's sociological sense of the term.) For it I should now like some other phrase, perhaps 'disciplinary matrix': 'disciplinary', because it is common to the practitioners of a specified discipline; 'matrix', because it consists of ordered elements which require individual specification. All of the objects of commitment described in my book as paradigms, parts of paradigms, or paradigmatic would find a place in the disciplinary matrix, but they would not be lumped together as paradigms, individually or collectively. Among them would be: shared symbolic generalizations, like '$f = ma$', or 'elements combine in constant proportion by weight'; shared models, whether metaphysical, like atomism, or heuristic, like the hydrodynamic model of the

[1] *This volume* p. 66.

electric circuit; shared values, like the emphasis on accuracy of prediction, discussed above; and other elements of the sort. Among the latter I would particularly emphasize concrete problem solutions, the sorts of standard examples of solved problems which scientists encounter first in student laboratories, in the problems at the ends of chapters in science texts, and on examinations. If I could, I would call these problem-solutions paradigms, for they are what led me to the choice of the term in the first place. Having lost control of the word, however, I shall henceforth describe them as exemplars.[1]

Ordinarily problem-solutions of this sort are viewed as mere applications of theory that has already been learned. The student does them for practice, to gain facility in the use of what he already knows. Undoubtedly that description is correct after enough problems have been done, but never, I think, at the start. Rather, doing problems is learning the language of a theory and acquiring the knowledge of nature embedded in that language. In mechanics, for example, many problems involve applications of Newton's Second Law, usually stated as '$f = ma$.' That symbolic expression is, however, a law-sketch rather than a law. It must be rewritten in a different symbolic form for each physical problem before logical and mathematical deduction are applied to it. For free fall it becomes $mg = \frac{md^2s}{dt^2}$; for the pendulum it is $mg \text{ Sin } \theta = -ml\frac{d^2\theta}{dt^2}$; for coupled harmonic oscillators it becomes two equations, the first of which may be written $m_1\frac{d^2s_1}{dt^2}+k_1 s_1 = k_2(d+s_2-s_1)$; and so on.

Lacking space to develop an argument, I shall simply assert that physicists share few rules, explicit or implicit, by which they make the transition from law-sketch to the specific symbolic forms demanded by individual

[1] This modification and almost everything else in the remainder of this paper is discussed in far more detail and with more evidence in my [1970]. I refer readers to it even for bibliographical references. One additional remark is, however, in place here. The change just outlined in my text deprives me of recourse to the phrases 'pre-paradigm period' and 'post-paradigm period' when describing the maturation of a scientific specialty. In retrospect that seems to me all to the good, for, in both senses of the term, paradigms have throughout been possessed by any scientific community, including the schools of what I previously called the 'pre-paradigm period'. My failure to see that point earlier has certainly helped to make a paradigm seem a quasi-mystical entity or property that, like charisma, transforms those infected by it. Note, however, as Section 3 indicates, that this alteration in terminology does not at all alter my description of the maturation process. The early stages in the development of most sciences are characterized by the presence of a number of competing schools. Later, usually in the aftermath of a notable scientific achievement, all or most of these schools vanish, a change which permits a far more powerful professional behaviour to the members of the remaining community. On this whole problem, Miss Masterman's remarks (*above*, pp. 70–72) seem to me very telling.

problems. Instead, exposure to a series of exemplary problem-solutions teaches them to see different physical situations as like each other; they are, if you will, seen in a Newtonian gestalt. Once students have acquired the ability to see a number of problem-situations in that way, they can write down *ad lib* the symbolic forms demanded by other such situations as they arise. Before that acquisition, however, Newton's Second Law was to them little or no more than a string of uninterpreted symbols. Though they shared it, they did not know what it meant and it therefore told them little about nature. What they had yet to learn was not, however, embodied in additional symbolic formulations. Rather it was gained by a process like ostension, the direct exposure to a series of situations each of which, they were told, were Newtonian.

Seeing problem-situations as like each other, as subjects for the application of similar techniques, is also an important part of normal scientific work. One example may both illustrate the point and drive it home. Galileo found that a ball rolling down an incline acquires just enough velocity to return it to the same vertical height on a second incline of any slope, and he learned to see that experimental situation as like the pendulum with a point-mass for a bob. Huyghens then solved the problem of the centre of oscillation of a physical pendulum by imagining that the extended body of the latter was composed of Galilean point-pendula, the bonds between which could be released at any point in the swing. After the bonds were released, the individual point-pendula would swing freely, but their collective centre of gravity, when each was at its highest point, would be only at the height from which the centre of gravity of the extended pendulum had begun to fall. Finally, Daniel Bernoulli, still with no aid from Newton's Laws, discovered how to make the flow of water from an orifice in a storage tank resemble Huyghens's pendulum. Determine the descent of the centre of gravity of the water in tank and jet during an infinitesimal period of time. Next imagine that each particle of water afterwards moves separately upward to the maximum height obtainable with the velocity it possessed at the end of the interval of descent. The ascent of the centre of gravity of the separate particles must then equal the descent of the centre of gravity of the water in tank and jet. From that view of the problem the long sought speed of efflux followed at once. These examples display what Miss Masterman has in mind when she speaks of a paradigm as fundamentally an artefact which transforms problems to puzzles and enables them to be solved even in the absence of an adequate body of theory.

Is it clear that we are back to language and its attachment to nature? Only one law was used in all of the preceding examples. Known as the Principle of *vis viva*, it was generally stated as 'Actual descent equals potential ascent'. Contemplating the examples is an essential part (though only part) of learning what the words in that law mean individually and collectively, or in learning how they attach to nature. Equally, it is part of learning how the world behaves. The two cannot be separated. The same double role is played by the textbook problems from which students learn, for example, to discover forces, masses, accelerations in nature and in the process find out what '$f = ma$' means and how it attaches to and legislates for nature. In none of these cases do the examples function alone, of course. The student must know mathematics, some logic, and above all natural language and the world to which it applies. But the latter pair has to a considerable extent been learned in the same way, by a series of ostensions which have taught him to see mother as always like herself and different from father and sister, which have taught him to see dogs as similar to each other and unlike cats, and so on. These learned similarity-dissimilarity relationships are ones that we all deploy every day, unproblematically, yet without being able to name the characteristics by which we make the identifications and discriminations. They are prior, that is, to a list of criteria which, joined in a symbolic generalization, would enable us to define our terms. Rather they are parts of a language-conditioned or language-correlated way of seeing the world. Until we have acquired them, we do not see a world at all.

For a more leisurely and developed account of this aspect of the language-theory parallel, I shall have to refer readers to the previously cited paper from which much in the last few paragraphs is abstracted. Before returning to the problem of theory-choice, however, I must at least state the point which that paper primarily aims to defend. When I speak of learning language and nature together by ostension, and particularly when I speak of learning to cluster the objects of perception into similarity sets without answering questions like, 'similar with respect to what?', I am not calling upon some mystic process to be covered by the label 'intuition' and thereafter left alone. On the contrary, the sort of process I have in mind can perfectly well be modelled on a computer and thus compared with the more familiar mode of learning which resorts to criteria rather than to a learned similarity relationship. I am currently in the early stages of such a comparison, hoping, among other things, to discover something about the circumstances under which each of the two strategies works more effectively. In both programmes the computer will be given a series of stimuli (modelled as ordered sets of integers) together with the name of the class from which each stimulus was selected. In the criterion-learning programme the machine is instructed to abstract criteria which will permit the classification of additional stimuli, and it may thereafter discard the original set from which it learned to do the job. In the similarity-learning programme, the machine is instead instructed to retain all stimuli and to classify each new one by a global comparison with the clustered exemplars it has already encountered. Both programmes will work, but they do not give identical results. They differ in many of the same ways and for many of the same reasons as case law and codified law.

One of my claims is, then, that we have too long ignored the manner in which knowledge of nature can be tacitly embodied in whole experiences without intervening abstraction of criteria or generalizations. Those experiences are presented to us during education and professional initiation by a generation which already knows what they are exemplars of. By assimilating a sufficient number of exemplars, we learn to recognize and work with the world our teachers already know. My main past applications of that claim have, of course, been to normal science and the manner in which it is altered by revolutions, but an additional application is worth noting here. Recognizing the cognitive function of examples may also remove the taint of irrationality from my earlier remarks about the decisions I described as ideologically based. Given examples of what a scientific theory does and being bound by shared values to keep doing science, one need not also have criteria in order to discover that something has gone wrong or to make choices in case of conflict. On the contrary, though I have as yet no hard evidence, I believe that one of the differences between my similarity- and criteria-programmes will be the special effectiveness with which the former deals with situations of this sort.

Against that background return finally to the problem of theory-choice and the recourse offered by translation. One of the things upon which the practice of normal science depends is a learned ability to group objects and situations into similarity classes which are primitive in the sense that the grouping is done without an answer to the question, 'similar with respect to what?' One aspect of every revolution is, then, that some of the similarity relations change. Objects which were grouped in the same set before are grouped in different sets afterwards and *vice versa*. Think of the sun, moon, Mars, and earth before and after Copernicus; of free fall, pendular, and planetary motion before and after Galileo; or of salts, alloys, and a sulphur-iron filing mix before and after Dalton. Since most objects within even the altered sets continue to be grouped together, the names of the sets are generally preserved. Nevertheless, the transfer of a subset can crucially affect the network of interrelations among sets. Transferring the metals from the set of compounds to the set of elements was part of a new theory of combustion, of acidity, and of the difference between physical and chemical combination. In short order, those changes had spread through all of chemistry. When such a redistribution of objects among similarity sets occurs, two men whose discourse had proceeded for some time with apparently full understanding may suddenly find themselves responding to the same stimulus with incompatible descriptions or generalizations. Just because neither can then say, 'I use the word element (or mixture, or planet, or unconstrained motion) in ways governed by such and such criteria', the source of the breakdown in their communication may be extraordinarily difficult to isolate and by-pass.

I do not claim that there is no recourse in such situations, but before asking what it is, let me emphasize just how deep differences of this sort go. They are not simply about names or language but equally and inseparably about nature. We cannot say with any assurance that the two men even see the same thing, possess the same data, but identify or interpret it differently. What they are responding to differently is stimuli, and stimuli receive much neural processing before anything is seen or any data are given to the senses. Since we now know (as Descartes did not) that the stimulus-sensation correlation is neither one-to-one nor independent of education, we may reasonably suspect that it varies somewhat from community to community, the variation being correlated with the corresponding differences in the language-nature interaction. The sorts of communication breakdowns now being considered are likely evidence that the men involved are processing certain stimuli differently, receiving different data from them, seeing different things or the same things differently. I think it likely myself that much or all of the clustering of stimuli into similarity sets takes place in the stimulus-to-sensation portion of our neural processing apparatus; that the educational programming of that apparatus takes place when we are presented with stimuli that we are told emanate from members of the same similarity class; and that, after programming has been completed, we recognize, say, cats and dogs (or pick out forces, masses, and constraints) because they (or the situations in which they appear) then do, for the first time, look like the examples we have seen before.

Nevertheless, there must be recourse. Though they have no direct access to it, the stimuli to which the participants in a communication breakdown respond are, under pain of solipsism, the same. So is their general neural apparatus, however different the programming. Furthermore, except in a small, if all-important, area of experience, the programming must be the same, for the men involved share a history (except the immediate past), a language, an everyday world, and most of a scientific one. Given what they share, they can find out much about how they differ.

At least they can do so if they have sufficient will, patience, and tolerance of threatening ambiguity, characteristics which, in matters of this sort, cannot be taken for granted. Indeed, the sorts of therapeutic efforts to which I now turn are rarely carried far by scientists.

First and foremost, men experiencing communication breakdown can discover by experiment—sometimes by thought-experiment, armchair science—the area within which it occurs. Often the linguistic centre of the difficulty will involve a set of terms, like element and compound, which both men deploy unproblematically but which it can now be seen they attach to nature in different ways. For each, these are terms in a basic vocabulary, at least in the sense that their normal intra-group use elicits no discussion, request for explication, or disagreement. Having discovered, however, that for inter-group discussion, these words are the locus of special difficulties, our men may resort to their shared everyday vocabularies in a further attempt to elucidate their troubles. Each may, that is, try to discover what the other would see and say when presented with a stimulus to which his visual and verbal response would be different. With time and skill, they may become very good predictors of each other's behaviour, something that the historian regularly learns to do (or should) when dealing with older scientific theories.

What the participants in a communication breakdown have then found is, of course, a way to translate each other's theory into his own language and simultaneously to describe the world in which that theory or language applies. Without at least preliminary steps in that direction, there would be no process that one were even attempted to describe as theory-*choice*. Arbitrary conversion (except that I doubt the existence of such a thing in any aspect of life) would be all that was involved. Note, however, that the possibility of translation does not make the term 'conversion' inappropriate. In the absence of a neutral language, the choice of a new theory is a decision to adopt a different native language and to deploy it in a correspondingly different world. That sort of transition is, however, not one which the terms 'choice' and 'decision' quite fit, though the reasons for wanting to apply them after the event are clear. Exploring an alternative theory by techniques like those outlined above, one is likely to find that one is already using it (as one suddenly notes that one is thinking in, not translating out of, a foreign language). At no point was one aware of having reached a decision, made a choice. That sort of change is, however, conversion, and the techniques which induce it may well be described as therapeutic, if only because, when they succeed, one learns one had been sick before. No wonder the techniques are resisted and the nature of the change disguised in later reports.

REFERENCES

Brown [1969]: 'The Electric Current in Early Nineteenth-Century French Physics', *Historical Studies in the Physical Sciences*, **1**, pp. 61–103.

Cavell [1969]: 'Must We Mean What We Say?', in *Must We Mean What We Say?*, pp. 1–42.

Gombrich [1960]: *Art and Illusion*, 1960.

Heilbron and Kuhn [1969]: 'The Genesis of the Bohr Atom', *Historical Studies in the Physical Sciences*, **1**, pp. 211–90.

Klein [1964]: 'Einstein and the Wave-Particle Duality', *The Natural Philosopher*, **3**, pp. 1–49.

Kuhn [1962*a*]: *The Structure of Scientific Revolutions*, 1962. [A second edition, revised and enlarged by a new chapter entitled 'Postscript 1969', is to be published as a Phoenix paperback by Chicago University Press in 1970.]

Kuhn [1962*b*]: 'The Function of Measurement in Modern Physical Science', *Isis*, **52**, pp. 161–93.

Kuhn [1964]: 'A Function for Thought Experiments', in Cohen and Taton (*eds.*): *Mélanges Alexandre Koyré*, Vol. 2, *L'aventure de l'esprit*, pp. 307–34.

Kuhn [1969]: 'Comment [on the relations between science and art]', *Comparative Studies in Philosophy and History*, **11**, pp. 403–12.

Kuhn [1970]: 'Second Thoughts on Paradigms', in Suppe (*ed.*): *The Structure of Scientific Theory*, 1970.

Nida [1964]: 'Linguistics and Ethnology in Translation-Problems', in Hymes (*ed.*): *Language and Culture in Society*, pp. 90–7.

Popper [1963]: *Conjectures and Refutations*, 1963.

Quine [1960]: *Word and Object*, 1960.

Raman and Forman [1969]: 'Why Was It Schrödinger Who Developed de Broglie's Ideas?', *Historical Studies in the Physical Sciences*, **1**, pp. 291–314.

Schagrin [1963]: 'Resistance to Ohm's Law', *American Journal of Physics*, **31**, pp. 536–7.

Scheffler [1967]: *Science and Subjectivity*, 1967.

Shapere [1964]: 'The Structure of Scientific Revolutions', *Philosophical Review*, **73**, pp. 383–94.

Shapere [1966]: 'Meaning and Scientific Change', in Colodny (*ed.*): *Mind and Cosmos: Essays in Contemporary Science and Philosophy*, 1966, pp. 41–85.

Toulmin [1967]: 'The Evolutionary Development of Natural Science', *American Scientist*, **55**, pp. 456–71.

Name Index*

* Italic page numbers denote the more important references.